Power Struggles

Power Struggles

Scientific Authority and the Creation of Practical Electricity Before Edison

Michael Brian Schiffer

3 1336 08378 0271

The MIT Press
Cambridge, Massachusetts
London, England

For information on quantity discounts, email special_sales@mitpress.mit.edu.

Set in Stone and Stone sans by SNP Best-set Typesetter Ltd., Hong Kong. Printed and bound in the United States of America.

Library of Congress Cataloging-in-Publication Data

Schiffer, Michael B.
Power struggles : scientific authority and the creation of practical electricity before edison / Michael Brian Schiffer.
p. cm.
Includes bibliographical references and index.
ISBN 978-0-262-19582-9 (hardcover : alk. paper)
1. Electric engineering—United States—History. I. Title.
TK23.S35 2008
621.30973—dc22

2007039865

10 9 8 7 6 5 4 3 2 1

In memory of my beloved father, Louie Schiffer (1918–2007)

Contents

Preface   ix
Acknowledgments   xi

1   Studying Technological Change   1

2   Technological Transitions   11

3   Electromagnetism Revealed   21

4   An American Physicist   31

5   Telegraphic Visions   41

6   Mechanical Electricity   49

7   The Blacksmith's Motor   63

8   The Chemistry Connection   75

9   A Peculiar Calling   91

10   Hard Times   105

11   It's a Blast   119

12   "What Hath God Wrought"   137

13   Magnetic Power Derailed   155

14   Humbug!   175

15   Action at a Distance   191

16   First Light   207

17   If at First You Don't Succeed . . .   221

18   A Thousand Points of Light   239

19   Machine-Age Electricity   255

20 A Beacon of Modernity   271

21 Enter Edison   283

22 New Light   299

Notes   317
References   385
Index   415

Preface

I was born in 1947, the same year that the transistor was invented and a century after Edison's birth. These coincidences, however, did not predispose me to study electrical things. Rather, the seeds of my scholarly interest were planted by child's play, for in the seventh grade a friend, Daniel Essin, introduced me to the wonders of electricity. In my parents' garage in south-central Los Angeles, I took apart old radios and televisions scavenged from junk shops and garbage cans, and read books and electronics magazines. I also took electric shop in junior high school, at a time when middle schools still valued technical education. In a word, I was a nerd, one determined to become an electrical engineer.

Career goals have a way of changing. In high school I became infatuated with chemistry. I started UCLA as a chemistry major, but the escalation of the Vietnam war caused me to rethink matters and change course. As a sophomore, I decided to major in anthropology, emphasizing archaeology, yet I was also smitten with ceramics in art courses I took as a senior. Although I almost applied to graduate school in art, I became an archaeologist. Archaeology, a challenging discipline, has held my interest over the decades.

I am, however, an archaeologist of a peculiar sort. Along with other behavioral archaeologists, I believe that our discipline possesses a special perspective on human society. For us, human life consists of the ceaseless and varied interactions between people and material things. And we may study these kinds of interactions in the deep past of the Paleolithic or in the present, and in any society—traditional, "developing," or industrial. To enhance such research, behavioralists have borrowed and crafted a suite of conceptual tools for studying technology and technology change.

Not until W. David Kingery joined the faculty at the University of Arizona in 1988 did I contemplate applying these tools to the history of electrical technology. Observing that I had become a collector of portable radios, Kingery suggested that I familiarize myself with a modern industrial technology to complement my expertise in hand-made pottery. This seemed like a smart idea that would allow me to combine my long-term archaeological interests in technological change with my enduring fascination with electricity. And so began a journey that commenced with portable radios, then moved to electric automobiles, and most recently engaged electrostatic technology of the eighteenth century. Conspicuously missing from my earlier projects was a treatment of the foundations of the electric light and power systems on which life in industrial societies so completely depends. In the spring of 2002 I set out to remedy that deficiency. *Power Struggles* presents the findings of my project on the pre-Edison electrical technologies of the nineteenth century. Regrettably, I had to be selective, because the variety of new technologies and the amount of information about them are simply overwhelming.

This book is aimed primarily at an audience of people who, like me, are curious about technological change and the beginnings of our modern electrical world. It is the sort of book that I would have enjoyed reading had someone else written it first. It assumes on the part of readers no prior expertise in electricity or studies of technological change. Although the first chapter, in which I present the framework that guided my research and writing, may seem overly theoretical to some, the reader can easily grasp the basic concepts and principles. Familiarity with these concepts and principles enriches understanding of the stories in the chapters that follow. When I have told these stories to friends, family members, and archaeology students—none of them interested in electrical history *per se*—I have been heartened by the responses.

Power Struggles may also appeal to historians of science and technology, to philosophers of science and technology, and to economic historians. Growing numbers of scholars in American studies, cultural studies, and science-technology-and-society studies may also find this work of interest. There is also a ready audience of electrical enthusiasts who are fascinated by the history of the technologies that captivate them. Finally, in my home disciplines of anthropology and archaeology, research on technological change, especially in industrial societies, has burgeoned during the past few decades, and so my colleagues may encounter something useful between these covers. Clearly, *Power Struggles* furnishes additional evidence, through a richly textured case study, that anthropologists and archaeologists are contributing potentially useful conceptual tools for studying technological change.

The intellectual fashion among many researchers today is to emphasize social and economic *processes* at the expense of discussing individuals and actual artifacts. This was an understandable reaction to internalist, antiquarian, and biographical works of past decades, but in my view the pendulum has swung too far. That is why, in this book, I fashion stories around the activities of real people and the technologies they helped to create. After all, as W. Bernard Carlson has shown in his biography of Elihu Thomson, social and economic processes are instantiated in the decisions and activities of specific individuals and groups. *Power Struggles* can, I hope, also show that it is possible to handle individuals, artifacts, and abstract processes in one work.

Acknowledgments

A project of this scope cannot be undertaken without contributions from many individuals and institutions. My greatest debt is to the people of the Smithsonian Institution, where I conducted much of the research intermittently from 2003 to 2007. For special thanks I single out Ronald Brashear and Kirsten van der Veen of the Dibner Library at the National Museum of American History (NMAH). I also mercilessly exploited the copying resources of the NMAH Library, and I thank the staff, particularly D. Chris Cottrill, for help and forbearance.

Beyond libraries and archives, the Smithsonian's greatest resource is its curators. I am deeply indebted to the following individuals in the NMAH who graciously shared their knowledge and research materials, showed me artifacts, commented on my ideas, and made me feel very much at home: Deborah Warner, Roger Sherman, Arthur Molella, Bernard S. Finn, Steven Turner, and Harold D. Wallace Jr. At the Joseph Henry Papers, Smithsonian Archives, Marc Rothenberg not only helped me to find relevant information but also generously shared his vast understanding of the Smithsonian's first Secretary, who looms large in this book. During his brief stay at the Smithsonian in 2005, W. Bernard Carlson and I had many exciting discussions about electrical history.

People at the following institutions also supplied useful information: Andria Fields and William C. Allen, Office of the Architect of the U.S. Capitol; Marjorie Ciarlante, U.S. National Archives; Maryline Simler, Société d'Histoire de la Poste et de France Télécom en Alsace, France; Laura Schiefer, Buffalo and Erie County Historical Society; Paul Carnahan, Vermont Historical Society; James Miller, Engineering Library, University of Maryland; Jodean R. Murdock, Rock Island Arsenal Museum, Illinois; Jack Kirby, Birmingham Museum and Art Gallery, Birmingham, England; Jane Insley, National Museum of Science and Industry, London. Materials were also furnished by the Saratoga County Historical Society, Yale University Library, Library of Congress, New York Public Library, University of Arizona Interlibrary Loan, and Marius Rensen.

For insightful comments on one or more chapters, I thank Linda Cordell, David Edgerton, Robert Friedel, Brian Hayden, Kacy L. Hollenback, Vincent M. LaMotta,

Philip K. Lundeberg, Catherine Roberts, Annette Schiffer, Jeremy Schiffer, James M. Skibo, Monica L. Smith, an anonymous reader, and especially Roger Sherman, Deborah Warner, and Bernard S. Finn, who pored over the entire manuscript. At Arizona, Janet Griffitts and Kacy L. Hollenback performed library errands.

Deborah Warner and Robert Friedel, historians and dear friends, have given me inspiration, guidance, and feedback throughout this project. Although differing greatly from what they might have produced, this book is nonetheless much better on account of their help. Another historian, Robert Post, has also given me much encouragement over the years. In addition, he assembled a file on Henry Paine and after retiring passed it on to Deborah Warner; she loaned it to me. I thank them both for their unstinting collegiality. Robert Post also gave me a file of images of nineteenth-century electrical technologies. Roger Sherman supplied NMAH catalog numbers for important artifacts.

Some support for this project was provided by the Department of Anthropology, University of Arizona, since 2004 through my Fred A. Riecker Distinguished Professorship. The Dibner Library of NMAH named me a Dibner Research Scholar in 2003, which helped defray research and living expenses. As a Lemelson Center Research Associate since 2005, I have also availed myself of sundry resources, including incomparable munchies; I thank Arthur Molella for giving me this appointment.

My editor at The MIT Press, Marguerite Avery, has been a great help; I especially appreciate her confidence in this project from the beginning. I thank the production staff, especially Paul Bethge, for work done well.

As always, I couldn't have completed this project without the help, encouragement, and companionship of my wife, Annette. She has been an incisive sounding board for the stories and theoretical arguments in *Power Struggles*. More than that, she is my best friend who nourishes me emotionally in times joyous and sad.

Michael Brian Schiffer
Alexandria, Virginia

1 Studying Technological Change

In the New York offices of J. Pierpont Morgan, impatient directors of the Edison Electric Light Company and members of the press assembled on September 4, 1882, to witness a long-delayed demonstration. For more than two years Thomas Alva Edison and his team had labored in Edison's "invention factory" at Menlo Park, New Jersey, and in the streets of lower Manhattan, to bring electricity to an area of newspaper and financial offices. The Pearl Street District—so named because Edison's generating station occupied a building at 257 Pearl Street—would be the first large-scale electrical system to bring light to a host of offices in a city.[1]

Edison had arranged for the station to begin producing power at 3 p.m. Confident that the extensively tested system was ready, he turned on the lights, erasing any doubts about whether it would work as promised. This dramatic event, Edison recounted years later, was the most thrilling of his life. On the following day, the *New York Herald* reported Edison's triumph as a successful "practical application" of his lighting system.[2]

In many histories, the lighting of the Pearl Street District is acclaimed as the beginning of the electrical age. Although it was a technological and symbolic milestone, privileging one high point among many in a lengthy torrent of technological changes is somewhat arbitrary. I could suggest instead that the first commercial application of generator-powered lights—in lighthouses during the early 1860s—was an equally momentous beginning because it had almost immediate social and technological consequences. But why stop there? The earliest commercial lighting systems predate 1860, and the first generators, whose descendants powered Edison's lights, date back to the 1830s. Clearly, in developing his light and power system, Edison made full use of scientific and technological resources that many dozens of people had already provided.

This book is about the scientists (then called natural philosophers) who came up with the first important effects, principles, and technologies; the people of various occupations and social status as well as institutions that sought to develop and bring electrical technologies to market; and the domestic, commercial, industrial, and governmental activities in which these technologies competed for applications.[3] By examining these pioneering technologies and enterprises, we can better appreciate Edison's achievements. As we shall see, his success represents, above all, the convergence of many favorable factors—financial, social, scientific, and technological—that enabled him to pursue his vision relentlessly to the point where his electrical system could be pronounced practical.

The words "practical" and "practicable," along with their negations, punctuate nineteenth-century discussions of motors, generators, lights, and the like. Seeing these terms used so often in so many ways aroused my curiosity. I also found them in recent books and articles in which authors invoke the word "impractical" to explain why a technology did not make it to market or reach consumers, and "practical" to explain a success. These explanations were unconvincing. Perhaps it is time to reconsider the meaning(s) of these terms, for they have been much abused and taken for granted by scholars past and present.

If we seek a definition of "practical" in the literatures of science and engineering, the search will be in vain, for the word has no formal technical meaning. From *Webster's Third New International Dictionary of the English Language* we learn that something practical is "available, usable, or valuable in practice or action: capable of being turned to use or account: USEFUL."[4] But *who* determines whether an invention is actually useful or capable of being useful, and useful for *what*? After all, the same technology could be labeled practical by one person and impractical by another, or could be regarded as useful for some applications and useless for others. Practicality, then, is not an inherent property of a technology, but is in large measure socially constructed: certain individuals or groups judge whether a given technology is practical—that is, appropriate for *their* purposes.

An exclusive focus on practicality's social dimension, however, can blind us to a technology's material nature—that is, how its hard parts work together while engaging people and other technologies in specific activities. To be considered practical by someone or some group, a technology must exhibit a modicum of functional competence—symbolic, utilitarian, and so forth—that depends on the design and performance of its material parts.

Thus, the meaning of "practical" has irreducible social *and* material dimensions.[5] We can recognize this duality by defining practicality as *a judgment that particular people render about a technology's fitness to perform specific functions in a given activity*. As I emphasize below, practicality judgments affect peoples' decisions about whether to invent, develop, bring to market, and acquire a given technology. These are the very decisions that produce technological change.

--∿∿∿--

A simple theoretical framework can help us to refine this concept of practicality, seek the social and material factors influencing practicality judgments, and assess the effects of these judgments on technology-related decision making. I present this framework at the outset to acquaint the reader with the most important ideas that structured the research for, and the writing of, this book.

Historians of technology, including social constructivists, teach us that technologies do not directly beget other technologies, for technological change is mediated by social and cultural processes, including the presence of supportive institutions.[6] Moreover, a new technology most often emerges not from an inspired genius but from the cumulative labors of many people possessing different skills and seeking different ends. Thus, a technological change is contextualized in relation to relevant social groups and institutions as well as in relation to sundry causal factors external to the technology itself. Above all, a technological change is believed to result from people making decisions according to the specific goals, interests, and values they have acquired as social beings.

Anthropologists supply perspectives that are eminently compatible with the history of technology.[7] Researchers begin by attending to the observable world of social life manifest in activities—patterned interactions among people and artifacts. This emphasis on actual behavior, whether termed "social practice," "skilled performance," or "action," means that we conceive of technological change as changes in human activities, such as inventing, developing, manufacturing, marketing, acquiring, and using artifacts.

Depending on our interests, activities can be aggregated at different organizational, temporal, and spatial scales. That is, we could focus on the manufacturing activities of a single artisan, or ask questions about technological systems that spanned continents and endured for centuries. Large-scale units are built up from observed activities or activities inferred from evidence surviving in the historical and archaeological records.[8]

Because technologies inhere in every activity, studies of technology merge with studies of virtually everything else that is human, including religion and recreation, medicine and magic, social organization and socialization, communication and economy, politics and travel, and science and art.[9] Thus, we can discern, for example, communication technologies, ritual technologies, transport technologies, and political technologies, and we can investigate their content, changes, and relationships over time. Accordingly, questions about technology imply questions about activities, social groups, and institutions—and vice versa. For example, in *The Material Life of Human Beings*, Schiffer and Miller show that inquiries into communication, even speech, must take into account participating technologies.[10]

Technological change can begin in any realm of activities. New religious beliefs and practices, for example, usually call forth new ritual technologies, such as architecture, musical instruments, and priestly vestments. The Reformation not only revised Christian beliefs and practices but also replaced a rich assemblage of ritual technologies with others more spartan and more consistent with Protestantism. These kinds of effects go both ways: the adoption of a new technology, wherever it originates in a society, may alter activities in other realms, often leading to unforeseen behavioral changes. Case in point: The transistor—invented in a communication context (AT&T's Bell Laboratories) in order to speed up telephone switching—came to be used in hearing aids, radios, televisions, and digital computers. Thus, the transistor affected not only manufacturing and economic activities but also social life, recreation at home, and office work.

Although this book mainly deals with the causes of technological change, sometimes the consequences of one change are the causes of another. For example, the spotty adoption of early battery-powered electric lights recalibrated expectations about artificial lighting in public places. This, in turn, stimulated inventors to develop steam-powered generators for furnishing cheaper electricity that might render electric lights suitable for a wider range of applications.

_⌒⌒⌒_

To round out the theoretical framework, I add ideas and terms employed by behavioral archaeologists.[11] An essential construct is *performance characteristic*, a behavioral capability or competence that enables a technology to take part in specific interactions with people and/or other technologies.[12] Thus, performance characteristics establish a technology's fitness to function in a given activity or activities. For example, an electrical generator used for lighting must be able to supply ample current at the correct voltage for its load, and must be robust enough to withstand long-term use. But performance characteristics go far beyond enabling electrical and mechanical interactions, and may make possible any kind of interaction—utilitarian, symbolic, even financial. Thus, an electric lighting system has a plethora of performance characteristics, including purchase price, installation costs, color and brightness of the lights, reliability, maintainability, operating expenses, risks to laborers and users, and—in the case of the earliest systems—the ability to symbolize technological prowess and modernity. In short, a performance characteristic is *any* capability or competence that can come into play in an activity-specific interaction. Given this expansive notion, we can explore the many factors that might have influenced the practicality judgments and consequent decisions contributing to any technological change.

I suggest that people's decisions take into account—explicitly or implicitly—a technology's *anticipated* performance characteristics in relation to a given activity's performance requirements. These forecasts may be projected on the basis of wishful thinking, technical reports, intuition, advertising, folk theory, scientific theory, word of mouth, or firsthand experience. Often a new technology's anticipated performance characteristics are contrasted with those of a technology it might replace. Thus, the expected performance characteristics of the first electric lights were compared to those of gaslights and oil lamps in applications such as lighthouses, factories, and city squares. When these comparisons were explicit in magazines, newspapers, or books, I was usually able to infer a technology's baseline performance requirements. Anticipated and actual performance characteristics were also inferred from historical materials, including museum specimens, and from findings in modern experiments and engineering science.[13]

_⌒⌒⌒_

Decisions take place during a technology's "life history," which can be segmented into three major processes—invention, commercialization, and adoption—each of which consists of specific activities.[14]

Invention is the creation of an idea or vision for a technology that has performance characteristics differing from those of other technologies. The idea may consist of a minor modification of an existing device or may be a vision breathtaking in its audacity, such as an undersea telegraph cable crossing the Atlantic.[15] Embodying the contributions of many people, inventions are usually materialized as prototypes, as models, or as descriptions and illustrations.

In commercialization, which includes research and development people refine the technology in the hope that it can be manufactured and made available to consumers; often they seek to sell it at a profit. Commercialization requires that a technology's champions—e.g., inventors, entrepreneurs, investors, and manufacturers—span a "developmental distance," which depends on resources such as time money, organization, labor, skill, tacit knowledge, raw materials, tools, and structures.[16] The promoters' wealth and income, loans and gifts, government grants and contracts, and stock sales can help to obtain labor and materials. However, creating tacit knowledge and skill entails experience and thus time. and forming an organization requires a suitable institutional framework, including permissive laws; these necessities cannot always be bought.[17]

Some developmental distances are short and follow well-traveled routes, and so the modest resource needs are easily forecast. For example, perfecting and marketing a new flavor of ice cream is a routine process in firms like Ben & Jerry's, which can exploit the findings of decades of food science along with the skills, expertise, and technologies in the company's own laboratories. Likewise, once long-lasting batteries came to market in the late 1830s, the modest requirements for electroplating copper became obvious, and so—in anticipation of avid consumers—many people commercialized this technology. In contrast, developmental distances for many novel and complex technologies may involve forays into uncharted technological territory, sometimes far beyond the borders of existing science and engineering science. The history of electrical technology is replete with people who underestimated the length of the journey, as in the cases of Cyrus Field's earliest Atlantic cables and Charles Page's electric locomotive. Edison also misjudged the developmental distance for creating his light and power system, but was able to acquire additional resources for finishing the job. Like many ambitious projects, both the Atlantic cable and Edison's system became nurseries for new engineering science.

When a developmental distance has been spanned and manufacturing is possible, the technology may become available for sale and adoption, where adoption is the acquisition and use of the technology by consumers. The latter may be individuals or groups (e.g., households, communities, churches, companies, and governments at every level). Consumers, of course, are the final arbiters of a technology's success, and many technologies do not succeed.[18]

A technology's life history may have a premature end: most inventions are never brought to market, and most new technologies fail to find many consumers.[19] The life history of a technology is moved forward (or not) through invention, commercialization, and adoption by the decisions of relevant players. By "player" I mean *role player* in the sociological sense—i.e., a person who performs a specific function in a specific social group. We should keep in mind that more than one person can play a specific role, and one person can play multiple roles. In studies of technological change, the most inclusive social group is composed of the people who take part in *all* of a technology's life-history activities, and thus players can include inventors, instrument makers, scientific authorities, patent attorneys, patent examiners, judges and juries, journalists, newspaper and magazine editors, entrepreneurs, investors, bankers, government officials, manufacturers, engineers, mechanics, corporate executives, wholesalers, retailers, consumers, and repairers. Players interact in specific activities with the technology itself or with representations of it, such as patent drawings, company prospectuses, marketing plans, legal documents, journal articles, advertisements, and magazine and newspaper accounts.[20] In the cases of the earliest electrical technologies, the only players were inventors, makers of scientific instruments, and consumers. In later decades the number of players multiplied, especially for complex technologies having appreciable developmental distances. (In the twentieth century, players included labor unions, environmentalists, community groups, consumerist organizations, health and safety advocates, and government regulators.)

Anticipated performance characteristics are assessed in relation to performance requirements on the basis of a player's activity-specific goals, interests, and values. Thus, a patent attorney evaluates an invention's patentability, whereas a manufacturer forecasts whether it will be purchased by specific consumer groups, perhaps defined on the basis of sex, gender, age, marital status, wealth, or ethnicity. Because different players usually cannot justify their judgments—especially negative ones—in the same terms (e.g., scientific and technical factors affecting patentability vs. assumptions about consumer-group preferences), and because the rationales for judgments formed intuitively cannot always be put into words, players often voice their views in terms of practicality.[21] In effect, a practicality judgment is the lowest common denominator of meaning that permits players to convey and justify their decisions to other players. By merely labeling a technology practical or impractical, a player can avoid furnishing a detailed and potentially incomprehensible rationale. For our purposes, a practicality statement is taken to be a judgment as to whether a technology's anticipated performance characteristics match the performance requirements of specific activities. When the match is poor, the player may state that the technology is impractical, and vice versa.

The decisions arising from practicality judgments can advance, retard, or truncate a technology's life history. Although we may be tempted to assume that all players have to agree on a technology's practicality for it to traverse a sizable developmental distance and achieve profitable adoptions, this is not always so. Sometimes an inventor possesses sufficient resources to develop, manufacture, and market a technology despite impracticality judgments from, for example, scientific authorities, journalists, and investors. More often a consensus on practicality is required from certain players because most inventors are not wealthy. The most consequential players often turn out to be investors, government officials, and manufacturers, who are able to supply or deny resources for commercialization. Scientific authorities, patent examiners, and others can also be consequential.

As players weigh in with their sometimes contradictory assessments of practicality, power struggles may ensue, resulting in lengthy and complex negotiations whose outcome is affected by the players' relative social power.[22] Thus, government officials who favor a certain invention can direct resources toward its development even though scientific authorities, investors, and manufacturers have declared the invention impractical. Likewise, scientific authorities having great social power can exercise more influence over the development process than lesser lights.

My task in constructing the stories I present in this book was to examine the life histories of specific electrical technologies, identifying relevant activities and players and their relative social power. I then inferred how players assessed practicality as they compared a technology's anticipated performance characteristics with those of competing technologies and with the performance requirements of specific activities. In narratives of a technology having a great developmental distance, I focused on players that affected flows of resources. I also considered many contextual factors *and* the technology's actual performance characteristics, particularly as viewed by consumers, in crafting a narrative that accounts for a technology's life history—long or short.

Although this theoretical framework has guided the research and writing of this book, it doesn't dominate the narratives.

_‿⌒⌒⌒‿_

Undertaking a study of the pre-Edison electrical technologies of the nineteenth century is appealing on several grounds.

First, these technologies are significant because they, along with new institutions, helped to lay the foundations of the modern industrial world.[23] Indeed, the period from 1800 to about 1880 witnessed the invention of the major hardware components of electric light and power systems: batteries, arc and incandescent lamps, motors, generators, transformers, and instruments for measuring current, tension, and resistance. And in responding to problems that arose during the construction and maintenance of telegraph systems, people began to fashion an abstract and quantitative engineering science. Also created were corporations for manufacturing electrical components and for constructing and operating large-scale electrical systems.

Second, the later changes that took place in electrical technologies were matched by—indeed, largely fostered by—changes in the institutional contexts for inventing, commercializing, and adopting inventions.[24] Most of the earliest electrical technologies were crafted by natural philosophers whose inventive activities were underwritten by personal wealth or by colleges and other private organizations. These men sought no patents, and their inventions were commercialized, if at all, by philosophical instrument makers. The market for their technologies was confined mainly to colleges, experimenters, hobbyists, and science lecturers. By 1880, however, there were many social, political, financial, and institutional incentives to invent electrical things, which drew in people of various backgrounds, occupations, and motives. There were also new business tools to supply resources for commercialization as well as competent manufacturing firms. And sometimes large markets emerged from long-latent needs, such as safer and brighter lights in factories. When Edison plunged into his light and power project, he was enmeshed in an essentially modern corporate capitalism and contending with many consequential players, including patent attorneys, Wall Street bankers, and city officials. Discerning the interplay between these many technological and non-technological factors is, for me, a fascinating challenge.

Third, there is the opportunity to examine in detail the "scientific authority" as a player in technological change. By "scientific authority" I mean a person represented as having pertinent expertise in science or engineering science.[25] As I familiarized myself with nineteenth-century technologies, I came to understand that the scientific authority—as distinct from the scientist as originator of principles and as inventor—*sometimes* consequentially influenced the life histories of electrical components and systems. For example, it was not uncommon for prospective investors or manufacturers to ask a scientific authority to evaluate a projected technology's technical feasibility—that is, its practicality in the sense of conforming to accepted scientific principles. However, such a judgment could be caught up in controversy if the scientific principles were underdeveloped, overgeneralized, or erroneous, and authorities sometimes interpreted the same principles differently or drew contrasting technological implications from different principles. Of course, personal, political, and social factors as well as institutional affiliations occasionally influenced evaluations and stoked the fires of conflict. In addition, scientific authorities sometimes went well beyond assessing technical feasibility by judging practicality on the basis of cost accounting and market forecasts, advancing opinions that were eminently disputable.[26] I believe that the scientific authority's involvement in technological change during the nineteenth century has been insufficiently examined.[27]

The foremost American electrical scientist during the period under consideration here was Joseph Henry. Exploiting principles and devices invented by Europeans, Henry built the first electromagnet capable of doing work in machines, fashioned an early electric motor, and crafted electromagnetic principles that contributed to the invention and commercialization of telegraphy. An internationally renowned physicist who held a professorship at Princeton and later became the first secretary of the Smithsonian Institution, Henry was often asked to assess technologies; indeed, he defined and personified the role of scientific authority in America. Accordingly, I attend closely to Henry's assessments of electrical technologies, including "humbug" inventions, motors, the telegraph, apparatus for lighting gas lamps in the national capitol's new dome, and lighthouse illumination. Because Henry's judgments were sometimes contested and did not always affect other players' decisions, these case studies bring interesting power struggles to light and help us grasp the varied effects that scientists had on technological change.

The scientific authority is often a consequential player in discussions of, and controversies surrounding, such cutting-edge technologies as genetically modified foods, nanotechnology, missile defense—and who can forget cold fusion? In the pages that follow, case studies of electrical technology—successes and failures—demonstrate that scientific authorities could become embroiled in heated controversies and power struggles. Indeed, the disputes of today are not anomalies but follow patterns of conflict that have centuries-long antecedents.

For those who believe that science is a body of secure knowledge that furnishes irrefutable answers, the specter of dueling scientific authorities is disquieting. But like other people, scientists have strong beliefs and passions, and science itself is an open-ended, long-term process that refines and sometimes replaces venerable generalizations. And so, in the short term, controversies that erupt in application-oriented contexts can get especially messy. Readers seeking insights into the influence of scientific authorities on technological change may find this very messiness interesting.[28]

2 Technological Transitions

The first age of electricity awaited neither Thomas Edison's genius and perspiration nor J. Pierpont Morgan's largess. During the middle of the eighteenth century, electrical technology had burst forth in hundreds of varieties. Among the most prolific contributors to this cornucopia of strange and wondrous things were Benjamin Franklin and his many friends and correspondents in Europe and America. In addition to the lightning rod, Franklin himself invented two electric motors, one of which rotated with no external connection; electric spiders that danced; the magic table, which could store an electric charge and shock the unwary; and the first "battery," which consisted of several Leyden jars—a kind of capacitor invented by Europeans in 1746—that could store a mighty charge. Other inventors created glass bulbs that gave off an eerie purplish light, self-ringing carillons with miniature bells, cannons and pistols that could fire with dramatic effect, and electric bandages.[1]

When Benjamin Franklin came up with an invention, he did not rush to patent it. In fact, he did not patent anything, believing that inventions should be made for the benefit of mankind: "That as we enjoy great Advantages from the Inventions of others, we should be glad of an Opportunity to serve others by any Invention of ours, and this we should do freely and generously."[2] This was not an unusual view at the time, at least not on the part of natural philosophers experimenting with electricity. Before 1837, all patents for electrical inventions issued in England and the United States could be counted on one hand. Certainly there was no basis for believing that sales of electrical things would ever yield great wealth.

Nonetheless, the lack of patent "protection" did not prevent electrical devices from being manufactured and sold. Established makers of philosophical and mathematical instruments kept abreast of the scientific literature, offering inventions that they believed might command demand. Their instrument catalogs listed dozens of electrical technologies, including some invented by Franklin; in comparison with microscopes and telescopes, most were inexpensive. Purchasers included experimenters, seeking in the laboratory new physical effects; college teachers and itinerant lecturers performing spellbinding demonstrations of electrical phenomena; physicians and self-appointed electrotherapists treating afflictions such as sore throats, hysteria, and hemorrhoids; and hobbyists and collectors of scientific apparatus. Occasionally rich collectors or even natural philosophers supported by a wealthy patron or museum hired instrument makers to build pricey one-off apparatus.

In this electrical world, so different from ours and even from the one that would take shape in the nineteenth century, explicit conflicts over practicality seldom surfaced. Perhaps one reason is that the players judged practicality *implicitly*. Inventors—mostly clerics, college teachers, and a few people of independent means like Franklin—took it for granted that their creations were practical; after all, they had invested their own time and sometimes money in crafting devices that were obviously useful in experiments or demonstrations. Instrument makers doubtless assumed that the new technologies reported in the scientific literature worked as described; after all, the inventor and the scientific authority were one. In any event, bringing to market these rather small-scale products of research required no infusions of outside capital; cranking out a new electrical thing was, for instrument makers, all in a day's work.

The authoritative judgments in favor of practicality, proffered by the scientist-inventor and affirmed by the instrument maker, were embodied in the artifacts that consumers could purchase. Perhaps having seen the same devices used in public lectures, consumers would have assumed that the instrument makers' offerings performed as expected. In the final analysis, very little was at stake financially in developing and commercializing eighteenth-century electrical inventions, and that, I suspect, is a major reason why practicality was not openly contested. This quiet rhetorical scene would be rudely disturbed after the late 1830s, when fortunes could be made or lost through investments in electrical technology and when the number of players increased dramatically.

_⁓⁓⁓_

Generators of the eighteenth century produced electricity by friction. In one early French design, the operator placed his hand against a rotating glass globe; other electrical machines had a small cushion covered with mercury amalgam that pressed against a glass cylinder or disk. The resultant charge on the glass was conveyed, through a metal conductor, to a Leyden jar or some other storage device, where it accumulated. The stored charge could be used later in lectures, in experiments on plants and animals, in medical treatments, and for other purposes. The technologies of frictional electricity enabled Franklin and others to set forth fundamental concepts and principles, including positive and negative charge, the notion of a circuit, and the distinction between insulators and conductors. After about 1820 these technologies and principles became known as "static electricity" and were covered in textbooks under the subject of electrostatics.

At the end of the eighteenth century, natural philosophers were struggling to coax new effects from electrostatic devices. Although gigantic frictional generators—one having two rotating glass disks 6 feet in diameter that yielded 24-inch sparks—bore limited scientific fruit, it was a discovery by an Italian physician using rather simple apparatus that reinvigorated electrical science and set it on a different course.

Luigi Galvani, a professor of obstetrics at the University of Bologna, had been experimenting at home with partially dissected frogs and other hapless critters. When he placed a metal object against an animal's crural nerve, its legs twitched. To account for this startling effect, Galvani offered an elaborate theory that proposed a new kind of electricity—"animal electricity"—that supposedly was generated in brains. Galvani published his findings in the early 1790s, and investigators throughout Europe began sacrificing animals so that they could repeat and extend his mystifying findings.[3] And a few people mounted strong challenges to the theory of animal electricity.

Galvani's most persistent critic was Alessandro Volta, a world-famous natural philosopher and a professor at the University of Pavia.[4] Already the author of several important electrical inventions, including an electric pistol, Volta disputed Galvani's claim to have discovered a special kind of electricity in animals. Instead, Volta insisted, the muscular movements were attributable to an imbalance in electrical fluid caused by external agents such as the metal items wielded by the experimenter. Frogs and other animals were simply sensitive electrometers, capable of reacting to minuscule charges.

Volta came to understand that working with dissected animals was in many senses messy, for creatures complicated an understanding of the underlying physical processes. With his exceptional inventive abilities, Volta forged new apparatus from common and inexpensive materials. Experiments with several metals eventually led Volta to formulate his "contact" theory. In an audacious claim that rivaled Galvani's, he suggested that merely placing two different metals or other conductors in contact with one another generated electricity.

In support of the contact theory Volta offered experiments with his "crown of cups" and his pile, the first electrochemical batteries of the modern era.[5] The crown of cups was just a line of salt-water-filled cups, usually of glass. They were connected by metal arcs—strips of silver and zinc joined in the middle—whose ends were inserted into the salt water. A device of greater force (or tension) could be made by chaining 40 or even 60 cups in a row. In testing this invention, Volta felt a shock simply by dipping a finger of each hand into cups some distance apart; the farther the cups were separated along the series, the stronger the shock. The pile, which soon acquired the moniker "voltaic pile," consisted of a stack of alternating silver and zinc disks, each pair separated by a conducting solution—salt-water-soaked pasteboard. In a tall stack, Volta found, each additional pair of disks strengthened the pile's ability to shock him. In 1799 Volta wrote a paper in which he discussed his contact theory, claimed that the metals represented conductive tissues in animals, and described his new devices. The paper was published in 1800 in the *Philosophical Transactions of the Royal Society of London*, a prestigious venue for reporting discoveries.[6] For many decades, battery-generated electricity would be called "voltaic electricity" or, in a curious twist, "galvanism."

_⌒⌒⌒⌒_

Controversies over the theories of animal electricity and contact electricity would persist for many decades, outliving their originators, but the electrochemical battery immediately became a valuable research tool and an important commercial product. Indeed, instrument makers, appreciating that experimenters could hardly wait to get their hands on Volta's inventions, brought them to market without delay. And they sold well. The voltaic (or galvanic) battery's main attraction to researchers was its novel electrical performance, which offered tantalizing possibilities for discoveries.

Seduced by the opportunities, researchers wanted to learn how a battery's construction affected its electrical performance, and so they varied factors such as the kinds of metals, nature of the conducting solutions, and size of the metal disks or plates. From these experiments came the basic design principles that battery builders applied when they wanted to optimize particular performance characteristics, such as intensity (in today's terms, tension, electromotive force, or voltage) or quantity (current). For example, a battery employing dilute acids instead of salt water generated a greater quantity of electricity, and some metal combinations yielded slightly higher tension; the latter findings were codified in tables that today are called the electromotive series. A few decades later, Michael Faraday introduced the term "electrolyte" for the conducting solution and "electrodes" for the metal components. Not surprisingly, instrument makers included many of the new battery designs in their catalogs.

An obvious move made by natural philosophers was to substitute galvanic batteries for electrostatic technology in routine experiments. In this way they could learn if, and in what ways, the effects of galvanism differed from those of frictional electricity. Although Faraday and others would eventually show that electricity was electricity regardless of how it was generated, galvanism and frictional electricity tended to have markedly different effects. In general, frictional generators produce very high tension and minute amounts of current, whereas batteries produce low tension and substantial current. Thus, galvanism failed to reproduce some old effects, such as creating long sparks, exciting light in partially evacuated glass globes, or causing someone's hair to stand on end. Yet the new technology yielded several well-known effects much more easily, such as melting metals and decomposing water. In fact, using a Leyden jar to produce the tiniest amounts of oxygen and hydrogen from the electrolysis of water had required thousands of discharges, an hours-long process. With galvanism, the same result could be had in minutes. In its ability to decompose compounds, some people discerned possibilities for creating new chemistry.

⌐∿∿∿⌐

One investigator who harnessed galvanism for chemical research was Humphry Davy. His studies were supported by the Royal Institution of Great Britain, which had been founded in London by the American expatriate Benjamin Thompson (later to be named Count Rumford).[7] Thompson intended the Royal Institution to be a museum where working-class people could be shown the potential of science to improve everyday technologies. However, it didn't turn out altogether as planned. Over the years, especially under Davy's leadership, the Royal Institution became a research powerhouse, particularly in chemistry and physics; its activities were supported mainly by expensive public lecture courses offered by distinguished men of science. Working-class people could hardly afford to enroll.

In Europe's rigidly stratified societies, natural philosophy had long provided a narrow path for upward social mobility. That Benjamin Franklin could rub elbows with French nobility was not a result of his work as a printer and publisher, but of his signal contributions to electrical science.[8] Humphry Davy also rose from obscurity to become a scientist respected worldwide.[9] Born into a family of very modest means, he was apprenticed in 1795 to a physician. After a three-year indenture, he assumed a post at the Pneumatic Institute, where he stayed for several years.

While experimenting with various gases and their possible effects on the human body (tests he often performed on himself), Davy made major discoveries, including the anesthetic effect of nitrous oxide, also known as laughing gas. Regrettably, surgeons ignored his findings. Nonetheless, publications resulting from his other experiments on gases helped establish Davy's scientific reputation.

In 1801, Benjamin Thompson was seeking someone to appoint to a lectureship at the Royal Institution. He turned to the gifted experimenter Davy, still in his early twenties. In 1802, Davy was named the Royal Institution's Professor of Chemistry. And there he would remain, conducting research in chemistry and electricity, and delivering entertaining lectures to well-heeled audiences.[10]

By the time Davy was appointed to the Royal Institution—shortly after Volta reported his pile and crown of cups—one could already purchase galvanic batteries. However, Davy and others anticipated that some experiments would demand more power than could be supplied by off-the-shelf batteries, which had at most a few dozen small zinc-copper "pairs" or "couples" (cells, in today's terminology).[11] A pair or a couple usually consisted of an electrolyte-filled glass jar or ceramic container, into which an electrode pair was immersed. (Most people today call even a single cell a "battery.")

Davy had several fairly large batteries built for the Royal Institution, one of which was composed of "24 plates of copper and zinc of 12 inches square, 100 plates of 6 inches, and 150 of 4 inches square."[12] Although puny in comparison with the behemoths built later, these batteries enabled Davy to carry out his most important work in chemistry. In experiments reported in 1806, he employed compounds of known composition to show that batteries could effect chemical decompositions. Confident in the technique, he next applied it to soda and potash. These were common substances that, though not previously decomposed, were believed to be compounds. In his first efforts, Davy applied battery current to soda and potash in aqueous solutions, but the result in both cases was the liberation of hydrogen and oxygen. The compounds had held fast, for only the water had decomposed.

Davy surmised that success might come if he kept out all water by fusing the potash and soda. He placed a quantity of potash in a platinum spoon and heated it with an alcohol lamp supplied with pure oxygen. A wire from the positive pole of the 100-plate battery was connected to the spoon, and a wire from the negative pole was dipped into the molten potash. The results were dazzling: "a most intense light was exhibited at the negative wire, and a column of flame . . . arose from the point of contact."[13] After reversing the polarity, Davy found that tiny globules formed on the spoon, floated to the top, and burned in the air. The experiment had liberated a previously unknown element: the lightweight and highly reactive metal potassium. But Davy was not done.

In order to receive credit for isolating a new element, Davy would have to collect a sample and determine its chemical properties. He obtained larger samples of potassium by using electricity both to melt and to decompose the potash. He also found that the metal could be kept in its elemental state by immersion in naphtha (a petroleum distillate), which facilitated characterization. Davy next isolated sodium from soda; however, this decomposition required more power, so Davy used both the 100-plate battery and the 150-plate one. And, in a feat of discovery that no individual would ever match, Davy isolated other new elements, including barium, calcium, strontium, magnesium, and silicon, using various modes of electrical decomposition. This work bolstered the theories of Davy and others that chemical affinities (bonds) were based on attractions between positively and negatively charged particles.

Davy's stunning discoveries showed unequivocally that galvanic decomposition was a practical tool for teaching and research, and other chemists agreed. On this foundation, entirely new industries that worked metals by electricity would arise, beginning in the late 1830s. In the meantime, some research projects demanded even larger batteries.

In 1808, John George Children, also in London, put together a huge battery, aiming to learn how a small number of enormous plates might affect chemical decomposition.[14] His battery had 20 cells of zinc and copper, each plate measuring 8 square feet. (In square inches, this dwarfed Davy's 250-cell battery by a factor approaching 12.) Experiments with the big battery failed to liberate barium from its compounds or to decompose other solid but poorly conducting substances; nonetheless it furnished information on how to design galvanic batteries for particular uses. Obviously, Children noted, decomposition of substances that conducted electricity poorly would require a battery with many more plates, capable of producing higher tension. In modern terms, Childrens's high-current, 20-cell battery produced at most 30 volts; its shock could barely be felt.[15]

Davy, who took part in the first experiments with Childrens's battery realized that it would be expensive to vastly increase the number of cells, so he turned to "a few zealous cultivators and patrons of science," doubtless arguing that their contributions would make possible significant discoveries.[16] Apparently his pitch was convincing.

In the laboratory of the Royal Institution, Davy built a battery containing 2,000 cells. The metal plates, each 32 square inches, were immersed in porcelain containers filled with a dilute solution of nitric and sulfuric acids. One rather unpleasant effect emerged at once: people who approached the monstrous battery received a shock through the stone floor. For insulation, Davy placed the porcelain containers on wooden supports, a move that not only prevented shocks but also improved the battery's performance. And perform it did, yielding "brilliant and impressive effects" that Davy cataloged at length in his chemistry text of 1812.[17]

It had been known since the electrostatic era that electricity passing through a thin metal wire caused it to heat up and eventually melt. What would happen if galvanism from a large battery were employed instead? Galvanism, it turned out, achieved these effects with unaccustomed speed, even on long wires. Discussing the light radiated by a white-hot platinum wire, 18 inches long and 1/30 of an inch in diameter, Davy wrote that "the brilliancy of the light was soon insupportable to the eye, and in a few seconds the metal fell fused into globules."[18]

In 1814, George Singer, another English researcher, described a fascinating variation on the heated-wire experiment: insert the fine platinum wire into the glass receiver of a vacuum pump, connect the wire to a battery, and then evacuate the air. The wire, Singer observed, will then attain a "glowing white heat."[19] We might be tempted to regard this little apparatus as the first incandescent lamp, but for Singer it was merely another way to show that galvanism could produce light. However, beginning in the middle of the nineteenth century, many inventors including Edison, would suppose that this technology could replace gas lights. And so they began their experiments on incandescent lamps with a platinum element in an evacuated glass globe.

Davy reported another effect that was even more dramatic: when two tiny rods or points of carbon, connected to a battery's poles, were briefly brought together and then separated slightly, there issued forth a light so bright that, in Davy's words, "sunshine compared with it appeared feeble."[20] Not surprisingly, exhibiting the dazzling carbon arc—the name it eventually acquired—became a staple demonstration in public lectures at the Royal Institution and elsewhere. The carbon arc, not the incandescent lamp, would become the technological basis of the first commercial electric lights.

The carbon arc also contributed to another effect: the production of very high heat. Davy tested the resistance of various materials to the dancing white flame by placing a small specimen at the top of the arc. He seems to have tried just about anything that was handy in the laboratory, including lime, quartz, and sapphire, and all melted immediately; fragments of diamond, however, evaporated. No other technology of that time could produce a heat so intense. This effect would underlie electric arc furnaces, which were commercialized at the end of the nineteenth century and became a successful industrial technology.[21]

With one fascinating exception, Davy himself derived no technological implications from the novel effects his research had revealed. They were, for Davy, discoveries that added new bricks to the edifice of natural philosophy and enhanced his reputation. However, one should not get the impression that Davy was unwilling to invent technologies for everyday life. Recall that Benjamin Thompson wanted the Royal Institution to show that new technologies could emerge from laboratory science. In fact, Davy invented several technologies that left the laboratory, the most significant of which was a "safety lamp" for coal miners.[22]

The safety lamp grew out of a "consulting" project Davy undertook at the invitation of a benevolent committee, set up by local clergymen and businessmen in Sunderland (in northeastern England) after a mine explosion in 1812 killed 92 men and boys. A mining engineer on the committee, John Buddle, believed that it was time to draft science for the job of figuring out the causes of coal mine explosions. After visiting several mines, Davy concluded that they were adequately ventilated. What was needed, he reasoned, was a lighting technology that would not ignite the inflammable gas escaping from the coal seams. He did experiments with phosphorus lights and with "the electrical light in close[d] vessels, but without success; and even had these degrees of light been sufficient, the processes for obtaining them, I found, would be too complicated and difficult for the miners."[23] Clearly, unfavorable performance characteristics of these lights rendered them unfit for this application.

Davy then turned his attention to the properties of the inflammable gas, which was known as "fire damp." (It was mostly methane.) His experiment found that its ignition temperature was surprising low. Next, using an electrostatic reaction vessel adapted from Volta's electric pistol, Davy determined the explosiveness of different ratios of air to fire damp. Integrating his findings, he concluded that the mixture of air and fire damp in a mine, if kept below the ignition temperature, would not explode, even when exposed to a flame. Thus, if a small-diameter metal tube or a fine wire mesh was placed around the lighted wick to cool the flame, the mine's atmosphere would not ignite. Applying this principle, Davy designed a number of oil-burning safety lamps, which he subjected "individually to practical tests" in the laboratory and, with John Buddle's help, in a coal mine. The results were more than gratifying.[24] Davy's safety lamp, which he did not patent, was judged practical by manufacturers, mine owners, and miners. Widely used during the nineteenth century, it was credited with reducing the risk of explosions. However, Davy's safety lamp did not eliminate disasters entirely, as miners were sometimes forced to work in ever more dangerous places.

_⌒⌒⌒_

Davy and fellow experimenters were aware that some of galvanism's dramatic effects might be turned into new technologies for creating artificial light and heat, but they did not highlight such possibilities. To understand their reticence, we need only consider the performance characteristics of the early galvanic batteries. Beyond the obvious expense of constructing them, huge batteries—indeed, all batteries—had a near-fatal performance deficiency: after a period of use, sometimes rather brief, their output dropped drastically. Power could be restored, but only after the battery was refurbished by replacing the zinc electrodes and refreshing the acid. (To reduce the rate of deterioration, some investigators suspended the electrodes above the battery, lowering them into the electrolyte only when needed.) I suspect that Davy and his contemporaries implicitly judged that a source of electric power so pricey, with effects so evanescent, could not be regarded as practical outside the scientific laboratory and the lecture hall.

Although producing heat and light required very large batteries, experimenters were able to create other effects with fewer and smaller cells. The most far-reaching of these effects is electromagnetism, the foundation of modern motors, generators, transformers, and countless other electrical things that make modern life possible. Let us now turn to the discovery of electromagnetism and some of the inventions that followed almost immediately in its wake.

3   Electromagnetism Revealed

The discovery of electromagnetism by Hans Christian Oersted is often attributed not to a clever experiment but to an accident. While manipulating a current-carrying wire next to a compass, the story goes, Oersted noticed that the needle moved. And presto, there was electromagnetism. This effect is so easy to produce that Oersted could have encountered it accidentally, but that is not how it happened.[1] Instead he set out with his long-held expectation of a relationship between electricity and magnetism, and with simple apparatus he found it.

Hans Christian Oersted (1777–1851) was born in Denmark, eldest son of an apothecary.[2] Overwhelmed by domestic duties, his parents placed him and a younger brother with a German couple. In 1794, having learned German, some French, and some Latin, Oersted returned to Copenhagen and entered the university. While earning a degree in pharmacy, he also studied philosophy and an array of physical sciences.

Oersted worked briefly in a pharmacy, then, in 1801, resumed his education by visiting major centers of scientific research in Germany. Not only did he witness firsthand the feverish activity occasioned by Volta's invention of the battery, he also learned about the latest philosophic fashion, *Naturphilosophie*, from some of its chief exponents. This trip was pivotal in forming Oersted's conceptual framework, for he would eventually wed *Naturphilosophie* to galvanism in electromagnetic research. He returned to Copenhagen in 1804 and two years later was appointed to a professorship in physics at his alma mater, where he gradually built a reputation as a careful experimenter.

Like all philosophical doctrines, *Naturphilosophie* came in many flavors. Unconcerned by these nuances and rejecting the anti-empiricism of its extreme adherents, Oersted adopted *Naturphilosophie*'s most fundamental premise: that underlying the various phenomena of the universe is a unitary, more basic force, and that the seemingly distinctive forces overtly manifest in chemical reactions, the motion of bodies, electricity, and magnetism are merely superficial.

Aesthetically and intuitively, the unity of forces, with its latent spiritualism, was a captivating conjecture. And it still is. Einstein devoted much of his adult life to forging theories for unifying the forces, and today's string theorists are driven by the same aim—and claim, not without strident detractors, to have completed Einstein's grand project.

In Oersted's time, one could summon in support of the unity conjecture several well-known relationships among the forces. Electricity, for example, produced heat, light, chemical changes, and mechanical action; it even affected living things. Through friction, mechanical force generated heat and, in the striking of a gunflint, gave off light. And some chemical reactions, such as combustion, yielded light and heat. On the other hand, gravity created only mechanical effects, and the purported effects of magnetism beyond attraction and repulsion were disputed. Moreover, some rather distinguished natural philosophers, including Charles Coulomb and Thomas Young, had ruled out a relationship between magnetism and electricity, even though it had been known since before Franklin's time that lightning could wreak havoc on a compass needle and sometimes reverse its polarity. Despite these inconsistencies, Oersted retained a fervent belief in the unity conjecture. As early as 1812 he had compared the effects of electricity and magnetism. Although acknowledging differences, he asserted that the "similarities between magnetism and electricity are so great that we need only remove the apparent contradictions in order to accept the identity of the forces in them."[3] He also suggested that investigators learn whether galvanism had any effect on a magnet.

As a philosophical speculation, the unity conjecture offered no empirical predictions for showing the identity of the forces common to electricity and magnetism. Investigators who wished to examine this issue in the laboratory had to be creative and cobble together appropriate apparatus. Although teaching duties and other responsibilities left Oersted scant time for original research, like many professors in that position today he occasionally combined research and teaching.

One day in April 1820, while lecturing to a group of advanced students, Oersted tried an experiment. He had already assembled apparatus: a galvanic battery and a fine platinum wire placed just above a compass. He expected that the wire, heated by the passage of galvanic current, would cause the compass needle to deflect. As predicted, when Oersted connected the wire to the battery, the needle moved slightly. Soon he learned that there was no need to heat the wire, for galvanic current alone produced the effect.[4]

Why hadn't Oersted's classroom experiment, involving the simplest manipulations of a few artifacts found in every physics laboratory, been done before? Investigators, I suggest, probably believed that any magnetic effect of electricity would be small, and so could not appear without a high-intensity discharge from an electrical machine or a Leyden jar. But, as Oersted showed in later experiments, only a galvanic battery—

even one with a single cell—could produce magnetism. And his attempts to discern any magnetic effects from the discharge of Leyden jars were utter failures. No doubt others had tried the same experiment, obtained the same result, and gave up.[5] Oersted concluded that it was not intensity but quantity—the ample flow of current—that generated magnetism.

After experimenting for several more months to refine his findings, Oersted published a brief paper that had a galvanizing effect on scientific men.[6] Like Volta's report on batteries, Oersted's announcement of electromagnetism set a new agenda for droves of enthusiastic researchers around the Western world. And why not? This was a fresh field that could be entered without lengthy preparation and required no expensive apparatus, and surely its first cultivators would discover additional effects and garner renown. Electromagnetism was so new and so startling that even established researchers, some with no experience in electricity, dropped what they were doing, bought or assembled galvanic batteries, and began experimenting.

⁓⁓⁓

Electromagnetism proved irresistible to several mathematical adepts in France and Germany who may have been looking for new physical phenomena to quantify. The most successful of these was the Frenchman André-Marie Ampère (1775–1836).

A self-taught natural philosopher and mathematician, Ampère eventually held a series of university posts and taught several subjects, including astronomy, philosophy, and mathematics. Although undeniably brilliant, Ampère's contributions before he took up electromagnetism had not elevated him to science's highest ranks. "Had he died before September [of 1820]," one historian has noted, "he would be a minor figure in the history of science."[7] His work on electromagnetism made Ampère a man respected throughout the scientific world, and in 1824 he was awarded a prestigious professorship at the Collège de France.

At a meeting of the Paris Academy of Sciences in September 1820, François Arago announced Oersted's achievement. Academicians, who typically reacted with reserve toward discoveries made outside France, were highly skeptical. And these French academicians had also been schooled in Coulomb's authoritative denial that electricity and magnetism were related. However, Ampère accepted Oersted's claim and embarked on experiments to establish the laws of electrodynamics—a term he applied to electricity moving in a conductor.

In only a few weeks, Ampère announced his first laws. And within a few years, his ponderous publications established the new science of electrodynamics, a science built around the theoretical premise that magnetism—including that of the earth and of permanent magnets—was merely electricity in motion.[8] Ampère, steeped in the tradition of French mathematical physics, which had blossomed in the mid eighteenth century after the belated embrace of Newtonian mechanics, eventually expressed his new science in the language of mathematics.

In the short run, Ampère's formulations had little effect on the development of electrical technology, for his writings were opaque to people untutored in trigonometry and calculus. Fortunately, researchers could achieve a qualitative appreciation of electrodynamic effects by copying Ampère's elegant devices and repeating his experiments.[9] One of these artifacts would be his greatest contribution to other inventors of electrical technology: the solenoid, a wire wound in a loose spiral, which Ampère named after the Greek word for pipe or tube.[10] The solenoid helped Ampère illustrate a number of effects. For example, by contracting when the current flowed, the solenoid showed that parallel wires carrying current in the same direction attract each other, and the overall magnetic force is increased. Ampère also noted that in exhibiting polarity a solenoid behaves just like an ordinary permanent magnet.

The solenoid, later called a coil (or helix), was the first of several crucial inventions, made by experimenters in three nations, that would result in an electromagnet capable of giving rise to motors, telegraphs, and countless other technologies. In the meantime, the solenoid served adequately to exhibit the intensification of magnetic force; it could also be used to magnetize compass needles.

The discoveries and demonstration devices of Oersted, Ampère, and other early workers led to the first instrument for indicating the strength of a galvanic current. Named by Ampère, the galvanometer was essentially a compass needle, pivoting on a vertical spindle that was placed in or near a coil of copper wire.[11] Depending on the strength of the current coursing through the coil, the needle deflected to a greater or lesser degree; whether it moved left or right was determined by the current's direction (figure 3.1).

Instrument makers, knowing that experimenters had lacked a convenient, reliable way to indicate current strength, seized the opportunity to bring a variety of galvanometers to market. Their faith in this instrument's marketability was justified, as the galvanometer became a staple in electrical researchers' laboratories. Late in the nineteenth century, one could buy sophisticated galvanometers graduated in the standard unit of current, fittingly named the ampere.

It is a commonplace in histories that Michael Faraday (1791–1867) invented the electric motor.[12] He did make a rotating electrical device, but it was not the first—and it was not a motor.

Faraday was born into humble circumstances near London, a blacksmith's son. He had little formal education, but was apprenticed for seven years to a bookbinder.

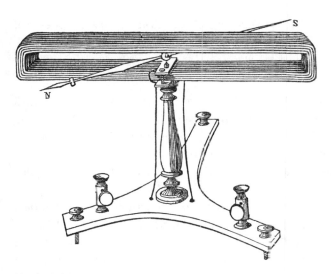

Figure 3.1
An early galvanometer. Source: Appleton 1866, figure 1180.

Faraday, who possessed immense intellectual curiosity, took advantage of the volumes in his midst to delve into countless subjects, including electricity. Fascinated by this mysterious force, he built his own electrostatic generator and conducted textbook experiments. In 1810, extending his scientific education and his social network, he joined the City Philosophical Society, a group of young men who assembled weekly to hear lectures on science and watch demonstrations. At one of these sessions, Faraday had his first exposure to a battery.

Although Faraday found electricity interesting, he became far more enamored with chemistry after reading Jane Marcet's *Conversations in Chemistry*. Like many other Londoners, Jane Marcet had been captivated by Humphry Davy's lectures at the Royal Institution. In *Conversations in Chemistry* she shared her enthusiasm for chemistry with a non-technical audience. Faraday came to believe that chemistry was the foundational science, capable of revealing deep truths about varied physical phenomena.

The paths of Faraday and Davy would soon cross. The opportunity came in the form of an unexpected act of kindness from one of the bookbinder's customers: a gift of tickets so that Faraday could attend Davy's lectures.

Faraday's bookbinding apprenticeship ended in October 1812. Rather than practice that trade, he took a job as an assistant to Davy, whose experiments with an explosive compound had caused blindness. Fortunately the blindness was temporary, and Davy was soon able to read Faraday's notes of his own lectures. Impressed, Davy added Faraday to the permanent staff of the Royal Institution. After accompanying Davy on a lengthy trip to the Continent, where he met many scientific luminaries, including

Ampère and Volta, Faraday returned to the Royal Institution to take up chemical research under Davy's tutelage.

‿‿‿‿

The many experiments that followed Oersted's discovery of electromagnetism led to a torrent of articles submitted to scientific journals. Bewildered by the contradictory theoretical and experimental claims, Richard Phillips, editor of the *Philosophical Magazine*, asked Faraday to make sense of it all. And so Faraday, in September 1821, took up the study of electromagnetism, aiming to build apparatus and conduct experiments, old and new.

Following Oersted, Ampère, and others, Faraday experimented with a compass needle and a current-carrying wire, carefully noting the needle's movements as he approached the wire from different directions. He affirmed that the current-carrying wire and the compass needle affected each other and that the forces around the wire were circular. To demonstrate this mutual interaction convincingly, he built a small device consisting of a wire crank placed vertically and supported at the top and the bottom.[13] With a powerful battery invented by Robert Hare, a professor of chemistry at the University of Pennsylvania, Faraday passed a large current through the crank. When he brought the pole of a magnet close to the crank, the latter revolved smartly until it hit the magnet. Were it not for this impediment, Faraday surmised, the crank would have continued to rotate.

Not long after the crank demonstration, Faraday commissioned an instrument maker named Newman to assemble a more complex two-part apparatus. In one part, a stationary wire was suspended in a container of mercury. A pencil-shaped magnet was inserted most of the way into the mercury and attached below to a ball-and-socket joint, enabling it to revolve freely. The other part was similar except that the magnet was fixed and the wire, dipping into the mercury, was able to rotate. When the circuit of the latter was completed, Faraday observed, "the wire revolved so rapidly round the pole that the eye could scarcely follow the motion."[14] With this elegant apparatus, Faraday called attention to a seemingly magical effect: a current-carrying wire would revolve continuously around a magnetic pole as long as the current flowed.

There is no inkling in Faraday's writings that he considered his invention to be anything but a device to exhibit electromagnetic rotation. It performed this function well, and so instrument makers offered it in their catalogs.[15] Decades later, however, writers on electricity began claiming that Faraday's invention was the first electric motor, an erroneous claim still repeated today.[16]

If converting electricity into motion—rotary or otherwise—is an electric motor's most basic performance characteristic, then Faraday was a latecomer to the invention, since artifacts with that capability had been built by Franklin and others nearly a century earlier. Their rotating devices effectively displayed certain electrostatic effects.

Further, if by "motor" we mean a technology that people at the time believed was capable, at least in principle, of doing work beyond impelling its own motion, then Faraday's rotating device does not qualify. Franklin suggested that one of his motors could be used as a spit to roast a large fowl, and in the 1770s James Ferguson made cardboard models of a grist mill and a water pump driven by electrostatic motors.[17] However, neither Faraday nor others envisioned such applications of his demonstration device. And, significantly, in the 1820s no one called it a motor or engine or prime mover.

The most telling objection is that the first electromagnetic motors, invented in the early 1830s, did not depend on a magnetic pole rotating continuously around a current-carrying wire (or vice versa). Rather, they were based on discontinuous electromagnetic effects created by alternating poles. Faraday's device was not their technological ancestor. In short, beyond serving admirably to illustrate electromagnetic rotations and to stimulate the invention of similar devices, Faraday's invention was a conceptual and technological dead end.[18]

Paradoxically, Ampère's much simpler solenoid was the starting point for developments that did eventuate in motors. Transforming Ampère's solenoid into an electromagnet capable of functioning in motors and other devices entailed, as we shall soon see, major design modifications. More than that, this transformation also required conceptual remodeling: inventors had to regard the electromagnet not as a self-contained display device but as one component in a more complex technology that could do work in the world.

⌁⌁⌁⌁⌁

Although William Sturgeon (1783–1850) did invent an early electric motor, he is mainly remembered for altering the solenoid and creating, according to many writers, the first electromagnet.[19] An Englishman, Sturgeon briefly apprenticed as a shoemaker, his father's sometime trade, but it was not to his liking. He spent nearly two decades in military service, acquiring mechanical skills, studying natural philosophy, and experimenting with electricity. After leaving the service, he became a bootmaker. However, in 1824 he embarked on a career as a science lecturer, which began at the Royal Military College of the East India Company. There Sturgeon modified the solenoid.

In surveying the various artifacts available for exhibiting electromagnetism, Sturgeon was struck by certain performance deficiencies. In particular, the devices were so small that they were difficult to use, and only people sitting in front rows could see their effects easily. And the batteries were large, unwieldy, and expensive to maintain.[20] His solution was to design "sufficiently large and powerful" electromagnetic devices that nonetheless could draw current from smaller, more manageable batteries.[21]

Sturgeon's radical redesign of electromagnetic apparatus was made possible by his discovery that inserting a bar of soft iron into a solenoid increased its magnetism dramatically. He also learned that when the current was cut off, the iron immediately lost its magnetism (in contrast to steel, which became a permanent magnet after passing through the solenoid). Taking advantage of these effects, Sturgeon incorporated an iron core into many ingenious devices of his own design.

In 1825, Sturgeon submitted a set of his new artifacts to the Society of Arts, boasting that by "acting on the principle of Powerful Magnetism and feeble Galvanism, [they] will, I trust, be found more eligible and efficient than any other [set] that has yet been brought before the public."[22] Society officials concurred, and awarded him 30 guineas and a large silver medal. By making this generous award, the society had affirmed Sturgeon's belief that his creations excelled in illustrating electromagnetic effects.

Among Sturgeon's assemblage of clever contrivances, all of which incorporated electromagnets of various shapes, one was a real eye-catcher. Not only was it powerful enough to support "a heavy bar of iron," it also had the familiar form of many ordinary permanent magnets.[23] This visually compelling design consisted of a soft iron rod, bent into the shape of a horseshoe, on which Sturgeon had wound a small coil of copper wire (figure 3.2). In every respect the coil itself was still a solenoid, made with bare wire, its turns somewhat separated. To insulate the solenoid from the iron, Sturgeon coated the core with shellac. The ends of the wires could be inserted into tiny cups of mercury, connected to the battery, when it was time for the electromagnet to perform. Though it was merely one among many in Sturgeon's award-winning

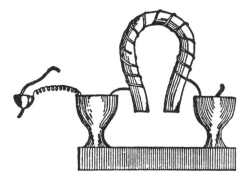

Figure 3.2
Sturgeon's horseshoe-shaped electromagnet. Source: Iles 1904, figure 32.

collection of 1825, this modified solenoid is singled out as the first electromagnet—a status not undeserved, because others took it as the starting point for their own experiments.

Instrument makers, believing that lecturers would find the performance characteristics of the horseshoe electromagnet irresistible, added it to their catalogs; Watkins and Hill, for example, offered it in 1828.[24] Sturgeon did not patent his electromagnets, for even at this late date patenting something sired in natural philosophy would have been regarded as unseemly. A few years later, however, that courteous culture receded into the background when it became clear that some electrical technologies, especially the telegraph, might prove highly profitable.

By buying or copying Sturgeon's horseshoe electromagnet, experimenters and lecturers also affirmed its practicality for their activities. This design was especially embraced because it afforded opportunities for spicing up demonstrations, both visually and acoustically. The lecturer placed a straight iron bar—sometimes called a keeper—across the horseshoe's open end. By suspending weights from the keeper, lecturers could readily materialize, even quantify, the magnetic force created by galvanism. When the current was disconnected, the keeper fell to the floor with an impressive thud. Soon experimenters in many nations would be in a race to see who could make the strongest electromagnet. An American, Joseph Henry, would be at the head of the pack.

4 An American Physicist

In some respects Joseph Henry's long and intensely productive life (1797–1878) echoed Benjamin Franklin's. Of humble origins, neither man attended college. Yet both men—brilliant, ambitious, energetic, and self-disciplined—attained a high level of learning. In their eras, Franklin and Henry became America's most celebrated physicists, their discoveries and inventions highly regarded throughout the scientific world. After achieving prominence in science, they served the nation for decades out of a deep sense of duty and left a lasting imprint on public policy and American life.

Unlike Franklin, Henry carried out research, not as an enthusiast made wealthy by his businesses, but as a teacher, a professor, and the first Secretary of the Smithsonian Institution. In these positions, Henry contributed greatly to defining and professionalizing the role of the scientist in America and to enhancing the scientist's power to influence technological development. Through Henry's efforts, scientists increasingly assessed the technical feasibility of inventions proposed by others.

Henry spent his youth in New York State. From his mother he received a strong dose of Presbyterian Calvinism that shaped his character and approach to his life's work. As he put it just four years before his death, "the certainty of having a just tribute paid to our memory after our departure is one of the most powerful inducements to purity of life and propriety of deportment."[1]

Henry received an elementary education in Albany and Galway New York.[2] He served an apprenticeship in Albany to a watchmaker and silversmith, but his passion was the theater. Nonetheless, through reading he discovered the allure of natural philosophy and determined to return to school. Although much older than the other students, he was admitted to the Albany Academy in 1819. Ostensibly a mere secondary school, the Albany Academy had a rigorous curriculum, ranging from Latin to trigonometry; it "was said to be a college in disguise."[3]

Before completing his coursework in 1822, Henry took a year off and worked as a tutor to earn money for tuition. The Academy's principal, T. Romeyn Beck, was so impressed by this able and hard-working young man that he hired him periodically as an assistant to lecture on chemistry. During the mid 1820s Henry also took assorted jobs as a surveyor, but being in the good graces of Beck paid off. In 1828 Henry became the Albany Academy's professor of mathematics and natural philosophy.

_⌒⌒⌒_

Like Ampère and Sturgeon, Henry began his experiments, not with grand visions of new machines built around electromagnets, but with the hope of creating new knowledge. Nonetheless, Henry did put an electromagnet to work in the commercial world. In the decades ahead the electromagnet, embodying Henry's design modifications, would become the main electrical component in countless technologies. It all began with Henry's alterations of Sturgeon's horseshoe electromagnet.[4]

Henry generally accorded others credit for their contributions to his projects, expecting that his would likewise be acknowledged. Accordingly, he recognized Sturgeon for showing that an iron core markedly increased the strength of an electromagnet, which could then be energized by a very small battery. Sturgeon's magnets were in Henry's words "more convenient" than earlier designs for research and lecturing.[5] Nonetheless, Henry believed he could achieve greater gains in strength by employing Professor Johann Schweigger's "galvanic multiplier," a galvanometer-type device that contained several turns of copper wire.[6] Like Ampère, Schweigger had found that multiple turns increased the magnetic effect.

To make his electromagnets, Joseph Henry adopted Sturgeon's soft iron core, adapted the multi-turn feature of Schweigger's multiplier, and added a few new wrinkles. Winding a solenoid-type coil with superimposed layers presented an obvious problem since the wires, pressing against each other, would produce short circuits. Henry found that by covering the wire with an insulation of silk thread, he could wind the coil in tight turns, which multiplied the magnetism. In one early horseshoe electromagnet, he used an iron rod one-half inch in diameter and 10 inches long, wound with 30 feet of insulated wire. Energized by a single galvanic cell with a zinc plate of just 2.5 square inches, it hoisted 14 pounds. He also found that by winding two separate coils on the core, the lifting power was doubled.

Curious to learn if there was a limit to how much magnetism a small battery could arouse in soft iron, Henry made a much larger horseshoe. Containing nine separate coils (each having 60 feet of wire) wound on a thick iron core, it was 9.5 inches high and weighed 21 pounds. With one galvanic cell whose zinc plate measured 0.4 square feet, the electromagnet held an astonishing 650 pounds, and with a larger battery 750 pounds. Henry cautiously boasted that this electromagnet was "probably . . . the most powerful magnet ever constructed."[7] Indeed it was. When using this giant electromagnet in classroom demonstrations, Henry suspended from its keeper 600 pounds of iron weights that, after the current ceased flowing, fell to the floor "with a great noise."[8]

Henry also explored the opposite extreme, fashioning a diminutive electromagnet weighing a mere 6 grains (about 0.39 gram) that lifted 420 times its own weight. Before that time, the strongest tiny permanent magnet was believed to be the one that Isaac Newton carried around on a ring. But this legendary magnet was no match for Henry's, for it suspended a mere 250 times its own weight. From this comparison Henry drew a far-reaching generalization: electromagnets can be made stronger than permanent magnets of steel.[9] Decades later, this generalization would be exploited in building the first dynamos.

At this time, Henry was trying to increase two strength-related performance characteristics: the absolute amount an electromagnet could lift and the ratio of that amount to its weight. In the years ahead, Henry and others would burnish their reputations as electromagnet designers by pushing these performance characteristics ever higher. Significantly, in the 1820s and the early 1830s no one was asking how much it would cost to purchase a battery powerful enough to sustain an electromagnet's force. One reason is that with the batteries of that time, the maximum force could be exerted for only a few minutes, at most. Even so, Henry was explicitly trying to get the strongest electromagnetic effect from the smallest—and thus least costly—battery.

_⌒⌒⌒_

Henry wasn't the only person making large electromagnets in the late 1820s. A Dutch researcher, Gerard Moll, had seen a Sturgeon-style electromagnet demonstrated in London by an instrument maker named Watkins; it had lifted 9 pounds. Using this magnet as a model, Moll made electromagnets of much greater strength.[10] He was quite successful: his most forceful magnet, an iron horseshoe weighing 26 pounds, lifted 154 pounds.[11] Thinking about applications, Moll suggested that his electromagnets could magnetize steel bars, and might also make it possible to conveniently remagnetize a ship's compass needle that had been affected by lightning.

In the fall of 1830, before Henry had written up his work on large electromagnets, Moll published a paper on his own experiments in a respected European journal. Unhappy that he had been scooped somewhat, Henry immediately wrote to Benjamin Silliman, a distinguished man of science at Yale College and editor of the *American Journal of Science and Arts* (known colloquially as *Silliman's Journal*).[12] He told Silliman, with whom he had socialized the previous spring in New Haven, about his new researches in electromagnetism, and inquired if a paper reporting his findings could appear in the next issue of the journal. What Henry did not know yet was that the next issue had been printed already and that it included a second paper on large electromagnets by Moll.

Silliman was sympathetic to Henry's plight. Like many American natural philosophers, Silliman believed—with ample justification—that American research was not taken seriously in Europe. Worse still, American discoveries and inventions were on occasion repeated and renamed by Europeans, who unfairly appropriated credit. Even so, in America there was little cutting-edge research in physics going on that was worthy of international attention. Thus, Silliman was especially eager to publish articles that broke new ground and might elevate the visibility of American researchers and add luster to his journal. In a letter delivering the disappointing news about Moll's new paper, Silliman promised Henry that his article would be published as an appendix in the very same issue.[13] This extraordinary move testifies to Silliman's admiration for Henry and to his commitment to promoting American science.

Henry's article describing the new electromagnets appeared in *Silliman's Journal* early in 1831. Anxious to establish scientific priority for his discoveries, Henry pointed out that he had carried out the reported experiments two years earlier and had published an even earlier work in 1828.[14] However, perhaps because Henry's paper had "been written in great haste" and rushed into print, he did not take care to distinguish his work from Moll's; yet the differences were substantial.[15] Moll, wanting to build stronger electromagnets, had merely scaled up Sturgeon's design, using ever larger iron cores and winding uninsulated wire, loosely, in one layer; and he showed no interest in strength-to-weight ratios. Not one feature of Moll's magnets, except gross size and the use of different metal wires, differed from Sturgeon's original design. Moreover, unlike Henry and even Sturgeon, Moll did not seek to economize on the battery; Moll's largest electromagnet—the one that held 154 pounds—required a galvanic cell whose zinc plate measured 11 square feet.

In these early years, the performance characteristic most important for research, and for dramatic demonstrations, was an electromagnet's lifting power. Readers could see that Henry's electromagnets performed much better than Moll's. An electromagnet that could suspend 750 pounds was the strongest magnet of any kind in the world. This feat in itself was probably sufficient to earn the admiration and respect of European researchers. And that it could be done with a small galvanic cell would have seemed nearly miraculous.

Clearly, by juxtaposing the Moll and Henry papers in the same issue, Silliman anticipated that Henry would become recognized as a first-rank researcher in electromagnetism. By the end of 1831, Henry's approach to designing electromagnets was being extolled in journals in England and Switzerland.[16] And in 1835 the French instrument maker Pixii offered a Henry-style electromagnet capable of lifting 200 kilograms for 100 francs, including battery.[17] Looking back from 1891, Franklin Pope justly remarked that the contents of Henry's paper "threw the scientific world of both hemispheres into a paroxysm of amazement."[18]

As word spread about Henry's dramatic "improvements" in electromagnets, other researchers and lecturers began to make them or have them made. Henry himself made magnets and sold them to colleagues at other colleges. Late in 1830 he offered to oversee construction of a large magnet for Benjamin Silliman at Yale. This electromagnet, which Henry promised to make for around $30–$35, would be capable of supporting 1,000–1,200 pounds—an appreciable increment above his previous record.[19] Silliman gave Henry the go-ahead, but stipulated that he would need the monster magnet before February 26, the last day of classes at the medical school.

Henry and Philip Ten Eyck (his colleague and sometime collaborator at the Albany Academy) began work on January 15, but missed the deadline. Despite the delay, Silliman was pleased when the magnet finally arrived. It greatly exceeded expectations, suspending 2,063 pounds with a galvanic cell containing just under 5 square feet of zinc. Although Yale's medical students would miss the magnet's awesome show of strength, Silliman could still use it in his lectures at the college. The plan was to employ a system of pulleys to suspend several men, adding one man at a time until the magnet let go. Although making this magnet ended up costing Henry around $45–$50 for materials and skilled labor, he kept his promise and billed Yale only $35. Gently protesting, Silliman sent him $40 and suggested he use the extra $5 to support new experiments.

The "Yale magnet," as it became known, was described in a brief note in *Silliman's Journal*.[20] It achieved international fame and, not incidentally contributed to Henry's growing stature in the scientific community. For decades the Yale magnet was exhibited in the National Museum of American History at the Smithsonian Institution, a symbol of electrical progress and national pride (figure 4.1).[21]

In building other large electromagnets, Henry made a number of design changes in insulation. For example, the magnet he made for Professor Parker Cleaveland at Bowdoin College had silk cloth between each layer of wire.[22] In addition, Henry had first covered the wire (about 1,000 feet) with a coating of shellac and mastic. The process of winding the coil on the 80-pound iron horseshoe "was a very tedious one and occupied myself and two other persons every evening for two weeks." Henry's out-of-pocket expenses were $46.13, but he charged Cleaveland only $45. As in the case of the Yale magnet, Henry donated his labor willingly because of the "additional knowledge and experience" he acquired along the way.

Why was Joseph Henry, teacher and researcher, commercializing his electromagnets without any interest in obtaining a patent or profit? Beyond gaining new 'knowledge and experience" (an obvious but, I believe, incomplete explanation), fabricating large, custom-made electromagnets for college professors was a generous act in support of science. After all, in the very early 1830s Henry alone possessed the requisite know-how to build large electromagnets. But surely Henry also knew that by bestowing favors on senior scholars, some of whom were in prestigious positions, he was building a social network and accumulating obligations. More than that, each electromagnet silently bespoke Henry's contributions to the science of electromagnetism as well as

Figure 4.1
The Yale electromagnet. Courtesy of Smithsonian Institution, negative 13346.

his technical expertise. These artifacts were, literally and symbolically, *Joseph Henry* electromagnets. Henry's sales, I suggest, were a kind of academic self-promotion—an investment in his future standing, not fundamentally different from publication, passing out reprints, and reading papers at conferences. That investment would soon return handsome dividends.

Not all of Henry's electromagnets were destined for college laboratories and lecture halls. In the spring of 1831, he took on a project for Allen Penfield and Timothy Taft, owners of an iron mine in the Crown Point area of New York State.[23] In their operation, magnets were used to separate iron-bearing minerals from the crushed ore. Hundreds of small permanent magnets, attached to a rotating wooden drum, captured the magnetic iron and allowed the remaining waste material to pass through. Over time, however, these magnets lost their strength and had to be revived by contact with another permanent magnet. Penfield and Taft believed that there must be a more effective remagnetizing process. Consulting with Amos Eaton of the Rensselaer School (later Rensselaer Polytechnic Institute) in Troy, they learned about Henry's electromagnets. They purchased from Henry an electromagnet that weighed, according to one estimate, about 3 pounds. With a two-cell Hare battery, it easily lifted a 100-pound anvil, and thus it had more than enough power to magnetize small pieces of steel. Some historians believe that Penfield and Taft's use of this electromagnet at their iron works "may very well have been the first industrial application of electricity," and I have encountered no earlier example.[24] However, it is not clear whether the proprietors of the iron works found the electromagnet indispensable; in fact, in late 1833 they sold it.[25] In the hands of the buyer, Thomas Davenport, this precious artifact would play a pivotal role in the invention of a rotary motor.

ˌⁿⁿⁿⁿˍ

Henry himself had invented a motor-like device, which he reported in *Silliman's Journal* in 1831.[26] Rather than producing rotary motion, its armature rocked like a teeter-totter. Two permanent bar magnets were oriented vertically on a small wooden platform (figure 4.2), with the north poles at the top. Between the bar magnets was a stand on which the armature—an electromagnet with a set of wires at each end—could rock like the beam of an early steam engine. Either pair of wires supplied current to the electromagnet when it dipped into a pair of tiny mercury-filled cups attached to a battery's electrodes. When the electromagnet was energized, its north pole was

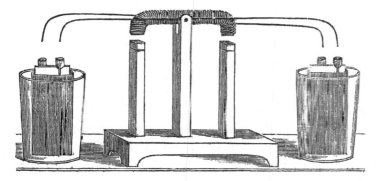

Figure 4.2
Henry's rocking-beam motor. Adapted from Henry 1831b: 342.

repelled by one bar magnet while its south pole was drawn toward the other. As the armature rocked, it withdrew one pair of wires from the cups and inserted the second pair on the other side. This immediately reversed the poles of the electromagnet, which then rocked in the other direction. The motion continued as long as the current flowed.

Henry regarded his little electric teeter-totter, which rocked about 75 times a minute, as "a philosophical toy," a phrase that meant at the time a device suitable for illustrating scientific principles.[27] It fulfilled that display function exceedingly well, exhibiting the principle that a machine's motion could be sustained when that very motion reversed the electromagnet's polarity. This would become the essential design principle of nearly all direct-current electric motors, rocking or rotary. In a few years, any mechanical arrangement that reversed an electromagnet's poles would be called a "pole changer."

Henry patented neither his rocking-beam motor nor his pole changer. Nonetheless, aware that his invention in its present form was incapable of doing any work, he cautiously staked out some technological territory: "... it is not impossible that the same principle, or some modification of it on a more extended scale, may hereafter be applied to some useful purpose."[28] Soon that very principle would animate an unimaginable variety of electric motors fashioned by people on both sides of the Atlantic, many of them having been inspired by Henry's demonstration device. Within 10 years, ingenious "electromagnetic engines" would in embryonic applications drive a toy train, turn wood on a lathe, and propel a boat upstream. Surveying this profusion of motors, Henry would insist that they were all merely variations on the scientific theme he had composed in 1831, and that he alone had invented the electric motor. This claim was consistent with Henry's definition of technology, expressed in an 1829 curriculum of the Albany Academy, as "the application of Natural Science to the Useful Arts."[29]

⎯⌒⌒⌒⎯

Joseph Henry did not think of himself as an inventor; few natural philosophers did. Yet physicists and chemists, among others, were prolific inventors. In fact, inventors abounded in the nineteenth century, a time when creative people, often not in the employ of large companies or other institutions (not initially, at least), contrived many new technologies—especially electrical ones.

After the War of Independence, many Americans, mostly mechanically inclined young men, came to view inventing as an activity that might yield riches. Although its rewards were potentially great, inventing was a risky endeavor that could lead to ridicule, disillusionment, and poverty. Evidently, inventors who lacked the backing of companies, colleges, or governments had to be gamblers, willing to sacrifice material comforts in the present for future rewards that might never come. The independent inventor—a social role more prevalent and better defined in the United States than elsewhere—would survive into modern times. Independent inventors would create important technologies, from xerography to the Polaroid camera, but the nineteenth century was their heyday.[30]

Henry occasionally received inquiries from inventors wanting to make electromagnets. In replying to a friend who had written on behalf of his son, Henry alluded to his recent publications on electromagnets, briefly mentioned the Yale magnet, and referred to electromagnetism as "a new department of science."[31] Only someone who had knowledge of electromagnetic principles, he cautioned, would be able to make such magnets. Henry hoped that his friend's son was not planning to make a perpetual-motion machine, since building one "on any mechanical or magnetic principle" was impossible. Knowledgeable inventors knew that working on perpetual motion was a sheer waste of time.

Henry believed strongly that invention, as the creation of new technology, was the straightforward application of scientific principles, and so he lacked patience for inventors who had not become steeped in the science relevant to their subject. During the nineteenth century, this view became a veritable ideology among scientists in their efforts to accrue social power and differentiate themselves from mechanics and inventors. Even *Scientific American*, which served an audience composed mainly of mechanics and inventors, adopted this ideology, repeating the science-begets-technology mantra as it debunked impossible schemes and championed a program of "scientific invention."

Although historians of science and technology as well as anthropologists soundly reject the notion that invention is simply applied science, this position, promoted by Henry and others, was widely adopted in the nineteenth century and survives today. That is why many people believed that a man of science with relevant expertise could determine if an invention's operation was technically feasible—i.e., consistent with known scientific principles. If the answer was no, then the invention could not work as claimed. On that basis, it was possible to rule out all purported perpetual-motion machines, no matter how ingenious. This role for scientific authority was not in fact a new one. In some European nations, people had been submitting their inventions for assessment by natural philosophers since the late seventeenth century. In the United States, scientific authority was vested in interested individuals or in committees appointed to sort out conflicts until the Patent Office was given statutory responsibility for evaluating inventions submitted for patents. However, in the middle of the nineteenth century the Patent Office often lacked the expertise needed to assess every application's technical feasibility.

In practice, judging technical feasibility could be far from straightforward. For example, scientific authorities might disagree about which principles to apply when assessing a given invention. Moreover, accepted principles sometimes turned out to be wrong. And what if an invention was based on principles not yet known to science? If any of these conditions prevailed, conflict might erupt among men of science. Indeed, while still at the Albany Academy, Henry had to overturn an accepted principle of electricity in the process of showing how to send current over long wires, as was required by the electric telegraph.

5   Telegraphic Visions

By 1830, "telegraph" was a familiar term to many people. Semaphore-type telegraphs had become an important political technology that helped several European governments acquire information rapidly from, and exercise power over, areas beyond the capital. France had installed a single, government-controlled telegraph network centered on Paris and used it during the Napoleonic wars.[1] Limited to line-of-sight transmission, semaphore-type telegraphs required many relay stations and personnel. Moreover, they worked slowly relative to the speed of electricity, and most shut down at night and in bad weather. These performance shortcomings helped to inspire visions of electrical telegraphs.

In Franklin's time, natural philosophers had discovered that static electricity could travel long distances through wires, apparently instantaneously. The first proposal for an electric telegraph appeared in 1753 in a letter published in *Scots' Magazine* signed simply "C.M."[2] This proposal was for an electrostatic telegraph that connected the communicating parties by a set of parallel wires, one for each letter of the alphabet. When a charge was sent through the wire labeled "d," for example, a ball at the receiving end would move, indicating the same letter. There is no evidence that C.M., whoever he was, built a telegraph on this plan.

In the 1780s, inventors in several European nations began to assemble prototype electrostatic telegraphs. A Frenchman named Claude Chappe came up with perhaps the most creative system, which used two modified clocks for selecting and displaying numerals.[3] On each clock's face he drew ten wedge-shaped zones and numbered them from 0 to 9. Beginning with perfectly synchronized clocks, the operator at the sending end discharged a Leyden jar when the second hand swept into the zone of the desired numeral. At the receiving end, the operator noted when the charge arrived (perhaps signaled by a bell or an electrometer) and recorded the numeral corresponding to the location of the second hand. Chappe's ingenious system did not go into operation, and he soon turned his inventive talents to mechanical-optical telegraphs.

The most ambitious electrostatic telegraph of the eighteenth century was built by Don Francisco Salvá, a Barcelona physician, who in this effort doubtless enjoyed the patronage of the Spanish crown.[4] Although sources differ on details, the telegraph seems to have required but one wire, some Leyden jars, and a large electrical machine. Extending from Madrid to Aranjuez, the line was 26 miles long and did transmit messages, even carrying news to the king.

Don Francisco's telegraph can be easily dismissed as an inconsequential royal indulgence. However, this is not an isolated case of a government investing in an electrical technology. In subsequent decades, political leaders in Western nations would cultivate certain electrical technologies, including lighting and the telegraph, their assessments of practicality sometimes far more favorable than those of other players. Indeed, inventors can sometimes convince officials of a technology's impending practicality for the government and thereby secure support for creating prototypes. Occasionally, as we shall soon see, governments take the initiative, seeking out and subsidizing inventors in the expectation that they can create a new technology to solve a specific problem, often social, political, or military.

For a variety of reasons, electrostatic telegraphs did not seriously challenge the mechanical systems. One important reason, suggested by John Fahie, a historian of early electric telegraphs, was that developing and installing any large-scale system of electric telegraphy—as opposed to a short experimental line—would have entailed a hefty capital investment.[5] Governments were unwilling to make that outlay for a still experimental and often fickle technology because they already had telegraphs that were, by contemporary standards, serviceable. Moreover, there was a dearth of private financial and entrepreneurial institutions capable of staking such a risky venture. Although electrostatic telegraphs continued to be invented in the early nineteenth century, none was developed very far and none went into regular operation.

_⌒⌒⌒_

As we have seen, after 1800 experimenters substituted galvanic batteries for frictional machines and Leyden jars in virtually every known electrical technology. In this context, it became a near certainty that someone, somewhere, would fashion a galvanic telegraph. Indeed, by 1810 several had been invented on the basis of electrochemical effects.[6] However, it was not scientific curiosity but the Bavarian government that helped to bring forth the most celebrated galvanic telegraph. The military value of telegraphy had been underscored by Napoléon's use of the French semaphore system to turn back an Austrian invasion of the Bavarian capital, Munich. Impressed, the Bavarian minister turned to his friend Samuel Sömmerring, a distinguished professor of anatomy and surgery, for help in creating a telegraph.[7]

Instead of recruiting electrical experimenters, Sömmerring himself plunged into this project. Late in the late summer of 1809, he exhibited an electrochemical telegraph to the Munich Academy of Sciences.[3] His apparatus consisted of 35 insulated copper wires bound into a cable, each wire corresponding to a numeral or a letter. Powered by a six-cell battery, the telegraph transmitted information by decomposing water. At each end of the telegraph was a water-filled trough containing 35 tiny gold pins attached to the 35 wires. By connecting a pair of wires to the battery at the sending end, Sömmerring could generate—depending on the polarity—hydrogen and oxygen at two pins on the receiving end, whose bubbles then indicated the corresponding letter or number. He also rigged up an alarm bell, triggered by a lever responding to the accumulation of gas, that was supposed to sound when a transmission was imminent.

In one version of his telegraph, Sömmerring sent messages over a 2,000-foot cable wound on glass spool, but noted that the decomposing action diminished and so reception was slowed. Consequently, most of his demonstration telegraphs, including those operated in the presence of government officials, had short cables.

Sömmerring tirelessly promoted his invention and even estimated the costs of constructing a full-scale version. Although he received numerous accolades for his handiwork, other people apparently judged his telegraph to be "complex and unpractical," and so Sömmerring was never commissioned to install a telegraph system.[9] Indeed, no electrochemical telegraph was ever adopted.

The technological legacy of Sömmerring's electrochemical telegraph was decidedly limited, for it merely underscored the decades-old finding that electricity could transmit information, more or less rapidly. Yet, like so many other inventions made by natural philosophers, this still-born telegraph advertised to his peers that Sömmerring was very clever.

___⌒⌒⌒___

After Oersted's momentous discovery, several researchers suggested that a telegraph might be based on electromagnetic effects, such as moving a needle at a distance. In 1820 Ampère himself envisioned a galvanic telegraph employing a wire for each letter and magnetic needles as the receiver; current would be directed through the appropriate wire by a keyboard.[10] This kind of telegraph, he believed, would make communication rapid and easy. However, Ampère took no steps to nurture his nascent telegraph; after all, it was merely a vision that grew out of his experiments. Nor did Ampère address the most obvious question that the possibility of an electromagnetic telegraph raised: how far might the transmission of information be extended? Would such a telegraph be fated by performance deficiencies to remain a scientific pipe dream, or might it become the basis of a nationwide communication system judged practical by governments or financiers? This question was largely held in abeyance until 1825, when an Englishman named Peter Barlow addressed it.

Barlow, an instructor at the Royal Military Academy at Woolwich and Fellow of the Royal Society of London, was already a famous mathematician and an expert on magnetism when he looked into the technical feasibility of an electromagnetic telegraph.[11] Indeed, in that same year, 1825, Barlow received the Copley Medal, the Royal Society's highest honor, for his contributions on magnetism. He had also achieved international recognition for devising a compass that was unaffected by the iron on ships. He radiated scientific authority.

Barlow began his experiments with the understanding that an earlier researcher, whom he does not name, had sent static electricity through a wire 4 miles long without any detectable loss of power. But, Barlow wondered, would this also hold true for galvanism? If it did, he reasoned, "then no question could be entertained of the [galvanic telegraph's] practicability and utility."[12] But should there be a diminution of power, he would formulate the law that described it.

Barlow performed his somewhat complex experiment outdoors with a Hare battery and 838 feet of copper wire wrapped around wooden posts that formed a large rectangle. Employing three galvanometers inserted at different points along the wire, he varied the circuit's length and recorded the needle's deflection in degrees. Barlow arrived at a measure of current strength by applying a trigonometric function (tangent) to the needle's angle of deflection, which he then compared to the circuit's length. On the basis of this mathematical analysis, Barlow proposed as his law that current strength was inversely proportional to the square of the distance.[13] This was a plausible result; after all, inverse-square laws were familiar in physics, for they described the force of gravity as well as magnetic and electrostatic attractions and repulsions.

Barlow's law had devastating implications for anyone who might have considered building an electromagnetic telegraph. Transmitted over a long distance, the current would be undetectable. Indeed, Barlow reported, "I found such a sensible diminution with only 200 feet of wire, as at once to convince me of the impracticability of the scheme." Apparently, this distinguished man of science had shown that the electromagnetic telegraph was a technical impossibility, a chimera not unlike perpetual motion.

Having established the relevant law and tendered his expert opinion, Barlow had evidently put the question to rest. However, scientific authority is often resisted by other scientists, who have many incentives—such as pride, prizes, prestige, and promotions—to be creative, to make new discoveries, and, especially, to reevaluate earlier work. Thus, a scientist's power to proscribe other scientists' research activities can dissipate in the blink of an eye. Barlow's law lasted a little longer, but it did temporarily curtail experiments with electromagnetic telegraphs.

The major challenge to Barlow's authority came from across the Atlantic. In the late 1820s, Joseph Henry and his colleague Philip Ten Eyck conducted research bearing directly on the feasibility of an electromagnetic telegraph. In one telling experiment, they measured the strength of current from a single galvanic cell as it coursed through circuits of 530 and 1,060 feet. They found that current strength was inversely proportional to distance—not to distance squared, as Barlow had claimed. Nonetheless, the researchers also learned that an electromagnet at the end of the 1,060-foot wire was energized only weakly by a single galvanic cell. Suspecting that they could overcome the electrical resistance of the long wire with a battery of greater intensity, they turned to a 25-cell Cruikshank trough whose total area of zinc plates was the same as the single cell. That did the trick. The high-intensity battery had more "projectile force" (as Henry termed it, using Hare's colorful expression). As a result, the current retained sufficient strength to produce appreciable electromagnetism.[14]

In further experiments, Henry and Ten Eyck showed that the number of turns of wire in an electromagnet affected the circuit's total resistance. Thus, to activate at a distance an electromagnet with a coil of many continuous turns, one had to employ a high-intensity battery. The need to match a circuit's total resistance with the battery's intensity was a fundamental discovery that would profoundly influence the design of all circuits containing electromagnets. Had Henry been aware of Ohm's Law, he might have cited it in support of his conclusions; however, Ohm's obscure formulations of 1827 were as yet little known to electrical researchers.[15] Henry published his findings in the 1831 paper that had been hurriedly written for *Silliman's Journal*. Perhaps because of his haste, Henry mentioned Barlow only once, without citation, noting that the new experiments, which showed the need for a high-intensity battery, were "directly applicable to Mr. Barlow's project of forming an electromagnetic telegraph."[16] This is a puzzling statement for many reasons, one of which is that Barlow's project was not to form such a telegraph but to assess its technical feasibility. This skewing of Barlow's goal may have resulted from careless wording.

But why didn't Henry directly contradict Barlow's claims and reevaluate Barlow's experiments? One could argue that Henry wanted to refute the great Barlow most delicately. After all, creating powerful enemies in science is not always a wise career move. As a young scientist seeking recognition, Henry might have expected his own experimental results, stated so plainly, to speak for themselves. His failure to confront Barlow by identifying weaknesses in his experiments may have been based on political calculation. Or he may have lacked the time to do a thorough critique; after all, in the same paper he had not engaged Moll's work either. In any event, the finding by Henry and Ten Eyck that an electromagnet could be energized at a long distance with a high-intensity battery, which raised hopes for telegraphy, did not go unnoticed; in fact, Barlow's law and its technological implications were now forgotten.

In the opinion of some who examined the history of the telegraph almost 30 years later, the discovery by Henry and Ten Eyck "was necessary to make the practical working of the electro-magnetic telegraph at considerable distances possible."[17] Also in retrospect, Henry himself claimed that, on the basis of this research, he "saw that the electric telegraph was now practicable."[18]

Henry had even stronger grounds for the aforementioned belief. In 1831 and 1832, he and his students had rigged up successively longer electromagnetic "telegraphs" on the campus of the Albany Academy. In one case they strung a wire many times around the inside of the assembly hall, where Henry lectured. The most ambitious of these lines reached more than a mile. Instead of using a compass as the receiver, Henry employed an intensity electromagnet of horseshoe shape that sounded a small bell. Between the poles of the electromagnet Henry placed one end of a permanent bar magnet, which could swing between the poles. When current passed through the electromagnet, the bar magnet, attracted to one pole of the electromagnet, pivoted smartly and struck the bell. By reversing the electromagnet's polarity, Henry could cause the bar magnet to move to the other pole, and this reset the receiver.[19]

Apparently, Henry did not use his apparatus to transmit information from place to place (there was no mention of a code); rather, he used it to illustrate the extension of electromagnetism's mechanical effects over some distance. It was a display device brimming with further technological possibilities, fully perceived by Henry, but it was not yet a telegraph. Because he had no interest in "abandoning my researches for the practical application of the telegraph," Henry did not develop a communication system ripe for commercialization.[20] He left that project to others, believing that "men of science" were disinclined "to secure to themselves the advantages of their discoveries by a patent."[21] Nonetheless, Henry's students were deeply impressed—and it is their recollections that testify to his handiwork, for Henry never published a word about these experiments until much later, when the invention of the telegraph was disputed in the courts.[22]

In June 1832, Henry received a surprising letter from John MacLean, vice president of the College of New Jersey (forerunner of Princeton University). Anticipating that the chair of Natural Philosophy might soon become vacant, MacLean inquired if Henry would accept the position were it offered. After deliberating for several days, Henry replied in a letter composed with the utmost care.[23] Acknowledging that there were advantages to being in Albany, Henry mentioned his annual salary at the Academy ($1,000) and complained about his current duties: "I am engaged on an average seven hours in a day, one half of the time in teaching the higher classes in Mathematics, and the other half in the drudgery of instructing a class of sixty boys in the elements of Arithmetic." In the next sentence, Henry gave his remarkably understated answer: "If I am not mistaken in the character of your college in the nature of the duties which will be required of me, I think I would be more pleasantly situated in Princeton than I am at present in Albany, and shall therefore accept the chair should your trustees see fit to offer it." Henry went on to admit that he was not a college graduate he also listed people, including Benjamin Silliman, who could testify to his "scientific character." Lastly, he mentioned the results of some preliminary experiments on "drawing sparks of electricity from a[n electro]magnet."

A few months later, Silliman, while passing through Albany, learned from Henry about Princeton's potential offer. Almost immediately he wrote a glowing letter in support of Henry's candidacy: "Mr. Henry [is] a young man of very uncommon talents and acquirements in the departments of knowledge to which he had devoted himself. He has a very Superior mind, active, inquisitive, inventive and ardent . . as a physical philosopher he has no superior in our country: certainly not among the young men. He has the important advantage of being an excellent practical mechanic and if placed in favorable circumstances I doubt not that he will add to the Science of physics and bring forward other discoveries besides the brilliant ones which have already made him extensively and advantageously known to the Scientific world."[24] Although expressing his sincere belief that this young man of science would distinguish himself at Princeton, Benjamin Silliman's exuberant praise also reveals a personal stake in helping advance the career of Joseph Henry—friend, protégé, and maker of the Yale electromagnet.

The job eventually opened up, and in late 1832, with Silliman's strong endorsement, Henry assumed his new post at Princeton.

6 Mechanical Electricity

In light of the unity conjecture and a growing roster of ways to convert one kind of force into another, natural philosophers believed that if electricity could produce magnetism then they might coax magnetism into producing electricity. Such a demonstration would be a triumph on a par with Oersted's great achievement, almost certainly reserving for the discoverer a place in the pantheon of science's immortals. Not surprisingly, many men reached for this ripe fruit, trying to materialize the predicted effect in numerous experiments, but none grasped it cleanly. As Joseph Henry put it, "at first sight it might be supposed that electrical effects could with equal facility be produced from magnetism; but such has not been found to be the case, for although the experiment has often been attempted, it has nearly as often failed."[1] Although scientists rarely report a negative result unless it undermines another scientist's claims, Michael Faraday was a significant exception who, after acknowledging failure after failure, reported in 1831 that he had at last plucked the luscious plum.

A blacksmith's son and a professor at the Royal Institution, Faraday kept a diary of his laboratory activities, its terse entries embellished by rough sketches. More than that, in published accounts of his projects Faraday sometimes described experiments that didn't work as expected. Thus, we can examine his trials on converting magnetism into electricity to learn how he eventually succeeded and perhaps to gain insight into why others failed.[2]

The seemingly simple effect that Faraday at last realized takes on special significance in the present study. That is because instrument makers, grasping its technological possibilities, immediately invented the first electromagnetic generators, whose descendants became the core technology of electric power systems. But during the 1830s inventors of these machines were not looking ahead to applications such as electric power and lighting; rather, they were trying to achieve with this new technology the same effects yielded by galvanism. For inventors, instrument makers, and potential consumers, a practical electromagnetic generator was one that could replace batteries in specific activities: scientific research, public displays of electricity, and electromedicine.

_⌒⌒⌒_

Like other believers in the unity conjecture, Faraday was smitten by the possibility of simply reversing the relationship Oersted had found. Thus, if a current-carrying wire produces magnetism as long as the current is flowing, then the steady force of a magnet ought to induce a steady electric current in a nearby coil. This obvious expectation was almost certainly the starting point—whether implicit or explicit—for Faraday and other researchers, but it turned out to be erroneous. Researchers who merely wrapped a coil of wire around a magnet (or electromagnet) and connected it to a galvanometer got no indication of continuous current.

During Faraday's desultory efforts of the 1820s, undertaken amidst hundreds of chemical experiments, he sought evidence for "magneto-electricity" without success. However, the pace of his research quickened in 1831 after he found out about the large electromagnets that Henry and Moll had made. Beyond perceiving that he had formidable competition in studying electromagnetism, Faraday learned from Henry's work that he could intensify an electromagnet's force by insulating the wire and winding the coil in layers of closely spaced turns. Henry's influence is also evident in the many trials Faraday carried out with coils having separate windings that could be connected in various series and parallel circuits. Using electromagnets in this research was a sensible choice because they could be made in various shapes and were very strong for their size.

On August 29, 1831, in an experiment that gave an inkling that success might be near, Faraday wound separate coils on an iron ring. First he connected one coil to a galvanometer. Then, upon connecting the other coil to the battery, he noticed something surprising: the galvanometer's needle deflected at precisely the instant that he made or broke contact with the battery. However, after oscillating briefly, the needle again became quiescent.[3] Because this fleeting electrical effect is easily produced with a small battery and one or two coils having a few hundred turns, other researchers must already have observed it. Henry, for one, noted that a spark appeared briefly at the point where he disconnected an electromagnet from its battery—his "sparks from a magnet." This phenomenon would later be termed "self-induction," an effect discovered by both Henry and Faraday.[4] However, being fixated on producing a *steady* current, Henry and others attached no significance to this apparently trivial effect. It is a testimonial to Faraday's scientific curiosity and persistence that he fastened on this transitory effect and explored it further. In one setup, he learned that the needle would also deflect when the second coil was moved toward or away from the first. In addition, Faraday found that thrusting a permanent bar magnet into a hollow coil, or removing it quickly, briefly stirred an attached galvanometer. In further experiments with permanent magnets and coils, he obtained a consistent result: only when the magnet was moved briskly in relation to the coil did the galvanometer's needle move. His provisional conclusion was that magnetism did produce electricity, but that it was only "momentary."[5] Perhaps because Faraday was still seeking a steady current, he did not yet report these findings.

Building on the rather puzzling results described above, Faraday embarked on a line of research incorporating apparatus employed by François Arago, a French electrophysicist, who had found in the mid 1820s that a rotating metal disk moved the needle of a nearby compass. Faraday apparently intuited from his earlier efforts—and from Arago's experiment—that electricity might be induced in a metal disk that rotated in the vicinity of a magnet. In looking for this effect, Faraday explicitly sought to "construct a new electrical machine."[6] By that he meant a contrivance that could, like an electrostatic generator, produce electricity continuously through mechanical motion. Generating electricity in this way would require, like an electrostatic generator, the application of an external motive power, such as a human arm. On October 28, 1831, after dozens of trials, Faraday finally found a configuration that worked. This date is sometimes taken as the beginning of the modern electrical age. He began with a copper disk, 12 inches in diameter and about 0.2 inch thick, mounted on an axle held fast in a frame. Next he placed the copper disk vertically between two small iron bars affixed to the poles of an enormous permanent magnet. Owned by the Royal Society, this magnet was probably the strongest in England at the time. Finally, he connected one wire of a galvanometer to the axle and held the other against the rim of the copper disk as it turned. What happened next is best described in Faraday's own words: ". . . the instant the plate [disk] moved, the galvanometer was influenced, and by revolving the plate quickly the needle could be deflected 90° or more. . . . Here therefore was demonstrated the production of a permanent current of electricity by ordinary magnets."[7]

After more experiments, Faraday offered generalizations about the process of turning magnetism into electricity. Substituting other objects for the rotating copper disk, he found that electricity was induced in any metal, whether it had the shape of a strip, a wire, a wire spiral, or a coil with an iron core—as long as the metal had a constant motion relative to the magnet's poles. That is, the moving metal had to cut across the "magnetic curves," which Faraday later termed "lines of magnetic forces."[8] Lines of force could be visualized by sprinkling iron filings on a piece of paper held immediately above a magnet.

Although Faraday's sensitive galvanometer indicated that current was always flowing in his machine, it was weak, for he could not reproduce common effects of electricity, including a sensation on the tongue: "Nor have I been able to heat a fine platina wire, or produce a spark, or convulse the limbs of a frog. I have failed also to produce any chemical effects by electricity thus evolved."[9] And despite several attempts modeled after his copper-disk device, Faraday was not able to build a "magneto" that could generate current perhaps ample to yield these effects.

Faraday presented his findings on "magneto-electricity," both successes and selected failures, in a long paper read to the Royal Society on November 24, 1831. Published the following year in the *Transactions of the Royal Society*, this report was reprinted in major German and French journals, and a synopsis appeared in *Silliman's Journal* in July 1832. Soon the community of natural philosophers on both sides of the Atlantic was abuzz with the exciting news that Faraday had completed the circle of magnetism and electricity that Oersted had entered.

In many publications, Faraday is said to have invented the first electromagnetic generator: his copper disk rotating between the poles of a magnet. For example, the historian L. Pearce Williams claims that "what Faraday had invented was the dynamo."[10] However, like his supposed electric motor that exhibited magnetic rotations, this apparatus merely materialized a new effect. Faraday did not suggest that it could be used in place of an electrostatic machine or galvanic batteries, for it was incapable of producing conventional effects. It was a tool of scientific research—a significant one, but so far nothing more.

Unlike Faraday's device for producing magnetic rotations, his proto-generator did embody the principles on whose basis others would quickly invent stunningly effective magneto-electric machines (later known simply as "magnetos"). Some of these artifacts would even be judged practical by experimenters, lecturers, instrument makers, and electrotherapists for replacing batteries in several activities. And nearly 30 years after this discovery, Faraday himself would oversee the first application of very large magnetos for lighthouse illumination; these installations employed the carbon arc reported by Faraday's mentor, Humphry Davy.

_⁓⁓⁓_

The first functioning magnetos were invented by Pixii et Fils, a distinguished French instrument-making firm that assembled apparatus for François Arago, André-Marie Ampère, and other elite men of science.[11] Capitalizing on a thorough understanding of Faraday's findings, Hippolyte Pixii and his son developed and marketed several magnetos in 1832.[12] Although the Pixiis may have been encouraged (perhaps even commissioned) by Jean Hachette and Ampère to undertake this project, the firm was apparently the creative force behind these designs.

The smallest machine, which sold for 180 francs, was remarkably simple.[13] On an iron bar, bent into the shape of the letter U, were wound two coils. In design this component was an electromagnet, but its function was reversed: it converted magnetism into electricity. This crucial component sat, open end up, on a short pedestal. Immediately above the coils, suspended from a wooden framework housing a crank-and-pulley mechanism, was a permanent magnet of horseshoe shape, open end down. When the crank turned the permanent magnet, the coils produced current, which flowed as long as the operator was willing or able to turn the crank.

Magnetos ordinarily generate alternating current (ac).[14] However, when alternating current is used to decompose water, hydrogen and oxygen are evolved at both electrodes; producing the gases separately requires direct current. Thus, inventors of magnetos, anxious to reproduce the major effects of galvanism, added commutators of varying designs for converting alternating current to direct current. And such was the case in Pixii's most elaborate machine, the price of which was a breathtaking 700 francs. This was equivalent to about $140, which approximated 6 months' pay for a common laborer in America at the time. Standing about 40 inches tall in its sturdy wooden frame, this impressive machine incorporated a large permanent magnet that rotated below the coils by means of a crank-and-gear mechanism.[15] The coils of copper wire with silk insulation were wound on a U-shaped piece of soft iron; both ends of each coil were capped with disks. This elegant design was the model for later generator coils and, especially, electromagnets. In Europe such coils were called "bobbins" because they resembled spools of thread.

Because the coils were wired in series, the job of the commutator, essentially an automatic switch designed by Ampère and called a "bascule" (French for seesaw), was periodically to reverse the current's polarity. The commutator was actuated by a cam attached to the magnet's shaft.[16] The generator's output was a pulsating direct current that was capable, in a decomposing cell, of evolving hydrogen at one pole and oxygen at the other.

Pixii magnetos were demonstrated before a commission of the Paris Academy of Sciences; impressed, this venerable body awarded the firm a gold medal worth 300 francs. Almost immediately, Pixii machines, judged practical for experiments and demonstrations, were acquired by the Paris Academy, the École Polytechnique, and the Collège de France. One was also bought by the Faculté de Médecine for use in electrotherapy. Aware of the potential medical market, the Pixii firm claimed in a pamphlet that its magnetos, by virtue of their superior performance, could replace electrostatic generators and batteries in medical treatments: they functioned "in all weather, without the use of acid, without preparation, and without any deterioration."[17]

The large Pixii magneto was greatly admired in the scientific community. Exploiting Faraday's scientific discovery, Pixii had crafted an instrument that, with the turn of a crank, dramatically created direct current capable of producing shocks, sparks, and decomposing water. In some applications, this machine and its immediate technological descendants could replace batteries as a source of electricity in laboratories and lecture halls. Consequently, the widely publicized Pixii product—not Faraday's—became the starting point for many magneto inventors.

In the mid 1860s, inventors replaced permanent magnets with electromagnets, creating dynamos. However, magnetos still found applications; for example, most small gasoline engines (e.g., those used to power lawnmowers) have a magneto that provides spark. In recent years magnetos have made a comeback as permanent magnets, spiced with rare earth elements, have become much stronger. Magnetos are even used today in high-tech wind turbines.

_⌒⌒⌒_

Other magneto inventors arrived at a different design by leaving the permanent magnet stationary and rotating the much smaller and lighter coils. Potentially as powerful as the large Pixii, the new machines were much more compact.[18] One of the earliest people to take this tack was Joseph Saxton (1799–1873), an American living in England.[19] The second child of eleven, he was born in the small village of Huntingdon, Pennsylvania. He received a common education, and like most of his contemporaries he did not attend college. During his teens and twenties he held assorted jobs, including laborer in his father's nail factory, watchmaker, and engraver. These experiences honed Saxton's drafting and metalworking skills and helped him to visualize the interaction of parts in complex machines.

The young Saxton showed an inclination to invent and became a prolific contributor of specialized gadgets, most of which were destined for commercial and industrial applications. It was for these skillfully constructed creations that he became well known among artisans and natural philosophers in Philadelphia. In all but name, Saxton was becoming a proficient mechanical engineer.

Like many young Americans imbued with an enthusiasm for science and the new technologies it might beget, Saxton traveled to England to enlarge his base of knowledge. He soon secured employment at the National Gallery of Practical Science in London, which on account of its address was often referred to as the Adelaide Gallery.[20] The brainchild of another Philadelphian, the inventor and entrepreneur Jacob Perkins, the Adelaide Gallery was a place where mechanics, inventors, and instrument makers could exhibit their wares in handsome displays. Admission cost one shilling, a price within reach of London's lower middle class, but exhibitors paid nothing. To enliven the displays, lecturers periodically discoursed on various topics of "practical science" and exhibited new technologies, including Perkins's steam gun. Perhaps the gallery's most famed lecturer was William Sturgeon.

With its eminent natural philosophers and affluent audiences, the Royal Institution relentlessly propounded the ideology that those who created scientific knowledge followed an ascetic calling, were oblivious to pecuniary gain, and laid the foundations of new technologies. Through their selfless efforts in the laboratory, natural philosophers believed, they had earned the right to be regarded as the only legitimate authorities on matters scientific. In contrast, the Adelaide Gallery fostered an appreciation for the essential contributions of practical men—inventors, artisans, and mechanics—to the creation of new technologies. Although most people understood at some level that "material progress" depended on synergies between makers of knowledge and makers of artifacts, an undeniable tension was appearing on both sides of the Atlantic as scientists began to define more sharply their social status and authority role in contrast to inventors and mechanics. In the United States, Joseph Henry would be in the vanguard of the scientists' movement.

In 1845, succumbing to competition and altered public tastes in entertainment, the Adelaide Gallery closed its doors, but not before Joseph Saxton made his mark on electrical technology. Among the exhibits that Saxton prepared for the gallery's opening on June 4, 1832, was a device for creating a spark from magnetism. It employed a horseshoe magnet around whose keeper Saxton had wound a coil. When Saxton abruptly separated the keeper from the magnet, a brilliant spark appeared between the ends of the coil's wires. Produced after much tinkering, this effect of magneto-electricity was an appropriate triumph for the gallery's inauguration. Not surprisingly, Saxton, who maintained a separate workshop at 22 Sussex Street, received a number of orders for his little gadget.[21] Soon he would make a magneto.

_____

Saxton and other early magneto makers were striving to craft machines that could reproduce well-known effects. Clearly, this interest reflected genuine scientific curiosity about whether magnetos were comparable in performance to electrostatic generators and batteries. Yet instrument makers were also thinking about potential markets, and so they strived to make products that could, at least in some applications, compete with and perhaps replace the older technologies. Thus, their development activities were guided by specific performance requirements, such as producing sparks, heating and melting wire, decomposing water, and giving shocks. Designs were compared on these concrete performance characteristics; at this early date, no one made the slightest reference to a magneto's "efficiency" or proposed its use in powering electric lights.[22]

When a new capability was attained, the instrument maker sometimes claimed that his magneto was an invention. However, other inventors sometimes disputed such claims vigorously. Neither then nor now do we have a way to objectively distinguish, in a sequence of related designs, between a trivial change and an invention. All changes are inventions, but for various economic, political, and legal reasons some come to be regarded as more or less original and important than others.

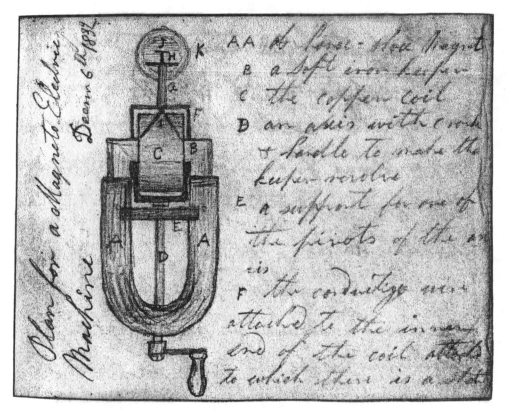

Figure 6.1
The first sketch of Saxton's magneto. Adapted from Smithsonian Institution Archives, RU 7056, Joseph Saxton Papers, Notebook #2, unpaginated.

On December 6, 1832, Saxton set down in his notebook a plan for a magneto (figure 6.1). His aim at first was merely to produce a more impressive spark by generating a continuous current. Eventually his machine was able to reproduce some of the effects that had eluded Faraday. Accounts of Saxton's magneto differ in design details, a consequence of his having made and sold it in different versions.[23] What follows is a generalized description of an early machine.

The core of Saxton's magneto was an elongated, U-shaped permanent magnet that was supported horizontally on brass pillars secured to a wooden base. Immediately adjacent to the magnet's poles, and firmly attached by a bracket to an axle, was a U-shaped keeper wrapped with coils (figure 6.2).[24] Revolving together by means of a horizontal crank-and-pulley arrangement, the ends of the coils swept by the magnet's poles in precise alignment. Current from the rotating coils was made available to the operator by means of a disk revolving in a cup of mercury.

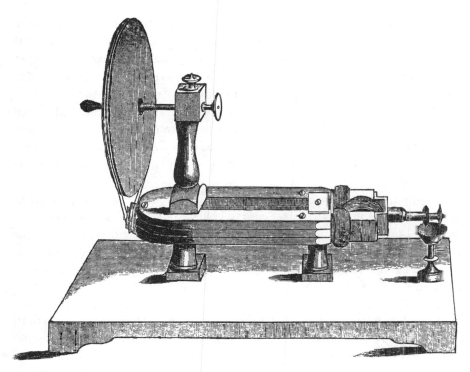

Figure 6.2
An early Saxton magneto. Adapted from *Mechanics Magazine*, volume 21, May 3, 1835.

As an instrument maker rather than a natural philosopher, Saxton did not immediately rush into print with a description of the invention. However, at Cambridge in June 1833, Saxton demonstrated his magneto before an appreciative audience at a meeting of the British Association for the Advancement of Science.

Saxton continued experimenting, trying to coax his magneto into reproducing the chemical effects of galvanism, and on August 25 he declared it done.[25] This formidable machine was put on display in the Adelaide Gallery, where members of the public could view it and sometimes see it in motion. A Saxton machine remained in the Adelaide Gallery for many years, but as Saxton tinkered with the design, periodically substituting newer versions, he did not publish a clear chronology of the modifications.[26] This undocumented updating of the machine rendered some of his priority claims problematic.

People who saw Saxton's machine in operation noted that it accomplished things that Faraday's copper-disk apparatus did not, including give small shocks, cause a platinum wire to get red-hot, produce sparks, decompose water, and even power a carbon arc.

Curiously, the *Literary Gazette*, in its issue dated August 1833, accused Saxton of performing experiments identical to those of the Pixii firm but failing to give the Frenchmen credit. In a later issue of the magazine, a somewhat exercised Saxton denied any prior familiarity with the Pixii magneto. I believe this claim to be credible in view of the very strong conceptual and technological continuity between Saxton's two inventions. As his original drawing (figure 6.1) makes clear, Saxton had turned the coil-wrapped keeper from his first invention into a rotating armature. In addition, although the Saxton and Pixii magnetos had similar electrical performance character-istics, they shared no important design features beyond placing permanent magnets and coils in relative motion.

In his reply to the *Literary Gazette's* charges, Saxton reduced his technological claims to an assertion that his magneto was superior to Pixii's. But was this true, and on what grounds could it be decided? In a dramatic solution to this quandary, the Adelaide Gallery hosted a public competition between the two magnetos. The instrument maker Francis Watkins brought along and operated a Pixii machine he had borrowed from the Count de Predevalle; Saxton demonstrated his own.[27] Among the luminaries who witnessed the contest were Michael Faraday and a chemist named John Frederic Daniell. The event was covered by the *Literary Gazette*, which lavished praise on both machines but—despite its original accusations—tilted toward Saxton's: "This splendid apparatus attracted the universal admiration of the scientific company present, not only from the beautiful and extraordinary effects produced by it, but also from its very superior mechanical arrangement."[28]

The next generation of magneto inventors would also opt for a stationary perma-nent magnet and rotating coils. The Pixii magneto was heavier and required more muscle power than the Saxton machine. By rotating the coils, Saxton and others made magnetos that were easier to use and easier to cart around while still producing similar electrical effects.[29]

An elegant piece of engineering, Saxton's magneto was not without performance deficiencies. Although it could decompose compounds, it lacked a commutator to convert alternating current into direct current. Thus, in comparison with a galvanic battery, or even the Pixii machine, Saxton's magneto was poorly suited for electro-chemistry. Sturgeon, experimenting on his own in 1834, installed a commutator on the Saxton magneto in the Adelaide Gallery and carried out successful experiments in decomposition.[30]

Not until 1835 did Sturgeon publish a description of his commutator, which he named a "Unio-directive Discharger."[31] By that time, he had begun to build his own magnetos, some with several rotating coils, hoping to replace the messy, inconvenient, and high-maintenance galvanic batteries in his lectures.[32] Sturgeon's commutator consisted of four wedge-shaped pieces, mounted on the axle. His description, without illustrations, was at best sketchy; we have to infer that the wedges were metal, insulated from each other and connected separately to the coils.[33] Pulsating direct current was drawn from two wires, one pressing against the rotating wedges and the other connected to the axle. With his commutator-equipped machines, Sturgeon could readily decompose compounds into separate products.

When Sturgeon published his modifications of the Saxton machine and the results of his electrochemical experiments, he also claimed credit for inventing a new magneto—credit that Saxton believed was undeserved. But the ensuing feud was minor in comparison with one that was brewing between Saxton and Edward Clarke (a competing instrument maker and an ally of Sturgeon).

Clarke's establishment was practically across the street from the Adelaide Gallery. Doubtless he had seen Saxton's magneto exhibited and been impressed by its ability to produce showy electrical effects. These performance characteristics inspired confidence that a similar machine, with suitable accessories, would be well received by natural philosophers and lecturers, but Clarke also had a much larger market in mind: electrotherapists. Clarke believed that his magneto would be "extensively used as a remedial agent in many diseases which will not yield to the ordinary modes of treatment."[34]

Since the 1740s, electrical technologies had been employed to treat sundry ailments. This was an obvious and potentially profitable application for any new source of electricity. After all, galvanic batteries had been turned to medical applications almost immediately after they were commercialized. Clarke's market forecasts were prescient: throughout the rest of the century, the current produced by magnetos, sometimes called "faradic" electricity, was harnessed for the healing arts.[35]

Able to build on the precedents of nearly a century of commercially successful electromedical technologies, Clarke homed in on several performance characteristics—compactness, ease of use, and portability—that could earn from electrotherapists the cachet of practicality. Electrostatic machines and galvanic batteries each had performance advantages and disadvantages for electrotherapy; his magneto, Clarke insisted, could in some cases replace either. To promote therapeutic uses, Clarke manufactured a pair of insulated directors that, pressing against moist sponges, could carry current to the patient's afflicted places. His magneto was also outfitted with a mahogany case that enhanced its portability for demonstrators and electrotherapists.

Taking advantage of the opportunity to advertise his wares, Clarke prepared a lengthy description of his magneto, generously illustrated with the accessories he sold for producing the various effects; it appeared in 1836 in the *London and Edinburgh Philosophical Magazine*.[36] Clarke made no mention of Saxton and claimed the invention as his own. When Saxton saw this article, he could not help but discern a marked

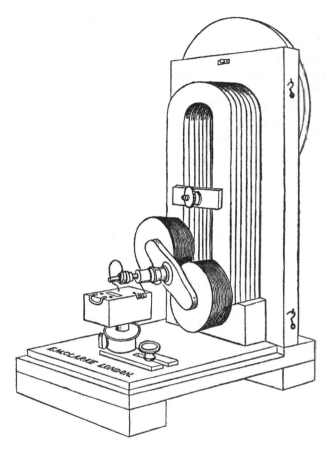

Figure 6.3
A Clarke magneto. Adapted from Clarke 1836: 262. Courtesy of Dibner Library, Smithsonian Institution.

resemblance between his machine and Clarke's (figure 6.3). The two magnetos differed mainly in the arrangement of their parts; in particular, Clarke's coils were turned 90° in relation to the magnet, which rendered the machine more compact.[37] In electrical performance the machines were similar, except that Clarke's had a commutator that produced direct current. Saxton responded immediately with a detailed account of his machine's design, which was published in the same journal as Clarke's "very disingenuous article."[38]

In Saxton's view, which Henry later seconded, the English instrument maker had perpetrated an act of "piracy."[39] The issue was not Clarke's manufacture and sale of the machine for profit; after all, in the tradition of natural-science instruments, it was not patented. Rather, the issue was Clarke's brazen appropriation of credit for Saxton's invention. But what, precisely, was Saxton's claim to originality, insofar as he was now aware of Pixii's priority in inventing the magneto?

In his response to Clarke, Saxton insisted that his real invention was the use of four coils, affixed to a cross-shaped bracket. Opposite coils were identical and wired in series, but the two sets were very different. The coils of one set had many turns of fine wire, which produced high tension; the others had fewer turns of thick wire and yielded high current. By adjusting a switch-like device situated between the coils and the commutator, Saxton could choose the kind of electricity the machine would generate. This was a significant development because each kind of electricity yielded a different suite of effects. Thus, with high current, the operator could rapidly decompose water and produce scintillating sparks, and with high tension the operator could deliver a smart shock. This, then, was Saxton's claim to originality.[40] Clarke disputed even that claim, insisting that he had added different coils to his machine several months before Saxton.[41] The dispute died quietly, and Saxton soon returned to the United States, where he enjoyed a distinguished career as a maker of precision instruments. Although this dispute had no significant consequences such conflicts over priority would assume great importance after the late 1830s, when the patenting of electrical inventions began in earnest. Establishing priority is a crucial factor in obtaining and defending patents. A strong patent position can affect the likelihood that potential investors and manufacturers will judge an invention favorably.

Regardless of who was the first to make a dual-electricity machine, both Saxton and Clarke had articulated a profound principle that would guide the construction of virtually all later electromagnetic generators. To wit, by varying the number of turns on the coil or coils, one could wind a "tension" armature or a "quantity" armature—or anything in between.[42] Applying this principle, inventors tailored their machines for particular applications. I infer that the machines Clarke made for electromedicine had two high-tension coils, whereas a multi-purpose machine, suitable for lecturers, most likely had interchangeable intensity and quantity coils.

Clarke's widely publicized magneto, which embodied design features invented by Saxton, Sturgeon, and Clarke himself, would become a model for other inventors. By the early 1840s, instrument makers on the Continent, in England, and in the United States were offering a variety of capable magnetos.[43] Though they differed in design, these machines all remained technologies for scientific research, displays of electrical effects, and electromedicine. Not until the 1840s would new kinds of magnetos be designed for applications in the commercial world.

Whereas Faraday, over the decades, seems to have garnered disproportionate credit for inventing the magneto, Pixii, Saxton, Sturgeon, and Clarke—the actual inventors—are known only to aficionados of electrical history. This kind of inequity is common in histories of science and technology. Indeed, when many people make inventions that contribute to creating an important technology, the more prestigious, better-known person is often named *the* inventor. The sociologist Robert K. Merton applied the term "Matthew effect" to the allocation of credit to famous scientists, those possessing disproportionate social power.[44]

Faraday's supposed invention of the magneto is a classic example of the Matthew effect. Not only was Faraday a member of the scientific elite, owing to his many real contributions to knowledge; he also held a prestigious position at the Royal Institution and was a member of the Royal Society. Moreover, he did, after all, formulate the generator's fundamental scientific principle. In contrast, the Pixii firm, Saxton, and Clarke were instrument makers, people of narrowly defined prestige in scholarly communities, and without advocates or hagiographers. As a consequence, their essential contributions to the magneto are easily overlooked. William Sturgeon, an inventor and a public lecturer, also had less prestige and social power than well-known natural philosophers had.

Yet Sturgeon chafed at his subordinate status. Sometimes tossing contemptuous barbs at the "philosophers," he nonetheless desperately wanted to be accepted as one of the scientific elite. Toward that end, Sturgeon offered his own theories and combatively disputed those of others, including Faraday. In June 1836, Sturgeon presented a theoretical talk at a meeting of the Royal Society. Though such presentations often were published in the *Philosophical Transactions*, Sturgeon's paper was rejected.[45] Evidently smarting from this slight, Sturgeon founded his own journal, *The Annals of Electricity, Magnetism, & Chemistry; and Guardian of Experimental Science*. It is generally regarded as the first periodical to be devoted mainly to electrical research. The first monthly issue came out in October 1836, and among the subscribers were Henry and Faraday. In the course of its 10-year run, *Sturgeon's Annals* (as it was often called) published many important papers, original and reprints, especially on electromagnetism. Prominent among the offerings was Sturgeon's own theoretical and experimental work. However, his ideas on electromagnetism, which developed in tandem with his inventions of captivating display devices and new magnetos, aroused no interest among prominent natural philosophers.

7 The Blacksmith's Motor

During the period 1832–1837, when magnetos were being developed in Europe that could function well in scientific research, lectures, and electromedicine, two Americans were trying to stake out for electrical technology entirely new commercial applications. Samuel Morse's electromagnetic telegraph succeeded brilliantly, but Thomas Davenport's rotary electric motor suffered a spectacular flop. This chapter and chapter 9 examine their earliest efforts, which eventuated in patents and the formation of companies.

Until the 1830s, the commercial world's involvement in the manufacture and sale of electrical technology was confined to the activities of instrument makers, who were allied with scientific experimenters in a symbiosis of sorts (such as the relationship between Faraday and Newman and that between Ampère, Hachette, and the Pixiis). These privately owned firms, often founded by one highly skilled person with an entrepreneurial flair, were sometimes passed down from generation to generation. As established businesses with successful products, financed with family resources and profits, instrument makers had little need to convince outsiders that their wares were practical before bringing them to market. They merely drew upon their own experience in judging when a new apparatus invented by a natural philosopher was likely to be of interest to other natural philosophers, to science lecturers, and to hobbyists. Although this intimate and comfortable system thrived in the nineteenth century and still exists today, the appreciable resources needed for commercializing new electrical technologies—motors and telegraphs were among the first—required new players and new forms of organization. By the late 1830s, no longer was the fate of an invention exclusively in the hands of natural philosophers, instrument makers, and traditional consumers. Now a multitude of entrepreneurs, capitalists, patent examiners, newspaper reporters, manufacturers, and new kinds of consumers—including governments—weighed in with practicality assessments that could affect a technology's life history.

It is hard to imagine two inventors more different in training, experience, and social position than Thomas Davenport and Samuel Morse. Davenport was a successful blacksmith in a Vermont village; Morse was a respected portrait painter and a professor of art in New York. As different as these two men were from each other, they were equally remote from traditional founts of scientific invention: neither was a natural philosopher or an electrical experimenter, nor did either have an association with an instrument maker. But Davenport and Morse had one thing in common: from encounters with electromagnets both formed vivid visions of technologies that could, respectively, replace existing prime movers and accelerate communication. And perhaps their inventions would bring them fame and fortune. As they attempted to realize the visions that had captured their attention and were consuming their lives, Davenport and Morse had to negotiate new kinds of practicality judgments with many other players, including Joseph Henry, the leading American authority on electromagnetism. Not being natural philosophers, Davenport and Morse were unfettered by the elitist ideology that discouraged direct participation in profit-making ventures based on scientific discoveries.

_____

Late in 1831, Henry completed an electromagnet for the J. and J. Rogers Iron Company in New York. Like the one he had sold to Penfield and Taft, this one was destined to magnetize small pieces of steel for an iron-ore separator. At some point in this project, Henry began to envision the possibility of processing iron ore using only electromagnets. But this was not a simple matter of substituting electromagnets for permanent magnets; Henry would have to redesign the iron-ore separator, no doubt an expensive and time-consuming endeavor. Working on this project without remuneration, Henry recognized that this use of his time and resources was not furthering his career in science, yet it might be lucrative.

In a letter to the Rogers firm, a conflicted Henry ruminated: "I have heretofore been perfectly free in giving to the public any Knowledge I might possess. . . . Although I have considered it almost below the dignity of science to ask pay for my Knowledge yet I now conceive that I have been rather too free for my own interest and that the subject of magnetism is worth something more to me than mere fame."[1] For the first and perhaps the only time in his career, Henry flirted with the possibility of obtaining a patent, but in the end he didn't make the move.[2] He appears to have decided that fame in science was, after all, a sufficient reward for his labors. From then on he watched as others drew upon his electromagnetic principles in fashioning new technologies, obtained patents, and formed companies with the expectation of realizing riches. And a few of these individuals, Morse among them, accumulated great wealth. Henry, however, would not be drawn into commercial ventures so long as income from the pursuit of scientific knowledge enabled him to support his family; fortunately, it did. As a Princeton professor and later as founding secretary of the Smithsonian Institution, Henry was ideally situated to promote an elitist ideology of science uncontaminated by pecuniary motives.

There is perhaps another reason why Henry sought no patents: his understanding of the patent system was peculiar.[3] Recall his belief that, in fashioning a new technology, an inventor was merely applying scientific principles.[4] Thus, according to Henry, if a principle had been made public, then no one could patent an invention based on that principle. In this view, Henry was dead wrong The American patent system, then as now, recognized as patentable any novel product or process having potential utility. Patent examiners also expected an invention to be consistent with scientific principles. Thus, any number of inventions based on established scientific principles might merit patents. Henry's erroneous understanding of patents also implied that mechanics and inventors operate at a level of intellectual activity somewhat beneath that of scientists pursuing original research.[5] Henry's mistaken views on patents and a growing elitism would color his dealings with both Davenport and Morse.

A native of Vermont, Thomas Davenport (1802–1851) was born into a poor farming family.[6] His father died when he was 10 years old, and his older brother Barzillai became his guardian. Thomas was indentured at age 14 to a blacksmithing firm in Williamstown, where for seven years he learned the metalworking craft. In exchange for his labor Davenport received, in addition to room and board, six weeks of rudimentary schooling each winter. Legend has it that the lad was sometimes seen reading a book while he pumped the bellows. One tool in the well-equipped shop, a trip hammer, impressed the young apprentice because it exploited gravity to shape horseshoes and nails.

After completing his indenture, Davenport set up his own business in the village of Brandon. He had no money, but his brother Barzillai was an established attorney in Brandon who doubtless vouched for his brother's character, and so Thomas was able to buy the shop and tools on credit. His occupation and independence gave Davenport a rural respectability that permitted him to marry into the relatively prosperous Goss family. His wife, Emily, would provide ideas, inspiration, and moral support for his adventures in inventing. For a decade Davenport made horseshoes, nails, and repaired the iron implements of everyday life. Then he heard about Henry's electromagnet at the Penfield and Taft iron works.

In the summer of 1833, Davenport journeyed the 25 miles to Crown Point, New York, to view and perhaps purchase the electrical wonder that could supposedly suspend a blacksmith's anvil. Finding no one at the Penfield works, he went on to Albany, hoping to quiz Joseph Henry. But Henry had gone off to Princeton, and Davenport returned home without result. On a second visit to Crown Point, Davenport persuaded his brother Oliver, a tin peddler, to accompany him, and they rode together in the peddler's cart. This time the proprietor was present. The price of the electromagnet (with a two-cell battery) was $75, far more than the brothers had with them. But the determined and resourceful Thomas would not be denied. He browbeat Oliver into auctioning off some of his goods and trading his horse for a sorrier one and some cash.[7]

Back home, Thomas methodically disassembled the prized electromagnet, and Emily took careful notes. Before the night was out, Thomas had forged a larger iron horseshoe and had wound onto it the copper wire from Henry's magnet. To insulate the layers of wire, he used pieces of silk that Emily cut from her wedding gown. According to Oliver's account, this electromagnet was more powerful than Henry's original.

Then Davenport had an epiphany: "Like a flash of lightening [*sic*] the thought occurred to me that here was an available power within the reach of man. If three pounds of iron and copper wire could suspend in the air 150 pounds, of iron, what would three thousand pounds suspend?"[8] But what could this power be used for? Davenport believed that "magnetic power" could replace other prime movers, especially steam.

Daily growing more important in an industrializing America, steam engines were already chugging away in the first locomotives and in a few factories.[9] And by the early 1830s steamboats were plying many rivers. On the well-traveled route on the Hudson River between Albany and New York City, nearly two dozen steamboats ferried freight and passengers during ice-free periods in 1833. The speedier boats could make the 150-mile trip in just under 10 hours, many times faster than travel by horse. According to one contemporary observer, the speed of steamboats was "extraordinary," and so too would be this technology's accelerating effects on trade.[10]

Whether or not Thomas Davenport had a firsthand acquaintance with steamboats, he was well aware of their reputation: steamboats were dangerous, their often flimsy boilers prone to explode. In one accident on the Hudson, perhaps known to Davenport, the boiler of the steamboat *Ohio* burst and five people died.[11] In actuality, steamboat travel might not have been more dangerous than other modes, but the accidents—featured prominently in the press—were spectacular and shaped attitudes. Magnetic power, Davenport claimed, would be "a valuable substitute for the murderous power of steam . . . no more aching hearts & desolate homes occasioned by the awful spectacle of hundreds & thousands of human beings annually hurled into eternity."[12] The power of electromagnetism "would no doubt eventually supersede that of steam"—and Davenport would build the engines that would hasten the transition.[13]

According to his hagiographers, Davenport invented his motor independently of the work of others. However, his early familiarity with *Silliman's Journal* suggests the likelihood that he had read about Henry's electric teeter-totter and adopted its basic operating principle: using the motor's own motion to operate a pole changer.[4] Davenport's first motors also employed wires dipping in and out of small cups of mercury. Yet Davenport did break new ground in America by inventing a *rotary* electric motor. (Many people in Europe were also devising rotary motors.)

Although a blacksmith, Davenport was not exactly an expert mechanic, much less an instrument maker, and so he often secured the assistance of others. In July 1834, with help from his neighbor Orange Smalley (a mechanic), Davenport succeeded in making a consequential electromagnetic engine.[15] Producing rotary motion at 30 revolutions per minute, it contained four electromagnets—two stationary and two in the armature. Unlike the devices of Faraday, Sturgeon, and others that merely *exhibited* rotary motion, this motor, Davenport believed, "opened up a wide field for philosophical research."[16] He admitted, however, that it wasn't very powerful and that 'the cost per diem of using it on a large scale would be enormous."[17] Nonetheless, Davenport expected that the expense of galvanic power would decrease dramatically, as had the expense of steam power in preceding decades.

Expounding his vision of putting steam power into eclipse, Davenport hoped to "no longer suffer the ridicule and derision of my friends & neighbors."[18] But that was wishful thinking. He became known to fellow villagers as the "perpetual-motion-man" whose electromagnetic engine produced only "musquitoe [*sic*] power."[19] Aware that perpetual motion was impossible, Davenport took umbrage at the implication that he was a charlatan trying to create something out of nothing.

Seeking to obtain the opinion and advice of a person more knowledgeable than his neighbors, Davenport toted his invention to Middlebury College and demonstrated it to Professor Edward Turner. Impressed, Turner urged Davenport to seek a patent and, moreover, volunteered to help prepare the application. Turner also gave Davenport a written statement that, according to the inventor, "expressed his entire confidence in the ultimate application of its power to propelling machinery."[20] Davenport fancied that he could use this certificate to entice a moneyed man to become a partner in the enterprise of building more powerful motors. He hoped to save lives by replacing steam engines, but was willing to accept wealth and fame if they came his way; perhaps then he could buy Emily another silk gown.

Turner's certificate did not open deep pockets, but Davenport was able to find local mechanics to help him make new motors of different designs. During the mid 1830s they also built a number of motor-driven models that suggested how electromagnetism might be used to power machinery. Decades later, H. S. Davenport recalled seeing in his uncle's blacksmith's shop an electrically driven "trip-hammer, a turning lathe and a machine for doubling, twisting and reeling cotton or silk, all at the same time."[21]

But these items of technological display, though attracting interest and spreading the blacksmith's notoriety, failed to attract backers. Appreciating that further development of electromagnetic power required more money than his neglected blacksmithing business could provide, Thomas Davenport decided to follow Professor Turner's advice and obtain a patent.

Though Davenport was a man of few words, some of his utterances had great persuasive power. After numerous entreaties, he convinced a few neighbors to finance a trip to Washington so he could apply for a patent. Before setting out for Washington, however, Davenport decided to bolster his case by getting additional endorsements from scientific authorities on his engine's technical feasibility.

From Edward Turner, at Middlebury College, Davenport received a letter of introduction to Amos Eaton, Principal of the Rensselaer Institute. Located in Troy, New York, the Rensselaer Institute had been founded to encourage technical training by the wealthy patroon Stephen Van Rensselaer, who many years earlier had hired Joseph Henry as a tutor. Turner's letter apprised Eaton of Davenport's invention, exulting that the blacksmith "will one day compete with Watt and Fulton in the glory of having added another prime mover of machinery to science and the arts."[22] Turner also urged Eaton to introduce Davenport to Stephen Van Rensselaer, and Eaton willingly complied.

Van Rensselaer apparently was not approached for money, and offered none, but at the end of June 1835 he dashed off a note introducing Davenport to Henry: "Mr Devenport [*sic*] of Vermont has exhibited in my Office a machine which evinces great ingenuity. He visits Princeton to submit to your inspection its operation and to obtain your opinion on the practicability of applying it to useful purposes . . . He is an intelligent unassuming Mechanic . . . destitute of Funds to make experiments on a large scale. I think he merits encouragement."[23] Davenport—by this time familiar with Benjamin Silliman's text on chemistry, which also described some of Henry's work—understood that America's foremost authority on electromagnetism would have to be consulted and his blessing sought.

When Davenport arrived in Princeton with his weighty endorsements, Henry was obliged to grant him a hearing. After observing the motor's operation, Henry supplied Davenport with a certificate. Fifteen years later, Davenport recalled that it spoke "highly of the novelty & originality of my invention." But, he noted, Henry had also discouraged him from attempting to scale up the motor to one horsepower, for fear that a public failure would label him a "humbug" and harm "the advancement of science & the arts."[24] While hardly a ringing endorsement of Davenport's vision, Henry's certificate attested to the technical feasibility of his rotary motor.

As to the broader practicality of electric motors, Henry was dubious. In a letter to Silliman, he expressed his unvarnished views: Davenport's "machine evinced much ingenuity but . . . I did not believe that electro magnetic power would be found sufficiently cheap for mechanical purposes." He also told Silliman that he had urged Davenport to "abandon the invention," for its only possible use was "in the way of exhibition as a curiosity." After all, Henry continued, any new motive power would have to be cheaper to use than steam, or more convenient in some way.

Henry's letter, in which he reiterated the view that his teeter-totter had established the basic design principle of the electromagnetic engine, also revealed his touchiness on the matter of priority. It appears that for Henry all similar inventions were uninteresting elaborations of his original philosophical toy: "The truth is that there is nothing new in the whole affair . . . it differs nothing in principle from the first one of the kind which I described in the Journal."[25] This patronizing proclamation utterly devalued the skill, mechanical insight, and technological knowledge that Davenport and his collaborators had brought to bear in inventing a rotary motor. Had the motor patent been disputed in court, Henry's assertion of priority might have been used against Davenport, probably to no avail. But it never came to that, because no commercial interests were at stake.

_~~~~~_

After leaving Princeton, Davenport continued south. At Philadelphia he met with Alexander Dallas Bache, a professor in the University of Pennsylvania, a great-grandson of Benjamin Franklin, and a close friend of Joseph Henry. At Bache's invitation, Davenport exhibited his motor in the Library of the Franklin Institute before a group of "scientific gentlemen."[26] Bache also gave Davenport a supportive certificate, writing that he *did* need to build a large machine in order to ascertain the cost of electromagnetic power.

Davenport arrived in Washington nearly broke. Lacking enough money to file a patent application *and* return to Brandon; he decided to forgo applying for a patent. On the trip back to Brandon, he managed to sell his motor to Stephen Van Rensselaer, who purchased it (for $30) for the Rensselaer Institute.

Meanwhile, an Albany newspaper reported that "the plan of the Brandon blacksmith will not work," and this story was repeated in the *New York Commercial Advertiser*.[27] According to Amos Eaton, this untimely report had the effect of discouraging potential investors. Eaton, after publishing a letter strongly in support of Davenport in the *Albany Daily Advertiser*, arranged for the inventor to give a public demonstration in Troy in mid October, which the professor narrated. It went very well, but investors did not line up to fund the project. Nonetheless, Davenport did receive an invitation from an interested member of the audience to visit Massachusetts. There he obtained a small amount of support from several sympathetic businessmen and got help in building a few more machines. One of the latter was a model train, pulled by an electromagnetic engine, that ran on a tiny circular track.

For two weeks in December, Davenport exhibited this model in the Marlborough Hotel in Boston. He was paid $12, which just covered his expenses. Though for Davenport it was a tolerable diversion, this trip failed to yield long-term financing.

In the summer of 1836, Davenport spent several weeks in Saratoga Springs, exhibiting the motor to paying audiences. There he met Ransom Cook, an artisan who owned a cabinetmaking shop. Davenport regaled Cook with stories of his projects and of the disappointments with which he had met. Cook, like Davenport, came to believe that electromagnetic engines had a great future, and the two craftsmen agreed to become partners. They worked together in Cook's shop through the autumn, making more motors (including one for the patent office) and preparing detailed drawings and specifications. They filed a patent application, but their timing was terrible. Before the patent could be issued, a fire broke out in the Patent Office, incinerating their paperwork and the precious model. Davenport and Cook quickly prepared a new model and new drawings (purportedly based on their latest experiments), which they delivered to the Patent Office in January 1837. On February 25, the government granted U.S. Patent no. 132 to Davenport for an "Electrical Motor."[28]

Davenport's patent was almost certainly the first U.S. patent issued for an electrical invention. And soon there was an English patent.[29] These patents marked a turning point in technological history, for in both nations they were followed by a flood of patents for electrical things, many of them incorporating the electromagnet.

⌐⌐⌐

The patent model of Davenport and Cook's motor (figure 7.1) has survived as a national treasure in the Smithsonian Institution.[30] I have examined the model and compared it with the patent drawings and specifications. It has two straight electromagnets, mounted together in cross fashion on a wooden disk that rotates on a vertical axle. In the wooden ring surrounding the electromagnets are two arc-shaped permanent magnets. Although there is no reason to doubt that this motor once worked, its mechanical instability suggests that it could not have run reliably without constant adjustment and repairs.

By this time Davenport and Cook had abandoned the use of mercury cups. In their place were four straight copper wires (the model has flat brass ones) that descended from the electromagnets. The bottoms of the wires made contact, at approximately 90°, with two small metal plates mounted on the motor's wooden base. The plates, connected to a battery, resemble a washer cut in half, with the halves slightly separated. In this configuration the electromagnets could be energized sequentially as the rotary motion brought the wires alternately in contact with each metal plate. The electrical continuity between the descending wires and the plates was tenuous. And

Figure 7.1
The patent model of Davenport's motor. Adapted from Davenport 1929: 145.

the wires' placement relative to the plates was mechanically weak and could be easily disturbed. As can be seen in figure 7.1, the brass wires are twisted, and thus the motor is inoperable.[31] These wires might have been disturbed when the motor was cleaned after the 1877 fire in the Patent Office, but in any case the design was deeply flawed.[32]

A second surviving Davenport motor in the Smithsonian has a very different arrangement for delivering power to the electromagnets. Resembling the Sturgeon-style commutator in the Clarke magneto, it has alternating conductive and insulating parts formed into a small cylinder that rotates on the motor's shaft. Pressing against the cylinder are two wires for making electrical contact. This motor is set on a circular wooden track about 2½ feet in diameter. If this is the same toy train that Davenport demonstrated in Boston in December 1835, it would suggest that he had already been experimenting with more reliable ways for powering the electromagnets. The toy train cannot be independently dated and may have been made after the patent model.[33]

The patent model (figure 7.1), unlike the patent drawing, has gears and a shaft for converting the axis of rotation from the vertical to the horizontal. There is also a slot in the base, directly below the horizontal shaft. The most likely explanation is that these seemingly extraneous features were parts of an accessory used in demonstrations. Most likely, a string attached to the shaft raised a weight suspended below the motor. If this is so, it casts doubt on the story that, after the fiery demise of the first patent model, Davenport and Cook built an entirely new motor incorporating the latest advances. The patent model may have been an older motor, one of several that Davenport had on hand for exhibition purposes.

_⌒⌒⌒_

Up to this point, Davenport had relied on ties of family and friendship to obtain funds for sustaining his inventive activities. In addition, through introductions and chance encounters, he had managed to expand his social network, drawing on people and resources outside his local community. And he had charged admission to view his seemingly magical motors in public exhibitions. Davenport seems to have possessed a soft-spoken charisma that exuded sincerity. Members of his extended social network came to believe that the earnest blacksmith could bring his project to a successful conclusion, or at least pursue it with unflagging determination and grit. People who still harbored lingering doubts that he might be a humbug could be convinced of the motor's technical feasibility through exhibitions and by written testimonials from scientific authorities.

But the possession of a patent opened a new world of opportunities for raising funds from strangers, most of whom Davenport would never meet. With a patent, he and Cook could form a joint-stock company (a corporation in all but name), sell shares, and use the proceeds to support their project to fashion more powerful motors. Considerable funds would now be needed to explore new dimensions of practicality, such as whether an electric motor could drive a full-scale machine. As Davenport and Cook attempted to scale up their motors, questions about performance characteristics such as reliability and costs of operation could be temporarily held in abeyance—or so they may have thought.

Before the early nineteenth century, the majority of joint-stock companies in England and America were formed for large-scale public works, such as the construction of roads and canals, that neither governments nor individuals were willing or able to undertake. In England, share prices and dividends of a few dozen companies were reported in periodicals such as the *Gentleman's Magazine*.

In America, in the first few decades of the nineteenth century, more and more speculative ventures were organized as joint-stock companies. In an increasingly common move, the patent was sold to the company, and the patent—along with the inventor's expertise and good will—became the company's major asset. The company then offered shares to the public, touting the invention's great promise, sometimes in newspaper advertisements or a prospectus. If the sale went well, the inventor was paid for the patent and the company began operation.

A joint-stock company was a curious creature that could not exist without state charter and state power to enforce its contracts. Its main attraction to investors was the promise of financial gain with limited liability: individual shareholders were not responsible for debts incurred by the company, and so the most a person could lose was the original investment. Enticed by lavish, sometimes outrageously misleading claims, people plunged; after all, the share price might rise, and some companies offered guaranteed dividends. In one growth industry after another, speculative bubbles inflated and burst.

The joint-stock company was—and still is—a magnet for con artists. In the middle of the nineteenth century, this form of business organization—despite its great potential for financial abuses—evolved into the bureaucratic corporation, becoming the institutional pillar of modern capitalism. After all, the joint-stock company made it possible to undertake risky projects, such as commercializing technologies having a great developmental distance, that apparently could not be financed in any other way.

_⏜⏜⏜_

In 1837, the same year in which Davenport obtained the motor patent, Joseph Henry took his first trip to Europe, where he visited some of the great luminaries of Western science.[34] In France he met Augustin Fresnel, whose lenses—designed on optical principles—were replacing the much less effective reflectors in French lighthouses. In the decades ahead, Henry himself would be evaluating and inventing new lighthouse technologies for the American government. In England, Henry spent much time with Michael Faraday, even attending some of his lectures at the Royal Institution. He also met J. F. Daniell (a chemist at King's College in London who had just invented a new battery), E. M. Clarke, Charles Babbage, and Charles Wheatstone. Wheatstone a professor of experimental physics at King's College, had wide-ranging interests in optics, acoustics, and electricity and was just then struggling to build an electromagnetic telegraph. Henry called to Wheatstone's attention the need for an intensity battery to achieve long-distance communication.

It did not escape Henry's notice that men of science such as Fresnel, Wheatstone, and Faraday often tackled technical matters of interest to industry and government. Indeed, Henry discovered that the opinions of scientific authorities were taken much more seriously in Europe than in America. In fact, inventions of overriding interest to the state had long been evaluated by fellows of the Royal Society of London and members of the Paris Academy of Sciences. However, the United States lacked a national scientific organization of comparable prestige (the National Academy of

Sciences was not established until 1860, and it did little for decades), and natural philosophers generally had low visibility and scant social power. In that power vacuum, charlatans, quacks, and humbugs rushed in, sometimes obtaining favorable publicity for egregiously defective inventions. Henry's friend Alexander Bache had become deeply concerned about this, arguing that science had to "put down quackery or quackery will put down science."[35] Influenced by Bache, Henry concluded that American men of science had a public duty to apply their specialist expertise in assessing inventions; this would benefit honest inventors as well as entrepreneurs and investors who might support their projects. With Bache's encouragement and collaboration, Henry determined to professionalize American science, increase its support, and enhance its prestige and its influence in public affairs.

One of Henry's first engagements in judging an invention's scientific validity occurred in 1838, when he published a piece that vigorously challenged a report by the Senate Committee on Naval Affairs. Henry Sherwood, a New York physician, had petitioned Congress seeking patronage for the invention of a "geometer" that purportedly employed the earth's magnetic field to make accurate determinations of latitude and longitude. The geometer was allegedly based on new principles of magnetism discovered by Dr. Sherwood. The Senate committee, relying on opinions of scientific men, evaluated the geometer favorably. Not only did Henry (an expert on magnetism) refute the supposed new science upon which the geometer rested, he also disputed the scientific qualifications of the outside reviewers. He argued that these individuals, lacking specialist expertise in magnetic science, were not qualified to pass judgment on Sherwood's claims. As for the invention itself, Henry merely pointed out that an invention based on "false principles" could "never give uniformly true results."[36]

The inventors of the electromagnet and the electric motor brought into human consciousness entirely new and marvelous phenomena. With little more than coils of insulated copper wire, pieces of soft iron, and a battery, it was possible to produce rotary motion or to lift weights exceeding a ton. These new technologies were exhibited in countless public lectures throughout the Western world, advertising that natural philosophy can drive human imagination and ingenuity. But not every demonstration came off precisely as planned. Large electromagnets quickly lost their maximum lifting power, and motors slowed down and eventually stopped.

The fault lay not with the electromagnets and the motors but with the batteries. Electromagnetic devices, with their heavy demands for current, established new performance requirements that existing batteries could not easily meet. The proliferation of electromagnetic devices in the early 1830s was a major stimulus for the development of new batteries.

To researchers schooled in the many disciplines of natural philosophy, it would have been obvious that the battery's ills could not be remedied by the application of electrical principles alone. Rather, the remedy required a heavy dose of chemical knowledge and experience. Beginning in the mid 1830s, a host of chemical researchers set out to solve the battery's persistent problems, for inventing a better battery might be a ticket to scientific acclaim.

The most serious shortcoming of batteries which came to be known as "polarization," was caused by the rapid buildup of hydrogen bubbles on the copper electrode. The greater the drain on the battery, the faster hydrogen accumulated. This nonconductive coating deeply depressed the battery's output. Several chemistry-savvy people solved the polarization problem by altering the galvanic cell's chemistry. The first of the so-called constant batteries to acquire the cachet of practicality from instrument makers and consumers was invented by John Frederic Daniell of King's College.[1]

Daniell (a friend and colleague of Charles Wheatstone, who was already developing a telegraph) was aware of the battery's limitations for powering electromagnetic technology. He wanted to make a battery "more efficient and convenient for all the purposes to which the common voltaic battery is usually applied."[2] Also motivating Daniell's investigations was Faraday's recent demonstration that a given "quantity of electricity" (i.e., amount of current) had a corresponding chemical effect in any electro-chemical reaction.[3] To verify and extend Faraday's findings, research that would have seemed timely to other natural philosophers, Daniell needed a battery that permitted experiments to be repeated exactly, time after time. Only under these conditions could he expect to make reliable measurements of electrochemical phenomena. In 1836, after lengthy experiments, Daniell claimed success and reported his findings.

More complex than the usual zinc-copper configuration, a Daniell cell held two liquid electrolytes in separate, nested containers. The inner one was a membrane that could pass current but not liquid; in the earliest version, Daniell employed a tube-shaped segment of ox gullet sealed at one end. Into this tube he placed dilute sulfuric acid and also one electrode—a zinc rod (figure 8.1). The outer container was a copper jar that served as the second electrode; it was filled with a saturated solution of copper sulfate. When the cell was assembled and operating, no hydrogen bubbles formed on the copper. Instead, this electrode gradually accrued a deposit of metallic copper, and so it remained highly conductive.

Figure 8.1
A Daniell battery. Adapted from Prescott 1877: 50.

When experimenting with his battery at higher temperatures, Daniell observed that the acid attacked the ox gullet. He substituted a tube made of porous earthenware (the kind of pottery used today in wine coolers), but he did not abandon the ox gullet for ordinary temperatures.[4]

Although current from a Daniell battery was relatively constant it could, for example, decompose water for 6 hours straight), so too were the maintenance activities. For example, apart from removing and cleaning the ox gullet after use, the sulfuric acid had to be topped off and eventually replaced, and the other electrolyte had to be replenished by adding copper sulfate crystals. Also, the zinc rod gradually wasted away, and so needed replacement from time to time. Batteries, constant and otherwise, had an insatiable appetite for zinc and fresh electrolyte, and this created an ongoing expense for generating electricity. Calculating this cost would soon occupy a number of distinguished men of science, and their pessimistic conclusions would diminish enthusiasm for using electricity to power motors and lights.

Instrument makers immediately brought Daniell's battery to market, and it found many buyers. Evidently, experimenters and lecturers considered the trouble and expense of maintaining Daniell cells to be bearable in light of their patent performance advantages. After all, a Daniell battery could sustain heavy-drain electromagnetic devices far longer than the older designs. Daniell himself believed that in large versions his battery opened up "the possible application of the extraordinary powers of voltaic currents to economical purposes."[5] On his visit to England, Joseph Henry saw the Daniell battery and thought highly of it.[6]

Publicity about the electric motor and about nascent electromagnetic telegraphs no doubt intimated that a large demand might soon emerge for even better constant batteries. In view of this anticipated demand, it is not surprising that several people with chemical knowledge and skills, including Robert Wilhelm Bunsen in Germany and William Grove in England, were soon designing alternatives to the Daniell cell that could also eliminate polarization. In the early 1840s instrument makers brought to market several new batteries (like Daniell's, named for their inventors). In subsequent years, the invention of constant batteries—many of them much simpler in construction than Daniell's—became a veritable cottage industry. Some of the new designs were patented, produced in factories, and enjoyed healthy sales during the rest of century.[7] Better batteries became an enabling technology, making made it possible for other electrical technologies to be judged practical for particular applications.

⎍⏜⏜⏜⎍

While people on both sides of the Atlantic were struggling to fashion prototype motors and telegraphs, Daniell's battery unexpectedly helped to sire an electrical technology that quickly became the basis of a prospering industry. The most widely used generic term for this family of technologies was "electrometallurgy"; the two main varieties were electrotyping (for making printing plates) and electroplating (for electrodepositing a thin layer of metal on a conductive object).[8]

Electrometallurgy rested on the discovery by Davy and others that a battery's current—as an agent of chemical decomposition—could liberate metals from their compounds. In the first few decades after this discovery, these effects were exhibited in lectures and described at length in textbooks, including George Singer's *Elements of Electricity and Electro-Chemistry* (1814). The simple apparatus needed to create this basic effect was common knowledge among electrophysicists and chemists. During the early 1830s, a handful of investigators reported the isolation of one or another metal in this way. Michael Faraday produced tin and silver from their fused chloride salts.[9] In 1831, trying to better understand electrical decomposition, an Italian named Carlo Matteuci used a 30-cell battery to release copper, silver, lead, and other metals.[10]

Although voltaic electricity could be employed to refine metals, no one at that time would have proposed using batteries commercially for extracting common metals from their ores. Because batteries consumed zinc and acid, refining copper or lead in this way would have been vastly more expensive than traditional technologies. But in 1834 Faraday suggested that electrical decomposition might be used in producing pricey metals: "The capability of decomposing fused chlorides, iodides, and other compounds . . . and the opportunity of collecting certain of the products, without any loss . . . render it probable that the voltaic battery may become a useful and even economical manufacturing instrument; for theory evidently indicates that an equivalent of a rare substance may be obtained at the expense of three or four equivalents of a very common body, namely, zinc: and practice seems thus far to justify the expectation."[11] With this comment, Faraday implied that natural philosophers were routinely using electrical decomposition in their laboratories to turn out sodium, potassium, and other rare elements. However, Faraday probably did not foresee that entire industries would one day be built upon the electrical refining of metals, rare and common.[12]

In the meantime, other possibilities arose for the commercial working of metals by electricity. Indeed, in the late 1830s several experimenters reconceptualized electrical decomposition as a process for both depositing and *shaping* metal. Electricity, for example, could be used to make and reproduce the copper plates that printed images in books and periodicals. Moreover, artisans could electrodeposit a layer of metal on virtually any object that could be made conductive. This sudden conceptual reorientation helped to unleash a flurry of discoveries, inventions, and trials, leading in just a few years to the adoption of electrometallurgy by the proprietors of hundreds of enterprises.[13] The immediate stimulus for this revolution—and the most likely explanation for its timing—was that many experimenters had just begun to use Daniell's constant battery. Recall that in a Daniell cell copper is continuously plated on the inside of the copper container. Experimenters noticed that when this copper layer was peeled away, it preserved an impression—in astonishingly fine detail—of virtually every irregularity on the electrode's surface.[14] Among the many who observed this effect, a few made an imaginative leap, foreseeing that it could become the basis of a commercial process. This belief seemed suddenly credible because a Daniell battery—and, later, others—could supply the abundant current, for long periods, needed to plate objects.

Two experimenters working independently, Thomas Spencer in England and Moritz Jacobi in Russia, simultaneously developed technologies of electrometallurgy and contributed to their lightning-fast commercialization. Yet, as Cyril Stanley Smith shows, others were working on similar projects in the years immediately after the appearance of the Daniell battery.[15] Debates about priority in this case serve no useful purpose because, as Spencer himself acknowledged—while nonetheless claiming credit—"scientific facts were all tending toward it," so rendering electrometallurgy inevitable.[16] Indeed they were, for the Daniell cell supplied both the immediate inspiration and the initial means to realize long-standing cultural imperatives in metal working, such as eliminating mercury from the gilding process. In this chapter I focus on Spencer as merely one of many electrometallurgical pioneers.[17]

_⌒⌒⌒_

A topic of growing fascination during the 1830s was the role of natural galvanism in the formation of metallic ores. Following up research by Andrew Crosse, who had shown that galvanism could create crystals, Spencer contrived a modified Daniell cell, employing plaster instead of pottery or ox gut to separate the two acids.[18] Observing his cell in action, Spencer grasped almost immediately that electricity could deposit copper on a surface of any shape. This effect, he believed, might be adapted to some metalworking industries. To increase the likelihood that tradesmen might judge the process practical, Spencer began talking to engravers, printers, and other potential adopters. Thus, the questions that guided his experiments were influenced not only by interests in natural philosophy and the principles of electrometallurgy but also by the desire to create technologies that would transfer readily from his laboratory to workshops and factories.

Spencer envisioned that electrometallurgy in industry would have two general applications: coating an object with a thin layer of metal and reproducing an object from a metallic mold. However, these applications imposed conflicting performance requirements on the technology. In the first, it was necessary that the metal layer, grown by galvanism, adhere tenaciously to the substrate; in the second, the electro-deposited metal had to come away from the mold cleanly and easily.

Spencer learned by accident that if he pre-treated an object with nitric acid, the deposited metal would not peel or flake off, and removing it mechanically was virtually impossible. Moreover, he contended that the surface of an object so treated promoted chemical bonding with the deposited copper. This process—later called "pickling"—was of signal importance.

To make a copper printing plate, Spencer had to deposit copper differentially across the plate's surface. He found (again by accident) that wherever an insulating material such as wax was put on the plate, it resisted the deposition of copper. After trials with various substances, Spencer came up with a workable "cement" recipe, consisting of "bees' wax, common resin, and a small portion of plaster of paris," that could be spread in a thin layer over a plate's entire surface.[19] When the cement was hard, Spencer could use an engraving tool to impart to the plate words, designs, or virtually anything he could draw. The plate was then subjected to galvanic action as an electrode inside the cell's copper sulfate container, and afterward treated to remove the remaining cement. Spencer found that the copper deposited in the engraved areas had sufficient relief to be usable in a printing press. He suggested that, with experience, engravers should be able to ply their trade on such plates "with great facility and precision."[20] And indeed they would.

Experimenting with a coin, Spencer also learned how to make ready-release molds for copying objects. He began by heating the coin, covering one face with beeswax, and then vigorously wiping most of the wax away. Presumably wax remained only in the pores, leaving the rest of the surface sufficiently conductive. After galvanic action had deposited a layer of copper, Spencer "applied the heat of spirit lamp to the back, when a sharp crackling noise took place, and I had the satisfaction of perceiving that the coin was completely loosened. In short, I had a most complete and perfect copper mould of one side of a half-penny."[21]

In a later exercise exploiting the same process, Spencer composed a brief text in an ornamental typeface and impressed it on a sheet of lead. The resulting copper plate, produced by electrodeposition and readily detached by heat, was "a most perfect specimen of stereotyping in copper, which had only to be mounted on a wooden block to be ready to print from."[22] Spencer also produced a plate from a wooden block pressed into lead, which was used to print an illustration accompanying an article in the *Liverpool Mercury*. Thinking ahead to the possibilities of copying things made of wood or clay, Spencer also devised ways to make the surface of a nonmetallic object conductive so that it, too, could acquire a layer of copper.

Spencer's attempts to electrodeposit gold, silver, and platinum did not achieve unqualified successes. Nonetheless, he was optimistic "that nearly all the metals will ultimately be brought under subjection to VOLTAIC ELECTRICITY."[23] To help bring about this happy result and, especially, to spur commercialization of electrometallurgy, Spencer reported his findings in ample detail. His first formal announcement took place at a meeting of the Liverpool Polytechnic Society on September 12, 1839. The following year, he detailed his experiments in a pamphlet, noting that printers of calico (a cotton cloth with brightly colored designs) were already using the process for making their copper plates. He believed that his discovery "now promises to be of much use to our manufacturers and the arts generally."[24]

Manufacturers could embrace this new technology without having to sort through conflicting opinions of scientific authorities. After all, electrical deposition of metals had been pioneered by both Davy and Faraday, two of the most esteemed natural philosophers of that time. And the writings and processes of Spencer, Jacobi, and others were there for all to read and replicate. Thus, electrometallurgy could be transferred from laboratories to the commercial world already possessing the cachet of technical feasibility. Many manufacturers and diverse consumer groups would soon affirm its practicality for their own activities.

⎯⌇⌇⌇⎯

Surely Spencer could have received patents for his fundamental processes, but he chose not to apply; in the tradition of Faraday and Henry, he regarded his work as scientific research pursued without pecuniary interest. Alfred Smee, author of the 1841 textbook *Elements of Electro-Metallurgy* and surgeon to London's upper crust, adopted a similar stance: "I determined to throw the laws and principles open for the benefit of the arts and manufactures of our great country." Noting with just a hint of envy that Russia, France, and Germany gave cash, medals, and pensions as rewards for important discoveries and inventions, Smee asked: "For what does he [the English man of science] labour? It is solely for the sake of science, and he is contented and satisfied with developing new scientific truths, or promoting the prosperity of our national manufactures."[25] When Smee wrote these words, Jacobi, whose research was financed by the Russian government, had already received its Demidov Award.[26] In his salaried position at the St. Petersburg Academy of Sciences, Jacobi did not have to worry about patenting.

In subsequent years, neither the benevolent cause of national prosperity nor the elitist ideology of natural philosophy deterred others from patenting electrometallurgical processes in England, on the Continent, and in the United States. This came about because new players in the commercial world regarded electrometallurgy as a potential basis of wealth. Although patents furnished scant protection against piracy, they were the institutional means for proclaiming novelty and also for declaring seriousness about turning scientific findings into a salable commodity, perhaps a profit-making enterprise. Many entrepreneurs, sizing up the performance characteristics of electrometallurgical processes, believed that they could compete successfully with—perhaps even quickly replace—many traditional metalworking practices. And so the technologies invented by men of science—Spencer, Jacobi, and others—were judged practical by manufacturers and entered the commercial world.

Many patents took the form of a recipe for electrodepositing a pure metal (e.g., gold, silver, or nickel) or a metal alloy (e.g., bronze or brass). Others specified techniques for duplicating and plating nonconductive objects—a wax or ceramic object, for example, was made conductive by brushing on powdered graphite. An important patent was awarded to George and Henry Elkington for electroplating gold or silver using a cyanide solution, a method inspired by the work of the chemist Karl Scheele.[27] Elkington, with factories in Birmingham and London, would become a large and prosperous manufacturer of electroplated objects, and also received royalties for licensing its patents. The contents of patented recipes testify that the development of electrometallurgy, like that of new kinds of batteries, depended on chemical expertise.[28]

Because copper could be so easily plated from a copper sulfate solution (I did this for a ninth-grade science project), many men and women took up the technology as a hobby, and for a few years it enjoyed favor as a "fashionable amusement." In 1860, after the fad had largely faded, Alexander Watt wrote: "Every one had his set of electrotyping apparatus, and his bath of sulphate of copper. Even among the fair sex would be found many a skilful manipulator, and in such hands, how could the art fail to give beautiful results!"[29] In addition to the satisfaction of mastering the process, the hobbyist could create striking objects able to astonish family and friends.

As already intimated, the working of metals other than copper involved much more chemical expertise. Moreover, in practice commercial firms often employed fairly involved chemistry-informed recipes, often using poisonous or dangerous compounds, to obtain products with the desired thickness, color, and surface characteristics, and to speed up the plating process.[30] For example, one recipe for plating gold, evolved from the Elkington process, required a heated solution of distilled water, gold chloride, sodium phosphate, sodium bisulfite, and potassium cyanide.[31] Not surprisingly, workers sometimes sustained injuries from the chemicals they handled.[32]

In addition to confecting many recipes for plating a certain metal on a certain substrate, electrometallurgists also developed new batteries and modified older designs. The new batteries, such as Smee's, typically were simpler than Daniell's, having only one liquid-filled compartment. Polarization could be eliminated in other ways, such as by amalgamating the hydrogen-attracting plates with mercury. Some of the new battery designs were patented, but others—such as Smee's—were not. In one of the better examples of a circle of technological innovation, Daniell's constant battery gave rise to electrometallurgy; in turn, electrometallurgy, with its seemingly insatiable demand for current, contributed to the creation of new constant batteries.

Figure 8.2
Commercial electroplating. Adapted from frontispiece of Watt 1860. Courtesy of Dibner Library, Smithsonian Institution.

Experimenters learned early on that electrodeposition could take place outside the battery in a separate container (an electrolytic cell) that held the plating liquid, the object to be plated, and often an electrode made of the plating metal that gradually went into solution.[33] In large enterprises, electrolytic cells mushroomed into vast tanks (figure 8.2).

Electrometallurgy made impressive inroads in the printing industry. Printing plates, which could contain both images engraved in copper and text set in lead type, had to be replaced periodically because the imprint became less distinct after each impression. In addition to enabling rapid and economical replacements, electrotyping had many attractive performance characteristics over mechanical processes of reproduction. J. W. Wilcox of Boston, who began using the process commercially in 1846, noted that "The advantages of this process are, 1st, its durability; the copper face of the type and illustrations lasting many times longer than type-metal, and 2d, the blackness of the impression taken from copper."[34] Another source noted that an electrotype plate "takes a sharper impression of the mould, and delivers the ink much more readily than type metal, besides being a cleaner process; it also takes up less ink, and consequently the printed pages dry more quickly."[35] For publishers and printers, however, the performance characteristic that mattered most was electrotyping's ability to lower printing costs for large runs because it was so inexpensive to replicate plates in quantity.

In concert with the phenomenal growth of the print media and their audiences, electrotyping was taken up quickly and developed further, especially in the United States. Wilcox used electrotype to print Daniel Davis's 1848 *Catalogue of Apparatus*, which he believed was the first book to be entirely electrotyped.[36] By the mid 1850s, *Harper's Magazine* was being printed with electrotype, and it was also employed, beginning in 1859, for publishing *Scientific American*. In that same decade, the U.S. Coast Survey was using the technology in house for copying maps.[37] By 1857, New York City had at least nine electrotyping firms; in 1874 there were more than 20.[38] Within two decades of its introduction, electrotyping had become throughout the Western world the industry-wide standard for making plates to print image-filled books, magazines, and newspapers. The widespread adoption of electrotyping helped to satisfy the insatiable demand for inexpensive printed works.

_⟋⟋⟋⟋⟍_

Led by Elkington, many manufacturers embraced electrodeposition for plating a thin—and inexpensive—layer of gold and silver on objects such as flatware, chalices, and candelabras. Traditional plating processes, such as soldering a layer of silver to a base metal or applying a mercury-gold amalgam and then boiling off the mercury, were expensive and labor-intensive, and the latter harmed workers. The perils of mercury poisoning were already well known (the venerable expression "mad as a hatter" comes from the effects of mercury poisoning on makers of felt hats), and thus many manufacturers eventually turned to electroplating.[39] The electroplating process could take many hours, depending in part on the desired thickness of the layer, but it did not require highly skilled labor or much supervision.

A person wanting to get started in the electroplating business or to add electroplating to a going concern needed little capital.[40] In the United States one could buy "Smee's Battery for Gilding and Silvering" at $2 per cell. Beginning in 1840, Edward Palmer in London sold six-cell and twelve-cell Daniell and Smee batteries starting at £2, 2s, as well as an "Electrotype Apparatus"—a modified Daniell cell—for 5s and up.[41] Supplies and accessories, including sundry chemicals and hardware, were relatively inexpensive in relation to the value that a thin layer of silver or gold added to an object of base metal. Even a small, family-owned jewelry shop could afford a rudimentary outfit that was easily expanded as sales and profits warranted. Moreover, beginning in the 1840s, innumerable manuals on electroplating were published in English, French, German, Russian, and Italian.

The ready availability and the affordability of electroplating technology hastened its adoption by enterprises large and small. Electroplating seemed a wise investment that promised profit because demand for prestige objects was growing along with increases in the size of the acquisitive middle class in industrializing nations Victorians bought facsimile symbols of wealth in large numbers because they furnished the same visual performance, in showrooms and in homes, as objects made from solid gold or silver, but were much cheaper. For example, from Richard & John Slack one could buy a dozen table forks in the "Fiddle Pattern," silver plated on nickel, for only £1, 10s, a fraction of the cost of sterling equivalents.[42] Because the salability of plated coffee pots, teapots, silverware, vases, and so forth was quickly shown by Elkington and other early adopters, manufacturers who had hesitated at first eventually embraced the new technology in order to stay competitive. Even some silversmiths added electroplating to their repertoire of metalworking technologies.[43]

Electroplating also found utilitarian applications, such as coating corrosion-prone iron with nickel or making seamless copper tubing. In photography, electroplating was sometimes used to lay down a thin. fine-grained, uniform silver or copper surface on a daguerreotype plate; daguerreotypes could also be duplicated by means of electroplating.[44] And in 1851, Paul Pretsch of Vienna invented an electrodeposition process for making plates to print photographs, but such processes did not see extensive use until the end of the century.[45]

The new technologies also gave rise to many specialty applications. In a factory in Lyons, France, copper wire was gilded and drawn thin for making lace; in Switzerland, watches were silvered.[46] Electroplaters also reproduced items such as coins, medallions, and seals, not to mention metallic versions of embroidery and wicker baskets. And, of course, the new technology made possible the "preservation," with a thin layer of metal, of perishable materials—leaves, twigs. flowers, fruits, vegetables, even insects.[47] When plating something organic, it was necessary to leave a small hole in the finished product so that the encased object could dry out completely.[48] To "preserve" objects such as fish, which in drying out might emit disagreeable scents, one could first make a plaster cast, then plate it.

Russians developed techniques for copying sculptures and bas-reliefs, and for forming and gold-plating large statues, some of which were emplaced at the St. Isaac Cathedral, at the Bolshoi Theater, and at the Hermitage. Drawing upon Russian expertise, the French firm of Christofle made for the Paris Opera house several copper statues standing more than 12 feet tall.[49] Between 1860 and 1865, Elkington fabricated 22 statues taller than 6 feet.[50] These examples suggest that the patronage of governments and wealthy institutions, including churches and museums, underwrote the perfection of processes for electroforming monumental objects.

In large-scale production of plated objects, electrodeposition took place in huge troughs filled with electrolyte (sometimes called "liquor") and connected to a battery of several large cells wired in series. Among the batteries in use were those of Grove, Smee, Daniell, LeClanché, and Bunsen, the maintenance of which still entailed replacing zinc electrodes and refreshing electrolytes. However, the commercialization of magnetos by instrument makers in the 1830s raised the possibility of an alternative source of current. After all, Clarke and others had shown that a magneto, like a galvanic cell, could produce electrochemical effects. More to the point, in 1836 Sturgeon reported to the Royal Society of London that he had "coated metals from metallic solutions" using a magneto.[51] Beginning in the early 1840s, inventors designed new kinds of magnetos for electrometallurgy.

We may surmise that magnetos showing commercial potential for these applications had to generate direct current in copious quantities, run reliably day after day, and, no doubt, operate more economically than batteries. Unless magnetos showed promise in meeting these performance requirements, few manufacturers were apt to regard them as practical. However, sometimes the mere promise of practicality is sufficient to lure a few customers into investing in a new technology. The first adoptions were made by Elkington, a large company that could afford to take some risks as it scaled up operations to meet rising demand for its silvered and gilded goods.

John Woolrich, a chemist and resident of Birmingham who became associated with Elkington, developed a series of magnetos of increasing power and durability which he patented beginning in 1842.[52] Woolrich's earliest magneto, reported in *Mechanic's Magazine*, used a stationary horseshoe magnet, two revolving coils, and a robust commutator.[53] In most respects this was essentially a Saxton-style magneto, but its few differences were telling, for it was not a scientific instrument or lecturer's tool; rather, it had been built for industrial use. Whereas every previous magneto had had a crank for rotating the coils by hand, Woolrich's machine had a pulley fixed on a shaft between two bearings. This machine was designed to be driven for long periods by a belt connected to a water wheel or a steam engine. In the early 1840s, steam engines were becoming more common, especially in large cities where waterpower might be in short supply. Some factories installed their own steam engines, and in later years some large commercial buildings made steam-generated power available to tenants by means of multi-story configurations of shafts and belts.

The increasing demand for current by the expanding electroplating industry inspired Woolrich to achieve greater power in his next magneto. This he accomplished by major design modifications and scaling up. The most radical design change in his 1844 machine stemmed from the appreciation that a set of identical magnetos could be placed radially on a single driveshaft. His first "compound" magneto had four huge

Figure 8.3
The largest Woolrich magneto, as used in electroplating. Adapted from King 1962b:355. Courtesy of Smithsonian Institution.

permanent magnets arrayed in a cruciform pattern around the armature. Eight coils were bolted between two disks that rotated on a common shaft between the magnets' poles. The shaft had lubricated bearings, a feature that was becoming more common on industrial machines of all kinds but had been lacking in magnetos of the 1830s. With its sturdy wooden frame, the magneto was nearly 6 feet tall and more than 2 feet deep; a later version had a sturdy all-metal frame. This robust machine remained in service for many years. It is said that Faraday inspected it and was delighted.[54] Remarkably, the machine still exists, carefully curated by the Birmingham Science Museum.

The third-generation Woolrich magneto (figure 8.3), apparently the last of the line, was an enormous all-metal machine with eight permanent magnets sandwiched between two disks. This model was used in the factories of Elkington (which now had hundreds of workers) and other companies.[55]

Woolrich's achievement is noteworthy for a number of reasons, not least of which is that his magnetos made possible the first industrial application of generator-produced electricity. Moreover, the compound magneto design he pioneered was the starting point for fashioning even larger magnetos to supply current for electric lights. Finally, there is the technological achievement itself, which spanned the considerable developmental distance between Saxton's display device and the hulking magnetos that worked effectively for years in electroplating factories. Woolrich obviously bridged this developmental distance in a series of steps, apparently supported by sales to

Elkington and other firms along the way. Perhaps future research will reveal the financial arrangements among Woolrich, the companies that manufactured his machines, and his customers.

Despite the many unknowns surrounding the commercialization of the first industrial magnetos, the Woolrich series calls our attention to a singular fact about the invention of many mid-century electrical technologies: their design involved little electrical engineering. In fact, Woolrich magnetos embodied only the simple electrical principles already materialized in the machines of Saxton, Sturgeon, and Clarke, with one exception: compounding furnished a smoother flow of direct current than the earlier magnetos. The real challenge in making a high-power industrial magneto was to scale up and ruggedize the parts—e.g., magnets, coils, commutator, shaft and bearings, and frame—so that it would operate over long periods with minimal wear and without shaking itself to pieces. Thus, designing magnetos that might be judged practical by prospective purchasers demanded expertise in *mechanical* engineering. Such expertise, of course, abounded in Birmingham, England's hotbed of heavy industry. Woolrich also hired local companies to cast the frames and make other parts for his all-metal magnetos.[56] Fabricating these machines was well beyond the capabilities of most philosophical instrument makers.

Elkington was not alone in trying out magnetos for electroplating. Manufacturers on both sides of the Atlantic, including Christofle in Paris, experimented with these generators.[57] On the eve of the American Civil War, George Beardslee of New York developed a compact magneto for electroplating that could be used on a smaller scale than the Woolrich machines; one model, which took up about a cubic foot, required only one-sixth of a horsepower.[58] The Beardslee magneto's novel design harked back to the first Pixii magnetos in that the coils were stationary and the magnets revolved (figure 8.4). In making the permanent magnets, V-shaped segments were cut sequentially around the steel disks. After being magnetized, the extremities of the disks exhibited alternating north and south poles. The disks were laid one upon the other and placed on a common shaft. Twelve coils were fixed rigidly to the machine's frame, six on each side of the compound disk magnet. Wires from each coil led to a switchboard on which the operator could key in any number of coils to make up the generator's output. A commutator of unique design (figure 8.4, upper left) converted alternating current from the coils into direct current. Beardslee magnetos were manufactured by Conrad Poppenhusen on Long Island in several sizes and sold at prices that reached $500.

One purchaser of Beardslee magnetos was the New York firm of L. L. & C. H. Smith, which claimed in 1861 that the Beardslee magneto "is more satisfactory in every respect than the battery. It is not only more cleanly and convenient, but it produces a better plate."[59] A few years later, however, after accumulating data about actual operating costs, Smith resumed using batteries. According to one account, "the

Figure 8.4
A Beardsley magneto. Adapted from *Scientific American* 5 (1861, n.s.), December 7.

machines did very good work, but the cost of steam power to drive them was greater than the cost of acids and metals for the batteries."[60] Apparently, in this case the long-term economies promised by magnetos did not materialize. But the relative costs of steam power and battery maintenance varied by installation, as did judgments about the most important performance requirements. Nonetheless, the vast majority of firms did not adopt magnetos, and each probably had its own reasons.

In trying to explain these very spotty adoptions, one cannot ignore the overarching economic factors that every firm would have considered: start-up costs, perhaps including a steam engine and its accessories, were high, and hard evidence of long-term savings relative to batteries was inconsistent or nonexistent. By the same token, several factors favored retention of the familiar and reliable battery, making the new technology—with its contestable performance characteristics—a hard sell to all but the most risk-tolerant manufacturers.[61] To wit, the battery was more flexible for handling jobs of varying sizes in small shops, could be maintained in house, and did not require a separate prime mover that might be off line when needed. Clearly, batteries gave firms virtually complete control over their source of electricity. For these reasons, I suspect that small firms almost uniformly would have decided against adopting magnetos. Tellingly, even Elkington did not entirely abandon batteries.[62] And manuals of electrometallurgy published from the 1860s through the 1880s gave extensive coverage to the use and maintenance of batteries, indicating that galvanic technology remained the dominant source of electric power in this varied and flourishing industry.[63]

Electrometallurgy was a technology that seemingly came out of nowhere around 1840 to become the foundation of a major industry in just a few years. Because Spencer, Jacobi, and other pioneers had already invented the basic processes, electrometallurgy arrived in the public domain essentially ready for application and elaboration; no one had to raise capital for traversing a large developmental distance. Also, the establishment of an electroplating company was not, at first, capital-intensive. In the early 1840s an aspiring electrometallurgist already could buy manuals that detailed fundamental principles and processes. Without having to issue stock and found a company, he could purchase a battery and supplies and, through practice, could acquire skills such as electro-gilding and electro-silvering. If his business was booming, he might reinvest profits in expansion. A few profitable and growing establishments, including Elkington, could even afford to try out new technologies for generating current, such as industrial-size magnetos.

In comparison with the commercialization of electric motors and telegraphs, electrometallurgy involved relatively few players and was seemingly free from competing interests and power struggles. Contributing to this uncommonly rapid transfer of technology from laboratory to shop floor were electrometallurgy's decisive performance advantages over many traditional metalworking practices. Also, this industry's prosperity was ensured by its ability to supply high-quality products to meet the demands of varied and growing markets.

Electrometallurgy's origins lay in physics and chemistry. This widely recognized pedigree helped to promote the Baconian notion, seldom previously instantiated, that science could beget technologies useful in contexts other than laboratories and lecture halls. The emergence in the public sphere of other astounding electrical technologies, notably the telegraph and its numerous offspring, also furnished incontrovertible evidence of Bacon's prescience. As a result, in both elite and popular cultures throughout Western societies, people came to believe that science is the fount of new and wondrous technologies. Soon *Scientific American* would enshrine that belief in its doctrine of scientific invention.

Although a strong case can be made that electrometallurgy was the first electrical technology to be successfully adopted for industrial activities, there are worthy contenders. Indeed, the constant battery also fostered invention of other electrical technologies that enjoyed significant adoptions beginning in the late 1830s, such as telegraphy. And electrical gadgetry soon came to be used in the detonation of explosives.

9 A Peculiar Calling

Peter Barlow was right on the mark when he stated in 1825 that the idea of an electromagnetic telegraph was "obvious."[1] After Joseph Henry showed that this technology could operate in principle, experimenters in England, Germany, the United States, and elsewhere began in earnest the quest to create functioning systems that might extend communication across hundreds, if not thousands of miles. In developing a railroad, steamship, or electromagnetic telegraph, however, the devil is in the details—in devising the myriad parts that must work together before a complex technological system can meet basic performance requirements.

The vision of the system itself may come early and often, but effective designs for specific components are far from clear at first. Moreover, the system's performance requirements change as it confronts more realistic operating environments and new players' differing expectations. To go from Henry's jury-rigged apparatus that could activate an electromagnet one mile away, to a telegraph that transmitted intelligence reliably over 100 or 1,000 miles, required inventors to come up with many new components, including sending and receiving devices.

The development process often engages the inventor in all-consuming campaigns to solve specific problems, often with the assistance of people having complementary skills. A common result is a series of invention cascades, each generating through trial-and-error a variety of artifact designs, one of which may solve the immediate problem.[2] During the 1830s and the 1840s, several groups worked out devilish details of electromagnetic telegraphs, including Charles Wheatstone and William Cooke in England, Karl Steinheil in Germany, and Samuel Morse and Alfred Vail in the United States.[3] These efforts eventuated in commercial systems, but not without large infusions of capital and the participation of new players, including government officials, investors, the press, and a mass of anonymous consumers.

This chapter introduces the story of how Samuel Morse, American art professor and painter, managed to piece together human and financial resources for transforming an obvious idea into a profitable telegraph system. It is indeed curious that a painter at the height of his artistic powers would invest so much effort in such a risky enterprise.[4] But Morse was a remarkable man who tried out many careers, including politician, anti-Catholic polemicist, and photographer, achieving exceptional success in several as he sought his true calling.

Because the electromagnetic telegraph recruited many inventors and, later, many companies building new lines, Morse became embroiled in priority disputes and patent-infringement litigation. In these adversarial settings where wealth and reputation are at stake, otherwise honorable people sometimes resort to exaggeration, misrepresentation, and worse. Morse was no exception. His son Edward, while praising his father as "a pure-hearted Christian gentleman, earnestly desirous of giving to every one his just due," also acknowledged that he was "jealous of his own good name and fame, and fighting valiantly . . . to maintain his rights; guilty sometimes of mistakes and errors of judgment; occasionally quick-tempered and testy."[5] Eruptions of ungentlemanly behavior would lead to lasting enmity between Morse and Henry. Both men rose from humble circumstances to the ranks of America's elite—one as inventor, the other as scientist and administrator—and both fought zealously to safeguard their hard-won reputations. My account of Morse's odyssey, in this chapter and in chapter 12, draws appreciably on Kenneth Silverman's biography *Lightning Man: The Accursed Life of Samuel F. B. Morse.*[6]

Samuel Finley Breese Morse (1791–1872) was born to Elizabeth Finley and Jedidiah Morse.[7] His father—a Congregationalist pastor in Charlestown, Massachusetts—was a well-known author of geography books who courted Benjamin Franklin and George Washington among his acquaintances. Two other Morse children survived to adulthood: Samuel's younger brothers Sidney and Richard. Although constantly struggling to make ends meet, the elder Morses placed a premium on educating their sons.

At Phillips Academy in Andover, Massachusetts, Samuel received a rigorous classical education. At home and at school he was trained to be "a Christian Gentleman—reverent, well mannered, and frugal, but aspiring to personal distinction."[8] The latter goal in particular would torment Morse throughout much of his life as he wrestled with conflicting impulses as to his proper vocation. Although he did not distinguish himself in his studies, at age 11 he began to show a gift for drawing, which his father acknowledged could be a useful amusement.

Morse entered Yale College in 1805. Sampling many subjects from geometry to French, he excelled in none, although science aroused some interest. He attended Benjamin Silliman's lectures, which included demonstrations of galvanic batteries, as well as those of Jeremiah Day, a natural philosopher also acquainted with electricity; and during one vacation he helped a tutor with electrical experiments. These exposures to electricity occurred more than 10 years before Oersted's discovery of electromagnetism, and so the young Morse learned mostly about electrostatics and the battery. While still at Yale, he took up portraiture, creating watercolor likenesses of faculty and other members of the New Haven community, which he sold for $1 each. He also became friendly with another Yale student, Henry Ellsworth, who decades later would become the U.S. Commissioner of Patents at a most propitious time for Morse.

After graduation in 1810, Morse, at his parents' insistence, began an apprenticeship with Farrand & Mallory, a Charlestown bookstore. But in evenings after work he indulged his passion to paint. The United States at the beginning of the nineteenth century was an artistic backwater, scarcely capable of properly nurturing or rewarding young talent. Aspiring painters usually had travel to Europe to view great collections and acquire instruction, and wealthy art collectors shopped abroad; Americans were not yet confident tastemakers. Morse became acquainted with Washington Allston, a young American painter trained in Rome, Paris, and London, who already enjoyed some recognition on both sides of the Atlantic. Morse yearned to study under Allston, perhaps even accompany him to London. Jedidiah and Elizabeth Morse resisted his plan but eventually relented, acknowledging that their son's future would not be in the book business. In 1811, Morse and Allston crossed the Atlantic.

In London, Morse was impressed by the esteem in which painters were held, in marked contrast to their lowly status in America. According to Silverman, "Londoners ranked painters with lords or barons . . . Indeed, in London art exhibitions were resorts of the fashionable, art was a constant subject of conversation, and no one was considered well educated who lacked an enthusiastic love of painting."[9] Morse's work earned him probationary status at the prestigious Royal Academy, which taught only anatomical drawing. Painting he did at home, gladly suffering Allston's relentless but constructive criticism of his canvases and learning the values of striving for excellence, attending to detail, and eschewing mere commercial success. Morse determined to master the demanding discipline of historical painting, creating magisterial scenes of events in American history that could, by stirring deep spiritual feelings, exalt the young republic.

In his spare time, Morse read classics, played piano, traveled in the south of England, attended the theater, and, partly by trading on his father's fame as a geographer, gradually gained entrée into London's literary and intellectual elite. Building such a social network was necessary for an artist whose survival as a professional would require patronage from the well-to-do.

During the War of 1812, Morse painted the monumental *Dying Hercules*. Included in a highly selective exhibition at the Royal Academy, the painting received a glowing notice in the *London Globe*. In preparation for painting his Hercules, Morse had fashioned a clay sculpture which itself was awarded a gold medal from the Adelphi Society of Arts. These accolades furnished Morse with the occasion to reassure his skeptical parents that he was making progress, achieving personal distinction. He sent them a newspaper clipping about his Hercules along with the gold medal, and not incidentally requested funds for a fourth year of study abroad. In turn, they counseled modesty in the wake of praise and told him of their straitened finances; they were, after all, now putting Sidney and Richard through Yale. Eventually they acquiesced, going deeper into debt to underwrite their son's final year in England.

After returning to America, Morse exhibited paintings in Boston and Philadelphia, gaining publicity but little else. Like many an American painter, he took to the road, securing the occasional commission for a $15 portrait.[10] This work was demeaning and barely covered expenses, but it had one benefit: while in Concord, New Hampshire, he met Lucretia Pickering Walker, and they were soon engaged.

Reminded by his parents that he would need a steady income to support a wife and family, Morse gave thought to abandoning art and briefly considered entering the ministry. He also dabbled in invention, in 1817 devising (with help from brother Sidney) a flexible piston that could be used in fire-engine pumps. The inability to cash in on the piston, despite its being brought to market by the Boston firm of Howard & Davis, tempered Morse's enthusiasm for the inventor's path, which "yields much vexation, labor, and expense, and no profit."[11] Nursing a nagging uncertainty about a permanent profession and the best route to personal distinction, Morse returned to painting.

_⌒⌒⌒_

Seeking new commissions for portraits, Morse headed to Charleston, South Carolina, where he enjoyed immediate success among the planter and merchant classes. Within three months he "had completed twenty-seven portraits and had commissions for forty more"—at $60 or $70 each.[12] Flush with cash and promise, he married Lucretia in 1818, then returned to Charleston to continue the lucrative career of portraying the elite. With his fame growing, Morse received a commission to paint President James Monroe in Washington. However, the economic downturn that began in 1819 drastically affected prices for southern crops and thus diminished his customer base. Morse returned to New England, where again he became an itinerant portrait painter.

The vision of a larger project began to take shape, one that would become Morse's most ambitious work to date. He decided to render on canvas the interior of the new chamber of the House of Representatives, and to depict all 67 members. Confident of the politicians' cooperation, Morse set to work, solving daunting problems of perspective he had not faced before. The breathtaking painting, patriotic but not historical, took 14 months to complete. It showed, in the cavernous, classically appointed chamber, a man in silhouette lighting the oil lamps of a huge lowered chandelier, surrounded by the congressmen.[13] The painting was 80 square feet in area and weighed 640 pounds. Morse had exhausted his savings in its creation, and exhibiting it in several cities failed to produce any profit. "This picture," he conceded, "has ruined me."[14] *The House of Representatives* was finished in 1822, two years after Oersted's discovery of electromagnetism—a phenomenon as yet unknown to Morse.

After another stint as an itinerant painter in New England, Morse moved to New York, which he believed was America's true capital. Demand increased for his portraits, and he even acquired a pupil. In 1824, beating out other famous artists, Morse won a $1,000 commission to paint the Marquis de Lafayette, aged hero of the War of Independence, who was touring the United States. While in Washington, painting the busy Lafayette when time allowed and watching debates in the House of Representatives over the unresolved election between John Quincy Adams and Andrew Jackson, Morse learned in a letter from his father that Lucretia, just weeks after giving birth, had died. The news arrived too late for Morse to return to New Haven for the funeral. Some say that this traumatic event indelibly inscribed in Morse's mind an appreciation for the need to communicate rapidly over long distances.

A disconsolate Morse eventually plunged back into painting, turning out portraits of such notables as William Cullen Bryant and DeWitt Clinton. Morse was soon elected an Associate of the American Academy of Fine Arts. The Academy, headed by the aged American painter John Trumbull, whose renown was based on his historical paintings of the War of Independence, nonetheless mainly promoted to its patrons the acquisition of European tastes and works. This emphasis, however, did not accord with Morse's vision of an American organization for the cultivation of American arts.

Prodded by students who found themselves literally locked out of the American Academy, Morse established the National Academy of Design, which he served as president for 15 years. This institution was dedicated to training American artists, and it succeeded brilliantly, nurturing the skills of Rembrandt Peale, Winslow Homer, and hundreds of others, notable and obscure. In delivering lectures on art at Columbia College and organizing the National Academy's exhibitions and other activities, Morse soon became "New York City's, perhaps the nation's, chief spokesman for American art."[15]

About the time of the National Academy's inaugural meeting in 1826, William Sturgeon was crafting and exhibiting electromagnets across the Atlantic. Morse became acquainted with Sturgeon-type electromagnets in 1826–27 while attending a course of lectures delivered by Professor James Freeman Dana at the New York Athenæum, where Morse also lectured on occasion. According to Morse, he and Dana became friends, and "the favorite topic of conversation between us was electro-magnetism, a science in which he was an enthusiast."[16] Upon Dana's death, in 1827, Morse came into possession of his friend's horseshoe-shaped electromagnet.

ᴫᴫᴫᴫ

Morse next planned a lengthy pilgrimage to Rome and Paris, intending to continue his art education. Several schemes to raise funds for the excursion failed to pan out, and he resorted to taking orders for copies of European masterpieces. Through this distasteful enterprise he raised $3,000.

After a 26-day crossing of the Atlantic on a sailing ship, a boat ride to France, and an arduous overland trip to Rome, Morse began his copying work and his studies in earnest. Traveling through Italy to view great art, ruins, and scenery, the New England Calvinist was appalled by the papal influence and what he regarded as the peoples' degenerate morals. Nonetheless, he expressed deep admiration for Italy's early architects, sculptors, and painters whose art, regrettably in Morse's view, served the Catholic Church. His own art, he believed, aimed at "promoting moral refinement and respect for republican ideals," but there was always the danger that it, too, might come to serve unworthy masters.[17]

Morse spent more than a year in Italy, then journeyed to Paris, where he pressed on with his copying work in the Louvre. After finishing the commissions, he embarked on another monumental painting, which would depict, in one scene, 37 masterpieces of the Louvre and several artists engaged in copying them. He spent 14 months on *Grand Gallery of the Louvre*, continuing despite a cholera epidemic.

On his return to America, Morse became Professor of Painting and Sculpture—the first such position in America—at the new University of the City of New York (later to be called New York University). This position did not include a salary. To make a living, Morse would have to sell his paintings and also charge students for instruction. And so he began to acquire commissions for works grand and otherwise. He eventually obtained living quarters and studio space in the university's new building on Washington Square in Greenwich Village.

In addition to his teaching and his painting (which produced a number of stellar works), Morse spent the next several years occupied with administration of the American Academy. He also found time to write political tracts about the purported Catholic infiltration of American institutions, honing the polemical skills he would one day turn against Joseph Henry. In 1836, still seeking his true calling, Morse ran for mayor of New York as the standard-bearer of the new Native American Democratic Association, which advocated draconian restrictions on immigration. He lost the election badly.

ᴫᴫᴫᴫ

The vision of an electromagnetic telegraph had come to Morse during his voyage home from Europe in October 1832. Another passenger on the packet ship *Sully* was Charles Jackson, a physician from Boston, who at dinner one night held forth on recent discoveries in electromagnetism. A captivated Morse, his interest in the subject rekindled, recalled later that "it occurred to me that by means of electricity, signs representing figures, letters or words, might be *legibly written down* at any distance."[18] In the weeks-long Atlantic crossing, Morse conjured up preliminary designs for an electromagnetic telegraph, freely sharing his ideas and sketches with others aboard ship. Later recollections of the *Sully*'s captain, William Pell, and several passengers—as well as a sketchbook—support the claim that Morse at this early date had conceived the rudiments of a telegraph system.[19]

Tinkering alone at NYU, he fashioned through trial and error the major parts, which eventuated in a functioning table-top telegraph. In addition to a battery and wires were two novel components: a sender, which he called a "port-rule," and a printing receiver or "register." Morse's sender was modeled after a printer's composing stick. Three feet long and made of wood, it was slotted lengthwise to receive flat pieces of metal "type," which at first Morse cast himself (figure 9.1, lower). In a mature version of the port-rule, the number and spacing of pointed, saw-like teeth and flat places along the top edge of the type indicated a digit from 1 to 9; these made up numbers that designated either a numeral or a word in a telegraph dictionary. A message could be constructed by placing a row of type into the slot. One end of a lever suspended above the port-rule rode the ups and downs of the type when the port-rule was cranked past it. The other end of the lever held a metal fork that dipped into and out of two cups of mercury, which opened and closed the circuit connected to an electromagnet in the register.[20]

Completed in September 1835, the first working version of Morse's receiver—called a "register" because it recorded the message with a pencil—was made in a wooden canvas-stretcher, an artifact that every painter had on hand (figure 9.1, upper).[21] The electromagnet's keeper, moving back and forth in response to the on-off current in the circuit, was attached to a triangular frame that held the pencil.[22] In turn, the pencil pressed upon a paper tape that was pulled along by a clockwork mechanism driven by a falling weight. The reciprocating movement of the frame imparted to the pencil the same motions made by the lever riding on the port-rule.[23] Thus, anyone who had a codebook could read a message on the tape.

By his own admission, Morse's telegraph as of 1837 "existed in so rude a form, that I felt a reluctance to have it seen."[24] Clearly it *was* rude by the standards of mechanical engineering in the early nineteenth century. Yet, taken on its own terms as a prototype, Morse's telegraph—constructed with limited resources—was a conceptual breakthrough that, surprisingly enough, worked. This was a significant achievement for a painter cum politician cum inventor who was neither an electrician nor a mechanic.

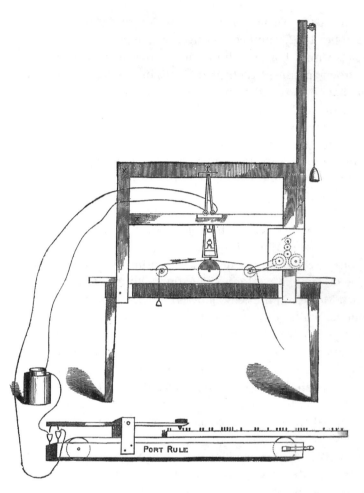

PORT RULE

Figure 9.1
Morse's original telegraph apparatus. Adapted from Prescott 1877: 422.

That said, it must be noted that the prototype could transmit through no more than about 40 feet of wire. In pondering this seemingly fatal constraint, Morse came up with a brilliant solution: "combining two or more circuits together . . . each with an independent battery, making use of the magnetism of the current on the first" to open and close the second circuit, and so on, seemingly without limit.[25] Experiments showed that the strength of current in a circuit, despite being attenuated by distance, could still effect a mechanical action sufficient to open and close another circuit. Indeed, Morse was supposedly fond of saying "If I can succeed in working a magnet ten miles, I can go round the globe."[26] "Relays" or "repeaters" became important links in commercial long-distance telegraphy in the decades ahead, making possible, for example, the 1861 line connecting New York and San Francisco, which preceded the transcontinental railroad by 8 years and ended the brief life of the Pony Express.

Other crucial solutions to the problem of limited distance were furnished during the winter of 1836–37, by Morse's NYU colleague Leonard Gale, who as a chemist had more than a passing familiarity with batteries and such.[27] Professor Gale, applying principles first enunciated in Henry's seminal 1831 paper on electromagnetism, identified the barriers to achieving greater transmission distances. Not only was the intensity of Morse's one-cell battery too low to impel current very far, the register's Sturgeon-style electromagnet also lacked sufficient turns of copper wire. By remedying both defects—using a 40-cell battery and winding hundreds of turns—the professors were able to send messages one-third of a mile. Clearly, Gale was now in a position to affirm the conformity of Morse's telegraph with known principles of electricity to potential backers.

In October 1837, confident that he had an invention of great importance Morse filed a caveat with the U.S. Patent Office in which he laid claim to an electromagnetic telegraph that recorded messages; after a delay requested by Morse to avoid jeopardizing his anticipated European filings, his friend Henry Ellsworth (now Commissioner of Patents) issued the patent in 1840.[28] Morse was eager to assert priority for his telegraph against European usurpers. Convinced that his idea had been stolen, and that all other electromagnetic telegraphs were derivative, he began a publicity campaign with an article in the *Journal of Commerce*. He claimed priority for his invention and asserted its superiority "over any of those proposed by the professors in Europe"[29] His brothers helped out by occasionally furnishing funds to the errant artist, now in the tenacious grip of his invention and painting little.

New York University's contribution to the telegraph did not end with Leonard Gale's advice on electrical principles. The labors of a recent graduate, Alfred Vail, would be indispensable for the project's eventual success. Vail's father owned the Speedwell Iron Works, a machine shop and foundry in Morristown, New Jersey. Morse appreciated that the younger Vail's mechanical know-how and access to his father's tools and workers could help improve the components. Having seen the telegraph operating and grasping its vast possibilities for technical refinement, Vail was eager to assist. The two men signed a contract that committed Vail to construct, at his own (meaning his father's) expense, a working model of the telegraph suitable for exhibition. In effect, the Vail family had become Morse's first big backer. Alfred Vail received in exchange for his labors, a 25 percent interest in the "Electro-Magnetic Telegraph" invention. He was also entitled to a 50 percent interest in any royalties accruing from the sale or licensing of European patents, for which he would construct the patent models and bear the application expenses. However, all patents would be taken out in Morse's name.[30] According to the painter, the telegraph had but one inventor.

By the middle of January 1838, Morse and Vail had sent a "pretty full letter" over 10 miles of wire. Writing about this achievement to his brother Sidney after a well-attended demonstration at Speedwell, Morse exulted: "The success is complete."[31] The intense and fruitful collaboration between Morse and Vail, who sometimes worked together at Speedwell, makes it difficult to discern which man was responsible for any particular design change. This kind of uncertainty is common in the history of complex technologies. The details are sometimes worked out through a sequence of suggestions and trials to which various participants contribute. The outcome is an invention without *an* inventor. It is clear, however, that Vail's contributions were substantial.

Morse was aware of the need to obtain endorsements from scientific authorities other than his colleague Gale. Moreover, he was now willing to call greater public attention to the project in the hope of attracting capital. And so Morse arranged to exhibit the telegraph before members of the Franklin Institute in Philadelphia. Reports of European electromagnetic telegraphs were appearing in American print media, adding urgency to his moves. Morse himself was now paying attention to relevant scientific publications, taking notes and digesting their implications for his project. Vail, too, was acquiring expertise in matters electrical.

During Morse and Vail's demonstration at the Franklin Institute, the telegraph performed perfectly, transmitting about 10 words per minute through 10 miles of spooled wire.[32] By this time Morse and Vail had modified the system slightly—for example, it now drew with a pen instead of a pencil—but it still employed the port-rule to encode messages. Members of the subcommittee assigned to examine the invention wrote a glowing report, noting that "the possibility of using telegraphs upon this plan in actual practice, is not to be doubted; though difficulties may be anticipated which could not be tested by the trials." One difficulty mentioned was Morse's plan to insert insulated telegraph wires into lead pipes placed underground: "more practical and economical means will probably be devised."[33] This reservation would turn out to be painfully prophetic.

The report of the subcommittee ended with a ringing endorsement, which emphatically urged the U.S. government to support a full-scale trial: "In conclusion, the committee beg to state their high gratification with the exhibition of Professor Morse's telegraph, and their hope that means may be given to him to subject it to the test of an actual experiment made between stations at a considerable distance from each other."[34] A copy of this report was sent to the Secretary of the Treasury. Months earlier, the Treasury Department had been directed by a House resolution to report on "the propriety of establishing a system of telegraphs for the United States."[35] Morse, aware of the Treasury Department's information-gathering efforts, alerted the Secretary that he would send along an account of his electromagnetic telegraph, and in due course he did.

Hoping to win favorable evaluations from government officials, which would increase the odds of getting federal support, Morse and Vail headed to Washington. During their stay of several weeks in February 1838, they successfully exhibited the 10-mile telegraph in the Capitol to a number of powerful men, including President Martin van Buren, members of the House Commerce Committee, and heads of executive-branch departments.[36] All sang the telegraph's praises; after all, this was an *American* telegraph, and it worked. Morse's creation, following in the wake of the steamboat, the cotton gin, the Blanchard lathe, and other important technologies, helped to focus attention on the young nation's growing genius for inventing, which perhaps merited greater governmental encouragement. One immediate result of the Capitol exhibition was a request to Morse by the House Commerce Committee (headed by Francis O. J. Smith, a lawyer and representative from Maine) for a detailed report on the telegraph. In his response, Morse suggested that a trial over a distance of more than 50 miles would indicate whether the telegraph could actually connect major cities. Building such an experimental system would, he estimated, cost around $26,000. Alluding to the Capitol demonstration, the committee's report remarked on the telegraph's "great and incalculable practical importance and usefulness to the country, and ultimately to the whole world" and recommended that the House of Representatives appropriate $30,000 for the experimental line.[37]

Representative Smith took a special interest in the telegraph, indicating a desire to partner with Morse. Indeed, in March of 1838, Morse, Smith, Alfred Vail, and Leonard Gale signed an agreement forming a company to commercialize the invention.[38] The role of each "proprietor" was specified in fairly general terms: Smith was to secure and market the foreign patents; Vail, at his own expense, would continue to perfect the invention and make prototypes; Gale would improve the telegraph's "philosophical and physical qualities or properties." Morse would help Smith exhibit the telegraph in Europe and supply the necessary apparatus made by Vail. (Once abroad however,

Morse would expand his role greatly, seeking to sell actual telegraph systems to European governments.) Shares or interests in the company were divided among the proprietors, with Morse retaining a bare majority of nine-sixteenths.[39] The agreement was also quite explicit about the company's intention to sell the telegraph to the U.S. government.[40]

What the agreement conspicuously lacked, like the Davenport-Cook venture being pursued at about the same time, was any mention of an organizational structure for making day-to-day decisions. The only nod in this direction was the problematic condition that "no partner could sell rights to the invention without consulting the others and having their consent."[41] Another shortcoming was the dearth of capital to underwrite the company's activities. However, any patents (none had been issued at the time of the agreement) would become company assets. In practice, Morse made most of the spending decisions and tapped Smith and Vail for reimbursements on the basis of detailed invoices. Clearly, this was not a company in the modern sense; rather, the men had merely agreed in a contract to work as partners toward a common goal, grounded in the shared belief that the telegraph was perfectible, marketable, and potentially profitable.

The company's first project was Morse and Smith's trip to obtain European patents, undertaken despite advice from Ellsworth that European nations would not be receptive to an American invention. They should have followed his advice. In England, their application was not seriously considered, perhaps because of opposition from Wheatstone. However, the visit did enable Morse to witness the coronation of Victoria at Westminster Abbey. In France, the reception was somewhat different: the men obtained a patent, but they were prevented from actually building a system because the French government maintained a monopoly on all telegraphs. Morse also carried on lengthy negotiations with a Russian diplomat but for naught.

Although the trip resulted in sales of neither patent rights nor telegraph systems, it had several less tangible benefits. The many telegraph demonstrations, including those before the Paris Academy of Sciences, members of the Royal Society of London, and both houses of the British Parliament, reinforced the impression, also fostered by Davenport's motor, that American inventors—like their European counterparts—were earnestly exploring uses of electricity in everyday life. Perhaps America was not so backward in applied science after all. Indeed, Morse's telegraph was effusively praised in France by important men of science, members of the nobility, and the press. The invention's technical competence, at least on a small scale, was strongly supported. And Morse, having seen the major European telegraph designs, was more convinced than ever that he alone had invented "the telegraph of a single circuit and a recording apparatus."[42] He was confident that his system would win out when governments at last bowed to what many technically savvy people regarded as the inevitability of electrical communication.

While lingering in Paris, awaiting appointments with French officials, Morse was approached by directors of the St. Germain railroad. They wondered if it might be possible to use the telegraph for regulating the movement of trains. After some deliberation, Morse presented them with a proposal for a costly project, which they declined to accept. In the decades ahead, however, railroad telegraphs became an important application, keeping track of trains on ever larger railroad networks in nearly every country. Morse also peddled the telegraph to the French military, proposing a system in which soldiers and horse-drawn wagons could carry equipment and lay wire. With telegraph wagons following front-line troops and reporting back to headquarters, commanders could act on timely intelligence from the battlefield. Such telegraphs did see extensive use in warfare, especially during the American Civil War, but in the late 1830s the French were not interested. His hope gone, Morse headed home.

In early 1837, with motor patent in hand, Thomas Davenport and Ransom Cook set up shop in Manhattan, a place they believed offered unparalleled prospects for raising money to perfect their motors. Shortly after their arrival, Edwin Williams, Secretary of the American Institute, an organization for promoting invention, approached Cook with a proposition: he and the inventors would form a joint-stock company. Davenport and Cook, rural blacksmith and cabinet maker, neophytes in the ways of big-city business, nonetheless seized the opportunity to parlay the motor patent into funding for their workshop.

Williams's lawyer, General Marvin, drew up a contract, which the inventors signed without benefit of legal counsel; it provided for the formation of "The Electro-Magnetic Association," with Davenport's patent as its major asset. The company was authorized to issue 3,000 shares of stock, priced at $100 each (figure 10.1). The company retained 1,000 shares for sale to minority shareholders, Williams received 100 shares for his services as promoter of the enterprise, and Davenport and Cook were issued 1,900 shares. Anticipating vigorous stock sales (the potential yield being $100,000 from minority shareholders), the contract also stated that the inventors would be paid, within a month of signing, $2,000 for expenses and $12,000 for their personal use. Other financial details of some importance were left unstated, such as whose shares Williams would be selling; in a way, that didn't matter, since he controlled the proceeds of the public sales. Unable to obtain a copy of the contract, and focused on building and exhibiting motors, Davenport and Cook left company matters in Williams's hands.[1]

The Electro-Magnetic Association was the first joint-stock company in America formed to develop an electrical invention. In principle, stock sales would supply the financial resources Davenport and Cook needed to make motors powerful enough to replace steam engines. The confident inventors rented space, bought equipment and supplies, hired helpers, and resumed their labors.

Believing that additional endorsements from scientific authorities would enhance the credibility of their enterprise—and thus spur stock sales—the men invited

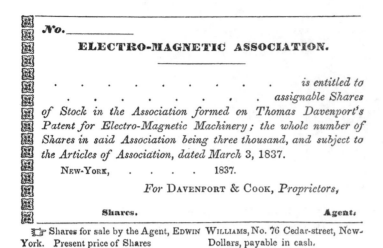

Figure 10.1
A facsimile stock certificate of the Electro-Magnetic Association. Source: Paine 1838: 33. Courtesy of Dibner Library, Smithsonian Institution.

Benjamin Silliman to examine their creations. Silliman was so impressed that he described the motors effusively in his own journal; the 8-page article, bearing a date only two days after Davenport's demonstration, was appended to the July 1837 issue. Two motors brought to New Haven performed well, operating, in Silliman's words, "with beautiful and surprising effect."[2] Silliman described Davenport as "more successful than any other person in the discovery of a galvanic machine of great simplicity and efficiency."[3]

There is some evidence that Silliman received an emolument for writing the aforementioned piece. A financial accounting of the Electro-Magnetic Association lists two curious payments for 1838: on May 11, "B. Silliman's dft . . . $53.53," and on August 9, "printing, (Silliman's Journal,) . . . $18.00."[4] There may be an innocent explanation for the check to Silliman, but on its face this entry suggests that he had a conflict of interest. In any event, that he accepted a subvention for publishing the article seems certain.

In reporting on Davenport's motor, Silliman as usual was calling attention to American discoveries and inventions. His effort succeeded: William Sturgeon, for one, published a two-page account in 1838; later that year he reprinted Silliman's article as well as two other short pieces on the American project.[5] Sturgeon, who as early as 1833 had built electromagnetic engines and displayed them powering model machines, was supportive of Davenport's efforts.

Silliman's glowing endorsement also advanced Davenport and Cook's agenda by addressing new issues of practicality. For example, he argued that scaling up the motor's parts and taking care to provide a proper battery should make it possible to increase the power "beyond any limit hitherto attained, and probably beyond any which can be *with certainty* assigned."[6] The actual rate of increase, Silliman emphasized, would have to be learned through further investigations "prosecuted with zeal, *aided by correct scientific knowledge, by mechanical skill,* and by *ample funds.*"[7] In this statement, Silliman came up just short of hawking stock in the Electro-Magnetic Association.

But Silliman wasn't done. The motor's motion could be sustained indefinitely, he maintained, simply by replacing the battery's exhausted acid and zinc electrodes. And, in a conclusion that differed sharply from Henry's, Silliman contended that "the power can be generated cheaply."[8] Despite the impeccable scientific credentials of both Henry and Silliman, their contradictory assessments were nothing more than unsupported opinions, for neither man had conducted any quantitative studies. In the coming decades, this dimension of practicality—the projected cost to users of galvanic power—would become ever more central and contentious.

_⌒⌒⌒_

Taking advantage of Silliman's encomium, the Electro-Magnetic Association included his article verbatim in a 94-page booklet promoting investment.[9] A masterly composition for its time, the prospectus built a seemingly unassailable case for buying shares in the company. Published in 1837, it began with a history of Davenport's success in obtaining rotary motion. It also recounted his many financial misfortunes. Stock sales were needed, it said, to defray the expenses of obtaining patents on the Continent (British patents were already in hand) and "to enable the proprietors to build a large working power; which, as soon as put in operation, will no doubt advance the stock several hundred per cent."[10] And two other curious claims were made in this section: that since coming to New York the inventors had increased the motor's power by a factor of exactly 528, and that by building five motors of different sizes they had shown "that the power of the machines increase[s] in a greater ratio than they are increased in weight."[11] Alluding to cost, it was noted that the materials consumed by the battery, being abundant, would make unlimited power available for human purposes.

Adding to the prospectus' scientific weight was a treatise by "Mrs. Somerville" on various electrical topics, including galvanism and magneto-electricity. The British scientist Mary Somerville was, according to a latter-day historian, "one of the foremost women of science of the nineteenth century."[12] In addition to her original contributions, she had popularized the findings of others, and so her surname might have been familiar to well-read Americans. Somerville's treatise had been excerpted from her influential book *On The Connexion of the Physical Sciences*, first published in 1834. Not surprisingly, the names of both of the Electro-Magnetic Association's scientific stars, Silliman and Somerville, were prominent on the prospectus' title page.

Articles about the motor project in seven newspapers, most in the New York area, were quoted extensively in the prospectus. All predicted that the electric motor would enjoy a rosy future. For example, the *New York Herald* presented the following claim: "The calculation is, that . . . an electromagnetic engine, constructed on this principle, will cost only one-tenth the expense of steam power, and only occupy one half of the space. There can be no doubt, in our mind, but the days of steam power, and animal power, and water power, are gone for ever."[13] For the *Baltimore Daily Gazette*, it was "obvious that the use of such a power—so safe, so cheap, so free from every thing that could cause annoyance in its use—will be, to Rail Roads, an acquisition of immense value."[14] Where did these optimistic forecasts come from? In responding to the reporters' obvious questions about operating costs, Davenport, Cook, and Williams doubtless quoted, with generous embellishment, Professor Silliman's unfounded claim that galvanic power could be produced "cheaply."

Newspaper articles also listed potential applications for electric motors. The *Baltimore Daily Gazette* suggested that they could be "applied more economically to a vast variety of manufacturing purposes than any other known power. For goldsmiths' and silversmiths' lathes, for silk and other reels, for cotton spindles, for an infinite variety of polishing purposes, for glass cutters, for ivory turner, &c., it is an invaluable power."[15] Purportedly, the article continued, a goldsmith who witnessed a motor in motion had offered to buy it for $25. The *New York Evening Star* claimed that galvanic power could even be used for "domestic purposes, such as churning, pumping, roasting jacks, washing machinery, &c."[16] This article went so far as to estimate that the patent rights for domestic applications might reach a value of $600,000. Acknowledging the inventors' need to find development funds, which could not be had from the government without special appropriation, the *New York Evening Star* counseled its readers to invest.

The newspapers' compelling endorsements of the invention were often packaged in prophetic prose. The *New York Evening Post* prognosticated as follows: "The application of this new principle of motion . . . is one of the most wonderful inventions of the age, and will hand down the name of its discoverer to future times along with those of Franklin and Fulton. It will furnish a power procurable, with the greatest cheapness and facility, in quantities to suit machines of any size, manageable with the greatest care, and free from any danger except such as necessarily belongs to rapid motion produced by any cause."[17] For the *New York Herald*, electromagnetic power "surpasses any discovery of ancient or modern times. The generalization of this principle . . . must and will create an entire revolution in all science, in all art, in all philosophy, and in all future civilization."[18]

Members of the press exposed to the performances of both motors and proprietors had been thoroughly enchanted. And they zealously conveyed to their readers an untempered enthusiasm for the cause of the electromagnetic engine. For most Americans, newspapers were a major source of information on happenings of general interest, such as inventions that heralded revolutions in everyday life. By virtue of this communication function, the press acquired an authoritative aura as well as the power to render judgments of practicality. Because of their favorable assessments and unabashed advocacy of the motor project, newspapers had become the willing mouthpiece of a stock promotion.[19]

Davenport and Cook built more motors and increased their power. A motor only 6 inches in diameter and weighing about 6 pounds reportedly raised 200 pounds a foot in less than a minute and supposedly could rotate at 1,000 rpm.[20] In claims published in *Silliman's Journal* and in *Sturgeon's Annals*, Davenport reported that he used a motor "to do all my drilling of iron and steel, to the [drill] size of one-fourth of an inch in diameter;" and that with a motor-powered lathe he turned hardwoods up to "three inches in diameter."[21] Neither of these tasks, however, required more than a fraction of a horsepower, which ordinarily would have been supplied by human hands or feet.

By the end of 1837, Davenport had set a more ambitious goal: to build a motor that could drive a Napier printing press.[22] Surely that feat would convince the world that the electromagnetic engine could become a practical technology for industry. After all, one or two horsepower sufficed to drive *individually* the vast majority of machines in workshops and factories that at present were powered by steam.

‿‿‿‿‿

Because newspapers often reprinted each others' articles, the exciting prospects of electric motors—the safe and supposedly inexpensive power of the future—eventually made their way to the far corners of America. Tantalized by the prospects of replacing other motive powers, proprietors of various firms wrote to Davenport and Cook inquiring if the motors could do their work. Some of the letters were addressed simply "Mr. Cook, New York City." That they were delivered at all suggests that the blacksmith and cabinet maker had become celebrities of a sort.

A man from the Massachusetts town of Medway wrote that he was looking for "an engine of three or four horse power" for his machine shop.[23] He wanted to know the purchase price and the daily operating expenses of Davenport and Cook's motor, and what dangers might attend its use. From Mobile, Alabama, one R. E. Redwood asked whether an electric motor's power was great enough and sufficiently steady to drive a lathe. Redwood also expressed an interest in the costs of purchase, installation, and operation.[24] The owner of a sugar mill in the West Indies wondered if an electric motor could replace the steam engines, water mills, and cattle that he used to express the juice from sugar cane.[25] A New Hampshire man asked whether Davenport and Cook could supply a machine for running an Adams Printing Press, thereby displacing a horse—provided, of course, that the electric motor's annual operating expenses were lower than "Horse Power" (figure 10.2).[26]

The letters to Davenport and Cook indicate that there was some felt need for motors of a few horsepower or less to replace animals and people as sources of power in relatively small factories and workshops. Proprietors of such enterprises, seeking a source of mechanical power, might choose the steam engine, but only with reluctance. Beyond the considerable first costs and the risk of boiler explosions, steam engines had serious performance deficiencies in these settings. For example, small engines used fuel very inefficiently, and the fact that they could not be turned on and off when used for intermittent tasks compounded their inefficiency. Of course they were also noisy and dirty, and they required constant attention.[27] The electric motor seemed like a gift from the gods—if the claims about its economy were true. Almost without exception, the aforementioned letters indicate that prospective consumers were seeking information not just about mechanical performance but also about cost-related characteristics. Hand-waving claims about cheap power might convince a potential investor now and then, but people in small businesses wanted real numbers based on experience. However, no one then could have assessed these financial performance characteristics reliably in terms specific enough to satisfy many would-be buyers.

Because of the wide range of potential uses for electric motors, estimating their operating costs would have been difficult. Certainly no single figure could have been relevant to all, or even many, applications. For example, a jeweler using an electric motor to drive saws, drills, and burnishing wheels might engage the motor only for short periods each day, thus placing little demand on the battery, whereas a printer might use the motor for many hours at a time. The jeweler might not calculate a dollar saving, but might instead relish relief from the exertions of hand- or foot-driven tools. The printer would have compared the costs of near-constant battery maintenance with those of hired hands and steam engines. Clearly, different users would have put varied demands on motors (and thus batteries) and, accordingly, weighted performance characteristics differently. Reliable data for making such assessments could be accumulated only after trials of electric motors in different activities. But that could not happen unless technically competent electric motors were available in sufficient numbers to attract a few pioneers willing to experiment with them, perhaps enamored by the motors' symbolic performance characteristics (the owner could possess *now* the power of the future).

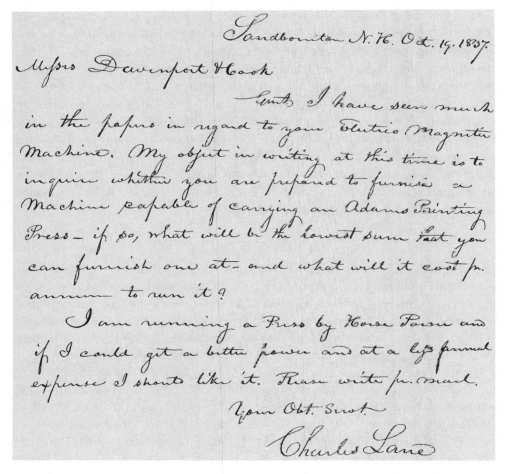

Figure 10.2
Lane's letter to Davenport and Cook, October 19, 1837. Source: Library of Congress, Thomas Davenport Papers, Box 1, Folder 8.

As of yet, the motors of Davenport and Cook, like other electromagnetic engines that instrument makers sold in the late 1830s and the early 1840s, were merely models or prototypes, useful only for exhibiting principles of electromagnetism—and the cleverness of their creators. Most such motors conspicuously lacked design refinements that mechanical engineers might have introduced, such as rigid frames and lubrication, which at the very least would have hinted at mechanical fitness and durability. And more development was needed to show that motors could drive heavy machinery. However, in the midst of the severe depression that followed the Panic of 1837, entrepreneurs and potential manufacturers of motors, perhaps discouraged by the specter of further development costs and an uncertain market, judged the electromagnetic engine impractical for their purposes. As a result, would-be buyers lacked the opportunity to evaluate the motor's practicality for themselves.

~~~~~~

In a very detailed letter to Davenport and Cook, Timothy Hudson of Medina, Ohio, posed ten questions, mostly about their motor's mechanical and financial performance characteristics. But Hudson's fifth question was "Has it the approbation of men of science generally?"[28] Apparently the opinions of scientific authorities did matter to at least some prospective consumers. The inventors did not answer this letter, perhaps because the judgments of men of science were becoming decidedly mixed.

Thomas Jones, a medical doctor and editor of the *Journal of the Franklin Institute*, had seen Davenport motors demonstrated in New York and Washington in 1837. (The Washington audience included President Martin van Buren and the heads of various executive departments.) Jones's upbeat assessment was that if they could be made more powerful such motors might compete favorably on an economic basis with the steam engine.[29] Several other men of science, however, weighed in with less favorable judgments. Robert Hare had witnessed a demonstration of a motor's lifting ability, but was evidently unimpressed. Surprisingly, he questioned whether the rotary motor was an advance over Joseph Henry's electric teeter-totter: "It does not . . . appear to me, that the practicability of the employment of electro-magnetic forces as a moving power, is more evident now, than at the period when Henry made his discovery."[30] Hare refrained from going public with his misgivings at that time, but he published them in an 1840 supplement to his chemistry lectures.

Other negative assessments reflected the elitist pure-science ideology cultivated during the Enlightenment by natural philosophers working out God's laws in nature and in the laboratory, supported by family wealth, colleges, and churches rather than by the fruits of invention. This ideology propounded the view that coupling the pursuit of wealth to scientific investigation was corrupting, almost blasphemous. Perhaps that is why so many men of science did not patent their inventions.

Silliman himself, when commenting on the Electro-Magnetic Association's prospectus, regretted "that this interesting application of electro-magnetism is attempted to be sustained by an appeal to the hope of immediate profit."[31] Yet he acknowledged that the joint-stock company was an interesting way to raise funds for further experiments.

After Joseph Henry, Charles Grafton Page was by far the most knowledgeable American experimenter working with electromagnetism. In an 1839 article about electric motors in *Silliman's Journal*, Page wrote: "It is much to be regretted, that in our country the invention should be a subject of mercenary speculation, when in reality it has no value except as an experiment, and that the public have been so far misled, as to withdraw that countenance and encouragement which the experiment really merits."[32] According to Page, the science of electromagnetism had been debased.

In a letter to Alexander Bache, Henry reflected on the prevalence of charlatanism in America, agreeing with Bache that "we must put down quackery or quackery will put down science."[33] In the very next sentence, Henry introduced a case in point: "the wonderful sensation produced in the country by magnetic machines. A company was formed in New York which succeeded in raising $12,500 . . . for experiments on the machine and after much puffing and the expenditure of the above mentioned sum the whole of course fell through."[34] Why did Henry equate the Davenport enterprise with quackery or charlatanism? The main reason, I suspect, is that he regarded his authoritative assessment of electric motors to be definitive. In his view, the enterprise was destined to fail because electric motors could not operate economically. No doubt he was also irritated because Davenport had pursued the project to build bigger motors despite Henry's counsel against it. For Henry, the true man of science strived only to increase knowledge, untainted by the vulgarity of money and speculation. Like Charles Page, Henry was concerned that the motor project had sullied the good name of electromagnetic science, which he had done so much to establish.[35]

Although sales of stock in the Electro-Magnetic Association were far from robust, Henry's obituary for Davenport's project was a bit premature. In 1838 a much briefer pamphlet was issued to accompany public exhibitions of the motor in Philadelphia. In addition to reprising Silliman's article yet again, the pamphlet touted the performance advantages of electromagnetic power over steam. For example, in a steam-powered factory multiple machines are driven by belts and shafts connected to one engine, which is always operating. In contrast, electric motors consume no power when at rest, and they can be installed on individual machines. As to financial performance, this pamphlet asserted that "the actual cost to run these machines is almost nominal, as the zinc [of the battery] is not destroyed, but merely held in solution, and can be converted into sulphate of zinc, sold to brass founders, or restored to its original state at a cost of only fifteen per cent., proving at once its immense advantages over steam in this particular." And unlike steam engines, electric motors did not require a tender, who might be "negligent and drunk."[36] The proceeds from these public lectures, the pamphlet proclaimed, redounded to the project's benefit. In a way that was true. Davenport had sold the rights to exhibit the motor in several eastern cities; most likely the exhibitors also had to buy the motors and batteries.[37]

In 1840, Davenport, who had been subsisting on meager fees from exhibitors, at last proved that an electromagnetic engine could power a printing press.[38] On a Napier press he published *The Electro-Magnet and Mechanics' Intelligencer*, which has been called "the pioneer electrical journal of America."[39] The motor that accomplished this feat, perhaps approaching one horsepower, was based on a new design that produced reciprocating rather than rotary motion. It had two hollow cylindrical electromagnets weighing 50 pounds each, into which were drawn iron rods, 2 feet long and 2½ inches in diameter, that were somehow linked to each other and to the press. Davenport's "perpendicular double helix engine" achieved 120 strokes per minute, running off 600 papers per hour.[40] To achieve this creditable performance, however, required a battery weighing 200 pounds. Davenport claimed that the expense of operating the press was 25 cents per day, but that figure is dubious. In any event, there was little demand for the journal, and it ceased publication after the second issue.

⁀⁀⁀⁀

The exhibition of the printing press attracted little attention, as the Electro-Magnetic Association had acquired a reputation as a stock swindle. Davenport had contributed to this state of affairs, not through dishonesty but by trying to get his proper share of receipts from the stock sales. But sales had not met expectations. A year after the contract was signed, only 131 shares had been sold, for a total of $8,265.15.[41] No doubt enthusiasm for new stocks had waned during the lengthy depression that followed the Panic of 1837.

The original terms of the contract were never fulfilled. Williams, after many entreaties, did release some funds to the inventors, but only to reimburse expenses. In February 1838, a disillusioned Ransom Cook sold his stock to Myrick Nelson, probably at a deep discount, and returned to Saratoga Springs. Believing that Edwin Williams had deceived his partners by keeping secret books, refusing to furnish the names of investors, and selling his own shares instead of the company's, Davenport and Nelson sued Williams. The document filed in New York Chancery Court by their lawyer, Elijah Paine, charged that Williams had perpetrated "a system of fraud and deception from the beginning."[42] Moreover, his deceit had shaken the public's confidence in the Electro-Magnetic Association, and thus adversely affected stock sales. Naturally, the lawsuit confirmed the worst fears of investors: word spread that the Association was a fraud, and its stock was soon worthless. Davenport and Nelson got no satisfaction, much less recompense, from this lawsuit; Williams was allegedly bankrupt.

As a result of Williams's shenanigans, or perhaps his incompetence, the Electro-Magnetic Association never became a real company. It had neither a chief executive nor a board of directors, although the latter was required by the original contract, and no meeting of stockholders was ever held. Nonetheless. the Association had been more than a slick stock promotion. It held the crucial patent as well as the allegiance of the indefatigable Davenport. Davenport was still plugging away on his motors, exhibiting them to anyone who visited his workshop, now located at 4 Little Green Street in Manhattan. The blacksmith perhaps harbored the hope that another Ransom Cook might, after seeing his ingenious creations, come to the rescue.

One visitor who stopped by to see the motor in motion was Samuel F. B. Morse, still struggling to commercialize his telegraph. Morse was sufficiently impressed to write a letter expressing "the high gratification with which I witnessed on Saturday last the operation of your beautiful and ingenious Electro Magnetic Engine. It performed more than was promised . . . it raised 523 pounds 6 feet in a minute." Morse further wrote: "The practicability of bringing into general use an engine on this principle as a motive power depends on several considerations, viz What is the first cost of construction? What are the consumption & cost of materials for generating the power in a given time? What is the amount of power obtained? Are the products of the decomposition in generating the electricity of little or great value?"[43] Morse's perceptive questions, which he believed could be answered by simple computations, probably went unanswered.

Sometime in late 1841 or early 1842, a man gave Davenport $3,000 in notes from an Ohio bank. He cashed the first note for $10, but the bank went bust before he could cash any more. Nearly destitute, his stock worthless, and no guardian angel in sight, Davenport retreated to Vermont and became a blacksmith once more. Although he occasionally invented a new electromagnetic device and sometimes treated invalids with electric shocks, Davenport's own health deteriorated, and he died in 1851 a few days shy of his 49th birthday.

Although Davenport failed to rescue humankind from the scourge of steam, he reached his goal of making more powerful motors and using one to drive a real machine (the Napier printing press mentioned above). However, no entrepreneur, financier, or manufacturer came calling with an offer to buy patent rights in order to set up a factory to produce motors. Without the support of such players who might develop and market Davenport's promising inventions, thereby enabling consumers to carry out experiments on perfected motors in real-world activities, steam was still safe. The electric motor, created in endless varieties during the next few decades, would remain largely a demonstration tool of science until moneyed players forecast that profits were in the offing.

Unlike many inventors, Davenport had hagiographers who perpetuated stories about his accomplishments. One of the most striking stories is that he, not Henry or Morse, invented the telegraph. According to recollections of family, friends, and other witnesses, Davenport built a prototype electromagnetic telegraph as early as 1833. The wires ran from Davenport's house to the workshop of Orange Smalley, where both men were working on the motor. There are sufficient witnesses to confer credibility on this story.[44] Missing from these accounts, however, is the recognition that Henry had built his apparatus on the campus of the Albany Academy several years earlier. Whether Davenport got the idea from word-of-mouth accounts of Henry's experiments may never be known. Nonetheless, Davenport, his eye squarely on applications, achieved what Henry apparently did not: he actually transmitted messages with an alphabetic code.[45]

Like Henry, Davenport did not publish his experiments on electrical communication, but stories spread about how Morse got the idea of using electromagnets for telegraphy after visiting Davenport and Cook's workshop in New York. This startling claim also falters on an anachronism: long before he met Davenport in 1837, Morse had already made prototype electromagnetic telegraphs.

Davenport's story has been told and retold by his descendants and by others. He has become an almost mythic figure whose accomplishments—real and imagined—are a source of pride to Vermonters. In a 1900 address to the Vermont Historical Society, one Davenport descendant said that Davenport's motor was more than an American invention, "it is a machine of independent Vermont origin; for Thomas Davenport was not only a Vermonter, but he practically completed his invention in Vermont. It was in a blacksmith shop in Brandon, under the shadow of the Green Mountains, that the great principle was discovered and laboriously worked out into concrete operative form."[46]

Thomas Davenport's story feeds into and feeds on the stereotype of the independent inventor's life, filled with triumphs and disappointments, often ending in defeat. Born into humble circumstances, Davenport was driven by the vision of improving the world's work by replacing a dangerous technology with a more benign one. Persevering stubbornly in the face of nearly constant financial woes and detractors, Davenport did make powerful motors. But, the story goes, Davenport was a half-century early; his invention could not become practical until there was a cheap source of electricity. Thus, his considerable technical achievements were not matched by achievements in the commercial realm; he died a poor man, his health broken by Herculean exertions, his dream of replacing steam unfulfilled.

Another stereotype of the independent inventor, a mirror image to that embodied by the Thomas Davenports of the world, is that of the person who succeeds brilliantly against great odds in turning an invention into a profitable product. Morse is one of the latter—his telegraph made him wealthy. But Morse also had years of desperation. However, he was more fortunate in receiving, eventually, favorable judgments of practicality from all consequential players, including consumers. After all, the telegraph was more than just an alternative to the mails for long-distance communication. The electromagnetic telegraph, freed from modes of transportation such as the boat, the coach, and the railroad, could transmit messages almost instantaneously. Such a technology was pregnant with military and commercial possibilities.

11   It's a Blast

The British flagship *Royal George* was more than 200 feet long and boasted more than 100 brass and iron cannons. Her masts—the main towering 114 feet above the deck—were the navy's tallest. Built at Woolwich between 1746 and 1756, she had served well, sinking the French warship *Superbe* and forcing another French ship aground. But one day in 1782, when the British were still fighting their wayward American colonies, the *Royal George* brought to George III's navy more than a measure of infamy.[1]

In the calm harbor at Spithead, near Portsmouth, the *Royal George*'s crew and hundreds of visitors—including wives, children, and a few working women—were enjoying a brief respite from war. In the midst of these amusements, the captain ordered the cannons moved to one side so that a pipe, ordinarily below the water line, could be repaired when the ship slightly tipped. In the meantime, another vessel made fast to the *Royal George* in order to unload its ample cargo of rum. As sailors rushed to help take on the precious brew, the list worsened and the ship also began to take on water. Ignoring warnings from a mere carpenter, the captain at first took no action. When he finally gave orders to right the ship, it was too late: in minutes, the *Royal George* capsized and slowly sank, drowning nearly 900 people. They were buried en masse in the Kingston churchyard, accompanied by no fewer than three generic monuments.[2]

The *Royal George* herself was also "memorialized." The Admiralty placed a marker buoy above the wreck to warn of the dangers lurking a mere 10 fathoms below. Hoping to salvage the *Royal George*, the Navy tried to raise her intact, but failed. Even attempts to retrieve her cannons and other valuable furnishings yielded little. And so the *Royal George* rested, decade after decade, her heavy oaken timbers yielding slowly to the elements. The marker buoy became a poignant reminder of naval incompetence.

This episode was at last resolved with the aid of electrical technology. In the first months of 1839, Colonel Charles William Pasley, a member of the Royal Engineers, reported to Faraday some experiments using galvanic electricity to detonate gunpowder underwater.[3] Franklin and others had shown that frictional electricity could ignite gunpowder, a finding that was repeated in many chemistry and physics texts. And in the early 1830s, Robert Hare had reported in *Silliman's Journal* that galvanic electricity could ignite gunpowder by heating a fine wire, an effect which he believed could be harnessed for commercial activities.[4] Prodded by one Moses Shaw, Hare devised, for the purpose of blasting rock in quarries, a fuse with a chemical primer. He also devised a new kind of battery, with 16 plates, made of zinc and copper, in a four-cell configuration. In the battery's resting position, the metal plates were fixed above a tray containing acid. To fire a charge, the operator lowered a lever that raised the tray, thereby immersing the plates in the electrolyte. The battery had enough current to detonate several charges simultaneously. Hare's reports claimed that his technology could also be employed "in all cases of blasting under water."[5]

Thus, by the time Pasley began work, Hare had shown that, in principle, electrically detonated explosions—a favorite stunt in public science lectures—might be put to use, even for underwater applications.[6] Convinced that electricity was the best agent for setting off submarine blasts, Pasley, consulting with Daniell, Wheatstone, and Faraday, worked on the technology. His system was constructed as follows: Inside a great charge of gunpowder, sealed in a waterproof wooden box, was a tiny sack of gunpowder containing the electrical fuse. The latter consisted of a thin platinum wire tucked into a small paper cartridge packed with detonating powder. Wires from the fuse were soldered to large copper wires that emerged from the top of the box. These were insulated with a coating of pitch, covered with a fabric tape, and coated again. When connected to a battery, the platinum wire got hot enough to set off the primer, which in turn detonated the entire contents of the box.

With the Admiralty's blessing, Pasley took on the *Royal George*. His first blasts, in August 1839, were carried out with a conventional fuse. The electrical fuse was tried in the following month, with his youngest son, then 7½ years old, closing the circuit. Confident in the electrical fuse, Pasley publicized his plans in advance, and so this test took place before a crowd that included Lord Durham and his son and "many other families of distinction."[7] The spectators were not disappointed.

And so began a campaign against the *Royal George* using Pasley's system of electrical detonation. With later explosions also publicized, the operation became an entertainment for the well heeled. Dozens of yachts stood by at a safe distance, their illustrious occupants hoping to glimpse a 50-foot fountain of water and assorted relics.

In addition to proving the utility of electrical fuses for underwater detonation, the clearing and salvage of the *Royal George* was a comprehensive early test of new technologies for working underwater. During the 26 months of work, blasts—sometimes using 2,400 pounds of gunpowder—were separated by periods of feverish salvage. Divers brought up everything from boilers to silverware. The divers, who also emplaced the explosives, wore rubber suits and metal diving helmets and received air through rubber hoses. When the salvage operation ceased, in the fall of 1843, the harbor had been cleared of debris and the mud mounds that had accumulated around the wreck. The Admiralty was at last able to declare the harbor safe, an act marked by removal of the ignominious marker buoy.[8]

~~~~~

Colonel Pasley's success in "salvaging" the *Royal George* revived interest in the use of galvanism to detonate mines, another application that Robert Hare had suggested.[9] The destruction of enemy vessels with submarine explosives had been a topic of desultory military interest at least since the American War of Independence. Even before the commercialization of constant batteries, experiments with galvanic detonation had taken place in several nations. Baron Pavel Schilling made copper conductors covered with tarred hemp that could be laid in water. Schilling's mines were detonated not by a hot platinum wire but by a carbon arc. Hoping to see his technology brought into use, Schilling demonstrated it before Tsar Alexander I. Instructed by Schilling, the tsar touched two wires together while gazing at a distant location where a mine had been emplaced. At the instant he closed the circuit, a conspicuous cloud of smoke rose from the explosion site. All were pleased with the result, but Schilling's system did not get adopted. Schemes for electrical detonation of undersea mines languished for the next two decades, a time of relative peace after Napoléon's defeat.

When European nations again took an interest in the technology, experiments were almost always initiated by governments, sometimes in secret. However, that was not the case in the United States. American military men were notoriously slow to promote technological change and especially reluctant to consider the offerings of outsiders. Even so, inventors and entrepreneurs commonly petitioned the U.S. government to buy their wares or furnish funds for projects.

The prevailing view in the U.S. Congress was that if private parties wanted to create new technologies, military or civilian, they should assume the risks and the costs.[10] If a technology had been commercialized, government agents could decide whether to adopt it. However, the United States made exceptions to this policy by funding a handful of projects in the 1840s and the 1850s, including Samuel Morse's telegraph, Charles Page's electric locomotive, and Samuel Colt's "Submarine Battery." In each case the inventor had to negotiate with new players, pressing officeholders and political appointees for appropriations. Although arguments for an invention's potential utility for government activities might carry weight, an official might also endorse a project if he could profit from it. Few inventors, I suspect, expected to be lobbying politicians.

~~~~~

The story of Colt's Submarine Battery was pieced together by the historian Philip Lundeberg, on whose monograph I draw for the following abbreviated account.[11] Colt (1814–1862), best known for the .45-caliber revolver that carries his name, was an ambitious inventor and entrepreneur from Hartford, Connecticut. While apprenticing in his father's dyeing firm, he began to think about deploying mines to defend American harbors. Such weapons had been invented before in the United States; even Robert Fulton, of steamboat fame, tried selling his all-mechanical contact mine to the government.[12] Although Fulton showed that underwater explosives could destroy a vessel, the government did not buy his technology. After all, the United States was plodding ahead on a conventional course to build stronger forts around harbors in response to events of the War of 1812, during which soldiers from the British fleet had burned down the Capitol and the White House.

After making his own gunpowder and covering copper wires with tar, Colt began experimenting with electrical detonation. In 1829, in a public demonstration, he managed to agitate not only a raft on Ware pond but also several drenched onlookers.[13] Colt temporarily set these experiments aside to work on the revolver. In the meantime, Robert Hare had published his findings on electrical detonation. In contrast to contact mines, an electrical mine could in principle be set off from afar by an observer monitoring the movement of ships in a harbor. Contact mines could not distinguish between friend and foe, but presumably a trained observer controlling electrical mines could do so (at least in daytime and clear weather).

Whereas Hare envisioned a network of mines that would complement coastal fortifications, Colt foresaw a radical system that would make forts obsolete. The plan for his Submarine Battery, which began to take shape in 1836, was to cover a harbor with a grid of mines connected by undersea cable to a "torpedo tower" holding a galvanic battery and switches.[14] After observing an enemy ship's position in the grid, a man in the tower could close the appropriate circuits, setting off mines near the vessel. The precise working of the system and the designs of the battery and the fuses were secrets that Colt disclosed selectively.

In 1841, as tensions flared between Britain and America over the boundary between Maine and Canada, Colt saw an opportunity to pitch his Submarine Battery to the U.S. government. Having been unsuccessful in an earlier attempt to sell his revolver to the Army Ordnance Office, which usually vetted military technologies peddled to Washington, Colt decided to try a more political strategy. He enlisted the aid of Samuel Southard, prominent Whig senator, acting vice president, and former Secretary of the Navy. Southard sent a letter to President John Tyler introducing him to Colt's system. Next, Colt himself wrote to the president requesting government funding for his invention, which, he insisted, could be deployed to defend a harbor the size of New York's for less than the cost of one steamship. Acknowledging Fulton's demonstration "that a certain quantity of Gunpowder discharged under the bottom of a ship would produce her instant destruction," Colt claimed that his invention could destroy a single ship or an entire fleet while allowing friendly vessels to pass safely. Colt's request for support to perfect the invention placed the following demands on the government: reimbursement for expenses estimated at $20,000, provision of other forms of assistance, and payment of an unspecified sum "as a premium for my secret."[15] Although Colt was granted a meeting with President Tyler and Navy Secretary George Badger, there was no pledge of support; the government, after all, was in terrible financial straits.

But Colt had another hand to play. Exploiting an invitation to visit Russia, he wrote to Senator Southard threatening to make the trip unless the government acted on his proposal.[16] This gambit resulted in a gentlemen's agreement that Colt would receive support from a $50,000 naval appropriation for weapons development. However, this gesture was nullified when all but one member of Tyler's cabinet, including Badger, resigned. Undaunted, Colt met with the new Navy Secretary, Abel Upshur, pushed his plan, and revealed its secrets. Upshur was sufficiently impressed to give Colt a $6,000 advance to show if his mines could indeed blow up a ship. This was a curious decision: Fulton had already proved the destructive power of underwater explosions, and Pasley and others had previously detonated underwater explosives electrically. With this money Colt could show only that he had mastered the technology sufficiently to do what was already known to be possible. Perhaps Colt and Upshur figured that a dramatic public exhibition would bolster support for further government subsidies.

In the meantime, Colt formed the Submarine Battery Company, selling stock to, among others, Senator Southard. Presumably the company would eventually manufacture components of the Submarine Battery and sell them to the government, and stockholders would benefit.

Aiming to perfect his system, Colt set up shop in laboratories at New York University, where the physicist John Draper and the artist-inventor Samuel Morse were employed. Colt and Morse had a common electrical interest: sending current long distances through insulated wires. Colt borrowed some of Morse's long wires and learned from him about Hare's blasting technology, which used a high-current, low-tension battery.

~~~~~~

Colt's first public demonstration took place in New York Harbor on July 4 1842. The location near Castle Garden on Lower Manhattan was apposite because of its proximity to a large press corps, which dutifully turned out and reported the event. From the deck of the vessel *North Carolina*, Colt set off the explosive charge below a slowly drifting hulk outfitted with makeshift masts and pirate insignia. According to the *New York Evening Post*, "the vessel was shattered into fragments, some of which were thrown two or three hundred feet in the air, and there was not a single piece left longer than a man could have carried in one hand."[17] Although Colt furnished an account of this spectacle to the Navy, high-ranking officers had not witnessed the demonstration. An official trial in Washington would still be required, and Secretary Upshur pressed for the funds.

However, changing political winds began to impede Colt's progress. Tensions with Britain subsided, and Senator Southard died. More ominously, former president John Quincy Adams, now a member of the House of Representatives, fomented opposition to Colt's project on both moral and practical grounds. Nonetheless, Colt prepared to exhibit his technology in Washington. Through Secretary Upshur's auspices he received sundry aid from the Navy; however, he kept a studied distance from naval officials, not ready to disclose his secrets to them unless his conditions were met. No doubt he also feared that, should hide-bound military men be too closely involved, they might place obstacles in his way. Although Colt continued to work political levers outside the normal channels of military procurement, this strategy was risky because it could alienate the very players who might be called upon to evaluate his invention.

On the Potomac River, adjacent to the Washington Arsenal, Colt had stationed his floating target, an old clam boat. Like the New York trial, this was a public affair, with 8,000 spectators on hand, including President Tyler and his entire cabinet aboard a steamer. Appreciating that this exhibition needed a new wrinkle (blowing up an old boat electrically was no longer a novelty), Colt had retired to Alexandria, about five miles south on the Potomac. From there he would set off the charge when alerted by a 24-gun salute. On August 20, 1842, all went as planned; the *New York Evening Post* reported that "the old craft was sent in ten millions of fragments five hundred feet into the air, and then fell into the water with a roar like that of Niagara."[18] The president and his entourage steamed down to Alexandria, invited Colt aboard, and gave him hearty congratulations. The inventor even received a bouquet of flowers from Tyler's daughter. Apparently, however, no one in this august party bothered to examine Colt's apparatus.

Setting off an explosion at this distance through an underwater cable was an engineering feat. Even the men of Capitol Hill took favorable note, although John Quincy Adams was still dead set against the entire concept, arguing that it was not "fair and honest warfare."[19] Adams's moral reservations notwithstanding, Congress appropriated $15,000 for continued tests of the Submarine Battery. The tests would show whether the system could meet the basic performance requirements, including the abilities to destroy the largest warships, remain functional after an explosion, operate at a distance beyond the range of large guns, and allow friendly vessels to pass safely. Colt, unready to conduct a trial that would resolve these issues, nonetheless wanted to blast another ship in New York Harbor, perhaps to keep public interest and pressure on Congress high. Upshur was unhappy with this plan, threatening to withhold funds; however, Colt, as a participant in the American Institute Fair, went through with it anyway. With cable borrowed from Morse, he managed to blow up a brig named *Volta*. In October 1842, a few weeks before this event, an Alexandria newspaper disclosed some mundane details of the Submarine Battery, perhaps with Colt's blessing. The explosive, contained in a sealed iron box, was ignited by a fuse consisting of a platinum wire embedded in the gunpowder; the current came from a battery of unspecified design.[20] These technologies were scarcely secrets worth keeping, but the use of Morse's miles-long submarine cable was something genuinely new. In recent times, Lundeberg, who showed that Morse and Colt were collaborating on cable design, reported that it consisted of strands of copper wire, "wrapped with cotton yarn impregnated with asphaltum and beeswax" and surrounded by a lead sheath.[21] The use of a similar cable in Morse's first long telegraph line would nearly cause his project's undoing.

In further experiments, Colt modified the electrical system by twisting the platinum wire into a tight coil, adding wires to the cable, and increasing the size of the battery.[22] Before he could carry out what he hoped would be the definitive test in Washington, Secretary Upshur, his supporter and the keeper of the invention's secrets, was killed (with four other federal officials) while witnessing an ordnance exhibition aboard the new steamer *Princeton*. Also present on deck during the notorious accident was President Tyler, who was not hurt.[23]

_⌒⌒⌒_

In the spring of 1844, Colt was at last ready to lay a minefield and exhibit the Submarine Battery's full capabilities. The culmination of his project took place near the Navy Yard on the Anacostia River, a tributary of the Potomac. As in his previous trials, Colt did not deign to include military observers; nonetheless, he did receive significant Navy materiel. The impending trial was widely publicized, attracting an armada of private vessels and dense crowds along the Anacostia's western shore. Excitement ran high as seemingly everyone in the capital rushed to arrive before the appointed time. President Tyler, members of his cabinet, and other officials were on hand, observing from the deck of a steamboat anchored nearby. The loss of Tyler's colleagues in the *Princeton* accident had not dampened the commander-in-chief's enthusiasm for watching exhibitions of new military technology. The target, an old 500-ton vessel named *Styx*, was sailed into the minefield and then abandoned to drift, propelled only by a breeze. Some distance away in his "torpedo tower," Colt fired a mine, but it went off ahead of the *Styx* and did no damage. He fired a second mine, and *Styx* again escaped. The third detonation scored a hit, though not exactly a bulls-eye. The *Styx*, her bow shattered, had been mortally wounded. With a large aft portion still intact above the water line, the old wreck resisted further efforts by Colt and others to blast her to smithereens.

As Lundeberg noted, the mysteries of this demonstration are the location of Colt's torpedo tower and how Colt determined which mines to fire.[24] And despite Lundeberg's sleuthing, the mysteries remain. Perhaps Colt's lack of candor on these matters and his failure to consult with military men contributed to a negative assessment published in the *Army and Navy Chronicle and Scientific Repository*: "As experiments, these, as many others have been, were very beautiful and striking, but in the practical application of this apparatus to purposes of war, we have no confidence."[25] On the contrary, Colt believed that the Submarine Battery had substantially met its performance requirements. But his apprehension only deepened in the weeks ahead, as the Secretary of the Navy and the Secretary of War responded to a House resolution calling for a professional assessment of the invention.

Secretary of War William Wilkins invited distinguished military and scientific authorities to weigh in on the Submarine Battery's originality and military value. Robert Hare replied that Colt's use of galvanism to detonate explosives was hardly new. Moreover, Hare expressed skepticism that it could arrest moving vessels. Joseph Henry's letter to Wilkins was no less damaging to Colt's cause: "The practicability of exploding gunpowder at a distance, in this way, was established by the experiments of Dr. Hare. . . . I do not think it in the least degree probable that Mr. Colt has added a single essential fact to the previously existing stock of knowledge on this subject."[26] These assessments seem unduly harsh, insofar as neither Hare nor Henry had inspected Colt's apparatus. Perhaps it was that very fact—they could not learn firsthand the details of what Colt had created—that occasioned their negative judgments.

Also perplexed by Colt's evasiveness, Colonel Joseph Totten, chief of the U.S. Army Corps of Engineers, represented the views of professional military men. Totten, of course, was not a disinterested party, for the Army Corps of Engineers had the responsibility for building the very coastal fortifications that Colt insisted his Submarine Battery made obsolete. Totten defended his turf, offering many objections to Colt's system, such as the torpedo tower's vulnerability to attack. His conclusion was relentlessly devastating: "The project of Mr. Colt, as a sole means of defence, is wholly undeserving of consideration; as an auxiliary, although it might in some situations be resorted to, it should in all, or nearly all such cases, be regarded as inferior to means that have long been known; and, even when resorted to . . . it may be applied without any indebtedness to Mr. Colt, either as an inventor, an improver, or an applier of the process."[27]

Secretary Wilkins, resting his case on the testimony of Hare, Henry, Totten, and other authorities, recommended against the government buying Colt's secret. On the advice of Representative Henry Murphy, Colt resorted to a desperate measure: he revealed some details of his technology to Henry Ellsworth, the Commissioner of

Patents. After studying Colt's draft application, Ellsworth affirmed that the invention was sufficiently novel to merit a patent. But this evaluation did not address the invention's usefulness under realistic conditions, which remained the major sticking point. In its final judgment, the House Committee on Naval Affairs did not recommend adoption of Colt's system.

Colt's ambitions had been thwarted in large measure by professional military men, whom he had avoided engaging in the first place. His suspicion that they would not give his invention a fair hearing had become a self-fulfilling prophecy. Colt failed to understand that these men, who were essential players in the process of developing and deploying military technologies, had to be won over, their favorable assessments assured. After lobbying members of Congress and the president, Colt did receive some funding, but in the end his politicking was in vain: Congress did not accept the proposal of a secretive inventor who lacked the endorsements of military and scientific authorities. Colt's tactic of retaining the supposed secrets of the Submarine Battery for use as a bargaining chip had backfired.

By attempting to perfect his technology with only minimal government support, without a detailed reimbursement schedule and no contract in the event that it met its performance goals, Colt was pursuing a high-risk strategy. (Sales of stock in the Submarine Battery Company might have provided additional funding, but this venture seems not to have raised much capital.) In the end, Colt was unable to parlay the success of his final Washington demonstration into a financial commitment from the government—his only targeted customer. Samuel Morse's federally supported project, which proved the technical competence of his telegraph, also failed to win a government contract. However, Morse had options for profiting from his invention that Colt lacked, as a large commercial market for telegraphy emerged immediately.

Colt never submitted a patent application for the Submarine Battery, and the secret of his targeting system died with him. But he did eventually receive lucrative government contracts for firearms, whose mode of action was painfully transparent.

American military officials took no interest in submarine weapons (then called "torpedoes") until the Civil War, when a Union vessel, the *U.S.S. Cairo*, became the world's first casualty of an underwater mine.[28] The *Cairo*, named for the town in southern Illinois, was one of seven ironclad warships that James Eads constructed in 1861 under a contract with the Union. The plan was that these vessels would choke off supplies to the Confederacy by taking control of the Mississippi River and its tributaries.

Costing nearly $90,000 and built in a few months, the *Cairo* bristled with 13 heavy guns and a howitzer, and was protected by nearly 100 tons of shatter-resistant iron armor. She was a side-wheeler, 175 feet long, driven by two large steam engines fed by five boilers. A floating fortress, the *Cairo* was also heavy and slow, and her flat underbelly of oak was unprotected. After several assignments, the *Cairo* was ordered to join a small fleet that would steam up the Yazoo River, a tributary of the Mississippi, to open a backdoor route for an assault on Vicksburg, Mississippi.

The Confederacy had many navigable rivers that Union forces might penetrate, but not nearly enough vessels for defense. Fortifications were built along rivers, but the Confederacy also invested heavily in perfecting torpedoes—both stationary mines and movable weapons. However, no Union ships had been lost to these technologies—indeed, no mine had been known to explode—before the Yazoo incursion. Because the federal government lacked comparable technologies, Union officers probably had little knowledge of the potential destructive power of these devious devices.[29] Few would have remembered Fulton's or Colt's dramatic demonstrations.

Men on the *Marmora*, the first Union vessel to navigate the Yazoo, spotted a number of floats and small scows, which they took as signs of a minefield. With a musket, one sailor "fired at one of the objects, which blew up with a tremendous explosion that rocked the boat from stem to stern."[30] Suddenly all aboard acquired a new respect for Confederate technology. A second mine detonated, with less effect, as the *Marmora* and other vessels headed downstream to deliver disturbing news about the minefield to the *Cairo* and the rest of the flotilla.

To hasten progress up the Yazoo, Commander Henry Walke sent the *Cairo*, the *Marmora*, and a ram to protect the two tinclads assigned to clear the mines. Lieutenant Commander Thomas Selfridge, captain of the *Cairo*, was put in charge of the expedition and urged to use the utmost caution. On December 12, 1862, the flotilla steamed up the river, stopping just short of what was believed to be the minefield. Several small boats were lowered from the *Marmora* and the *Cairo* so that their men could liberate the mines from their moorings. In the meantime, the *Cairo* had drifted slightly toward shore. Selfridge, putting the engines in reverse, ordered the *Marmora* to proceed. Adding to the confusion of big ironclads maneuvering in the river was fire from a distant battery.

As the lumbering *Cairo* began to back up, preparing to fire on the shore positions, it set off a mine. According to recollections of a *Cairo* crew member, First Class Boy Yost, the mine "exploded under our starboard bow, a few feet from the center and some 35 or 40 feet from the bow proper just under our provision store room, which crushed in the bottom of the boat so that the water rushed in like the roar of Niagara. In 5 minutes the forward part of the Hold was full of water and the forward part of the gundeck was flooded."[31] Perceiving that his boat was doomed, Selfridge headed her to shore. Twelve minutes after the blast, the Yazoo River swallowed the *Cairo*, but not before all men aboard—more than 100—had reached safety.

The surprising loss of an ironclad to a Confederate mine was big news. On the next day, both the *Chicago Tribune* and *New York Times* reported the debacle in front-page articles.[32] These accounts made clear that the mine—a large bottle (called a demijohn) filled with gunpowder—had a mechanical detonator. When the taut mooring rope was tugged, it moved a match-like friction device in the neck that ignited the primer, which in turn set off the mine. The mine's mode of construction was inferred from other mines in the area that men of the *Marmora* had disabled and inspected. Despite these contemporary accounts, a rumor arose that the *Cairo* had been sunk by an electrical mine. That rumor was perpetuated in an exhaustive history of the Confederate Navy published in 1887.[33] It persisted into modern times. Indeed, the *Cairo's* twentieth-century excavator, Edwin Bearss, claimed that the mine was exploded by the spark from a galvanic cell, actuated from the shore.[34] More recently, however, John Wideman dug deeply into archives and set the record straight.[35]

The mechanical mine that destroyed the *Cairo* was one of many torpedo designs that the Confederacy deployed against the Union Navy. As Wideman shows, the rebels made up for deficiencies in conventional weapons and materiel with materials at hand, ingenuity, and more than a dash of daring. Among the other mechanical mines were keg torpedoes, raft torpedoes, drift torpedoes, and spar torpedoes, the latter held at the end of a ram.[36] The most sinister torpedo of all was made from an irregularly shaped container of cast iron, which was filled with gunpowder and coated with coal dust. A few of these "coal torpedoes," surreptitiously planted on Union coal barges, made their way into ships' boilers. In all, Confederate torpedoes destroyed 29 Union vessels and damaged 11 others.[37]

_⁓⁓⁓_

Although the *Cairo* did not fall victim to an electrical mine, the Confederacy did develop such weapons at an early date. In March 1862, Union soldiers entering Columbus, Kentucky, came across several iron casks containing grapeshot and gunpowder. These mines, which had been buried along the riverbank, were connected by wires to firing stations in Columbus.[38] Many electrical mines were jury-rigged affairs, built by mechanics hoping to land the Confederate government's bounty for sinking a Union vessel. However, a Virginian named Matthew Maury (1806–1873) spent much of the war perfecting electrical mines and firing systems.[39]

Matthew Maury had been Superintendent of the U.S. Naval Observatory in Washington. In that capacity he had digested reams of data on wind and currents from countless ships' logs, and constructed charts that enabled navigators to make much speedier crossings of the Atlantic, saving shippers untold millions of dollars. At the outbreak of the Civil War, Maury relinquished his federal post in order to assist the Confederacy at the rank of Commander. He was sent to England to obtain materials denied the South by Union blockades and to work on perfecting electrical mines; he also served as the Confederacy's informal ambassador to England.

Maury designed a minefield that echoed Colt's system. Two men, observing from different positions, would pinpoint the location of a target vessel. Communicating by telegraph, they could then detonate the nearest mine. After spending 2½ years in England, Maury was at last ready to deploy his system. On May 2, 1865, he landed in Havana, only to learn that General Lee had surrendered and the war was over.[40] His system went into storage in Havana.

The *Cairo's* demise changed naval warfare forever, not only in America but also in the rest of the maritime world. Immoral or not, mines—many of them electrical—would eclipse shore batteries as a major defense against invading vessels, just as Colt had foreseen. However, in virtually every nation it was governments, not independent inventors, that financed their development, often in secret programs.[41]

Awakened from its slumbers by the Confederacy's prowess in mine warfare, the United States entered this technological arena, one of its naval officers conceding that "military and naval engineers seem to have agreed that the most useful form of defensive torpedo is that which is exploded by an electric battery, and their attention has recently been particularly directed towards perfecting the apparatus by which its explosion may be controlled."[42] In 1869, the federal government established the Naval Torpedo Station, a facility for developing torpedoes (mines and movable devices), at Newport, Rhode Island. In 1874, one of the Torpedo Station's publications declared that electricity was "almost exclusively used as a firing agent."[43] Samuel Colt died the same year as the *Cairo's* demise and so did not witness the government's embrace of his brainchild.

Beginning in the 1860s, inventors in England, Germany, the United States, and elsewhere devised magnetos that could be used in place of batteries for detonating mines and for other applications requiring only a brief surge of electricity. A typical magneto was operated by a plunger or crank that engaged a gear which, in turn, rotated a magnetic armature between the poles of stationary coils; a spark ignited the primer. Among the inventors of "torpedo exploders" were Siemens-Halske (Germany), Charles Wheatstone (England), and, in the United States, George Beardslee and H. J. Smith.[44]

Frederick Abel, who worked for decades on electrical fuses and power sources in the employ of the British government, outlined the performance advantages of electrical mines (in contrast to mechanical ones): "They may be placed in position with absolute safety to the operators, and rendered active or passive at any moment from the shore . . . they can be fixed at any depth beneath the surface . . . [and can] be removed with as much safety as attended their application."[45] Abel also oversaw development of electrical detonators for guns and cannons, which were extensively deployed by Her Majesty's Navy. He was knighted for his contributions to the arts of war.

In the early 1830s, when Robert Hare was writing about electrical blasting, massive engineering projects were being undertaken in Europe, the Americas, and elsewhere to hasten travel, trade, and communication and to serve manufacturing. The earth's surface was being transformed on an unprecedented scale by the construction of canals, railroads, and harbors along with the expansion of mines and quarries for supplying countless materials needed by burgeoning industries. Perhaps the best known of the mid-century projects were the Suez Canal and the American Transcontinental Railroad, both completed in 1869.

Electrical blasting contributed to the success of some projects because it enabled large explosions to be set off at a safe distance, sometimes simultaneously. Safety was an important consideration because this era witnessed the adoption of explosives much more powerful than gunpowder. including nitroglycerin and its less temperamental derivatives dualin and dynamite.

An especially ambitious American project was the railroad tunnel through Hoosac Mountain in northwestern Massachusetts.[45] Although a railroad completed in 1842 already connected Boston to the Hudson River at a point near Albany many merchants and businessmen favored building a second line further to the north. With gentler grades and fewer sharp curves, it would accommodate longer trains, reducing shipping costs and giving Boston an economic advantage over New York in capturing more of the expanding western trade fostered by the Erie Canal.[47] When the tunnel was proposed to the Massachusetts legislature in 1851, the projectors—the Troy and Boston Railroad Company—estimated that it could be completed in 4 years at a cost of less than $2 million. The audacious project's feasibility and rosy financial forecasts were questioned. Oliver Wendell Holmes even penned a poem intimating the tunnel's impossibility:

When the first locomotive's wheel
Rolls through the Hoosac Tunnel's bore—
Till then, let Cumming blaze away,
And Miller's saints blow up the globe;
But when you see that blessed day,
Then order your ascension robe.[48]

Despite detractors and political opposition, the legislature was enticed by the projected economic benefits, and agreed in 1854 to the loan.

As with many grand projects, the early time and cost estimates proved to be wildly optimistic. In fact, the tunnel (nearly 5 miles long) was not put into service until 1875. It might have been completed sooner had it not been for on-again off-again political support and work stoppages occasioned by the railroad company's insolvency, which led to a state takeover. The taxpayers of Massachusetts eventually picked up most of the tunnel's tab, more than $14 million.[49]

A pioneering project on the scale of the Hoosac Tunnel entails an appreciable developmental distance. In this case, the massive cost overruns were borne by the state, which could afford to fund experiments with new technologies. Building the tunnel was a continuous learning exercise involving ventilation, drilling, blasting, and the hauling of waste rock. Technologies employed at the beginning of the project were obsolete by its end, replaced by others that had been created in the interim. Mistakes made along the way, such as the early use of a boring machine that got stuck and was abandoned in the shaft, imparted valuable lessons to civil engineers.

With the failure of the boring machine, holes for inserting black powder and blasting caps had to be drilled by hand. The blasts were set off by a ribbon of black powder or tape fuses, ignited at a distance. Using these traditional technologies, progress was painfully slow. The tunnel could advance no more than 60 feet per month. At that rate, the tunnel, though being bored from both eastern and western ends, might not be done for decades. However, a Massachusetts inventor named Charles Burleigh designed a pneumatic drill whose bit pounded, turned slightly, and pounded again, ad infinitum; it could penetrate rock to a depth of more than 3 feet. Project engineers enthusiastically adopted the Burleigh drill, and supplied it with outside air from compressors driven by steam power on the west and water power on the east. Mechanics mounted four Burleigh drills, each weighing 500 pounds, on a carriage running on tracks that could be easily positioned against the working face. These machines proved to be effective and relatively easy to maintain, and so hastened progress. They did, however, produce a deafening din.

After the Civil War, as the project got back on track (again) under a new general contractor (the Canadian firm of Walter & Francis Shanly), engineers began trying out new explosives to speed up the work. Nitroglycerin, whose invention dated to 1848, was the most powerful explosive known, but it was liable to go off if handled carelessly. Nonetheless, it could shatter a lot more rock than black powder, and so after trials in 1866 it was adopted for the Hoosac Tunnel—the first major use of this explosive in the United States. To avoid having to transport nitroglycerin over long distances, a local chemist, George Mowbray, was hired to operate a nitroglycerin factory on the construction site, which supplied in total half a million pounds of the oily substance. There were still accidents, but nitroglycerin was responsible for only a small fraction of the nearly 200 men killed during the project's duration.

Most casualties in blasting of any kind were caused by "hanging fire" when the tape fuse failed to ignite the charge.[50] In that event, the miners would warily return to find out what went wrong, sometimes arriving just in time for a tardy explosion. Electrical detonation promised to solve this problem, and was adopted along with nitroglycerin. The holes made by the Burleigh drills received a container of nitroglycerin and an electrical fuse invented by Charles Browne of North Adams, Massachusetts, and manufactured locally. The fuse consisted of a hollow wooden cylinder containing, in a tiny wooden plug, a small amount of a fulminating chemical that served as the primer. Into this powder were inserted two copper wires, held firmly in place with a tiny gap between their ends. Another explosive material filled the remaining space in the wooden cylinder.

The fuses were connected in series and the wire run to a source of electricity of sufficient tension to span the gap between the wires' tips. The generator was described in *Scientific American* as "a round case about 15 inches in diameter and 4 inches thick, resembling the case of a clock, and having a small crank protruding from its upper surface by turning which a current of electricity is generated."[51] This box could have easily accommodated any of the available electrical blasters. Indeed, the article stated that the box contained a magneto. However, *Scribner's Monthly* reported that the power source was an "electrical machine," which perhaps at that time still meant an electrostatic generator; a small one would have fit nicely in that round case.[52] Also, in several nations, including the United States and Austria, frictional generators—which give great spark—had been commercialized for detonating mines and torpedoes, and could have been easily adapted to terrestrial blasting.[53] It is also possible, perhaps likely, that project engineers were trying out various generators. Visitors might have observed different machines in use, depending on when and where they stopped by. In any event, electrical detonation had been coupled with the use of nitroglycerin to accelerate the project.

On Thanksgiving 1873, 600 railroad officials, engineers, state legislators, and other dignitaries gathered to witness the final blast along with a gaggle of newspaper reporters. When the circuit was closed at 3:05 p.m., Hoosac Mountain rumbled as 160 pounds of nitroglycerin—more than double the usual dose—violently dislodged the last 13 feet of rock separating the tunnel's segments. After the dust had settled, State Senator Johnson, who chaired the tunnel committee that year, was the first person to walk through.[54] Although the segments lined up beautifully—offset by only ½ inch horizontally and 1½ inches vertically—much work (e.g., completing the brick lining and laying the permanent track) remained to be done before the tunnel would be ready for trains. The first trains did not pass through the tunnel until early 1875.

Although the Hoosac was not the longest tunnel in the world when completed (that record was held by the Mount Cenis tunnel through the Alps, nearly 8 miles long), it was for decades the longest in the western hemisphere. In its time the Hoosac Tunnel was held up as a supreme engineering achievement testifying to the can-do spirit of American civil engineering. Newspapers around the country carried stories of the tunnel's impending completion, presenting capsules of the project's history, emphasizing the tremendous obstacles that workers had to overcome, describing the new technologies, and inspiring in readers a sense of wonder at this nearly superhuman feat. The *New York Times* proclaimed it "one of the greatest engineering works of the age."[55] The Hoosac Tunnel became an instant American icon, exemplifying what the historian David Nye calls "American technological sublime."[56]

Committed to fostering commerce through new transportation and communication technologies, promoters seldom pronounced a project too outlandish to contemplate; and sometimes they convinced capitalists or a government that the huge investment would pay off handsomely. The northern rail route did yield many of the promised economic benefits.[57] Nonetheless, Massachusetts eventually sold the line at a loss. The Hoosac Tunnel, said by some to be haunted by the ghosts of those who died building it, remains in use today, silently affirming the usefulness of electrical blasting and the efficacy of state subsidies for infrastructure.[58]

_⟋⟋⟋⟋⟍_

The Hoosac Tunnel was one of many earth-modifying projects that exploited electrical detonation. Civil and mining engineers increasingly adopted electrical blasting, believing that its performance advantages over conventional methods were decisive. An electrical fuse was clearly safer because it did not "hang fire" or set off a blast prematurely. According to one source, "probably nine tenths of the accidents which occur in the use of powder in mining arise from difficulties inseparable from the use of the common tape fuse."[59] Blasting would remain dangerous, but the electrical fuse could eliminate the major source of accidents and thus fatalities.

Electrical detonation also allowed more than one blast to be set off simultaneously, an impossibility with other methods. This lesson had been learned in 1843 during the construction of the Southeastern Railway in England, which connected Dover and London. Standing in the way of this project was Round Down cliff, about 400 feet tall. To overcome this obstacle, the company electrically detonated at once three charges of gunpowder, totaling 18,000 pounds. The result was a thundering cascade of chalk rock, estimated to weigh more than a million tons. This spectacular feat was also a financial triumph, for the company saved a year's labor and £10,000.[60] Perhaps the largest simultaneous blast of that era took place in 1876 at Hell Gate in New York's East River, where a large reef imperiled shipping. In a single blast, engineers detonated nearly 50,000 pounds of explosives with 4,427 charges.[61] Electrical detonation was clearly the technology of choice for removing large masses of material with simultaneous blasts.

Although electrical blasting was safe, reliable, and highly effective in comparison with the simple tape fuse and blasting cap, it was a complex technology that involved not only fuses themselves but also wires, perhaps switches, and a source of electricity that had to be maintained by a knowledgeable person. And it was more expensive. These factors posed a significant barrier to widespread adoption, especially on small projects. Nonetheless, makers of fuses and electrical accessories were convinced that once electrical blasting was tried, few would give it up. Companies increasingly turned to electrical blasting—especially during the 1860s and the 1870s, when more components (including various fuses and generators) were commercialized; however, adoption was far from universal.[62] One authority in 1883 noted with regret that the old method of detonation was "still very extensively practised in mining and blasting operations, and [was] likely to continue so on account of its comparatively simple and inexpensive nature."[63] Traditional fuses, their reliability somewhat improved, survived alongside electrical detonation into the twentieth century.[64] That more dangerous methods remained in use permits us to infer that some companies placed little value on the lives of laborers, many of whom were immigrants, powerless to affect the adoption of detonators.

In March 1839, after spending nearly a year in England and France, his hopes for telegraph sales repeatedly buoyed and dashed, Samuel Morse returned to America by steamship. The discouraged inventor was broke and falling deeply into debt, his affairs in disarray. A possibility remained that the House of Representatives would act on the Commerce Committee's recommendation to fund an experimental telegraph line, but in the meantime Morse might need a new vocation.

With an eye toward the future, Morse sought a meeting with Joseph Henry. Capitalists, he may have reasoned, were likely to consult Henry when seeking answers to questions about the practicality of a full-scale telegraph. In a very deferential letter, Morse inquired: "Have you met with any facts in your experiments, thus far, that would lead you to think that my mode of Telegraphic communication will prove impracticable?" Henry's reply was all that the inventor could have desired: not only did he invite Morse to visit; he also asserted that "science is now ripe for this application and . . . there are no difficulties in the way but such as ingenuity and enterprise may obviate."[1] However, Henry did suggest that telegraphy over longer distances might require more power.

Bearing a series of electrical questions, Morse soon called on Henry at Princeton. The quality of the questions indicates that Morse had boned up somewhat on electrical principles. Concerned as ever about long-distance telegraphy, he asked: "Have you any reason to think that magnetism can not be induced in soft iron at the distance of 100 miles or more?" Henry responded with an encouraging "no."[2]

While waiting for the telegraph to find buyers, Morse embarked on a new career. During his lengthy stay in Paris, he had visited the studio of Louis Daguerre, inventor of a remarkable new technology, photography, which was all the rage in the French capital. In the Daguerre system, positive images were formed directly on copper plates coated with a light-sensitive silver salt. Even in photography's first year, 1839, the detailed images were simply stunning; Morse could scarcely believe his eyes. Photography soon became an art in itself, and also offered new possibilities for the painter. Morse grasped at once that a daguerreotype image could capture a scene that might be painted later at leisure, allowing a rendering more faithful than a hasty sketch.

Morse bought one of the first copies of Daguerre's photography manual and had a camera made according to the Frenchman's design. His first results were mediocre, so he enrolled in a course taught by one of Daguerre's associates. He also began experimenting with John William Draper, a professor at New York University whose interests were in optics and photochemistry. Morse and Draper laboriously perfected their techniques and set up a studio on the roof of the university building where, with benefit of sunlight, they practiced portraiture, a challenging application because it required long exposures. After many trials, they succeeded in reducing exposure times from 15 minutes to 2 minutes or less.

His collaboration with Draper having ended amicably, Morse built a studio atop a building owned by brother Sidney, who was now prospering quite nicely. Morse was able to get out of debt by selling daguerreotypes and enrolling people in his photography lessons, which emphasized composition as well as chemical processes. Among his students was Mathew Brady, who would become famous for photographing Civil War carnage. Enjoying success as the major practitioner of the Daguerre method in America, Morse surely wondered—with his painting neglected and the telegraph a money pit—whether photography was, after all, his true calling.

⎯⎯⎯⎯⎯

Apart from Morse's contacts with Henry, the telegraph project was essentially dormant. Consultation with the company's proprietors was exceedingly difficult, as they were scattered in New York, Philadelphia, and New Orleans.[3] Morse became preoccupied again with anti-Catholic politics and lost another mayoral race by a devastating margin. Most dispiriting of all, Wheatstone and Cooke, who had established a 13-mile telegraph line along the Great Western Railway in England, invited Morse to secure for them an American patent.[4] The terms were generous—a half interest—but the Englishmen could not have begun to grasp how deeply their offer offended Morse. French inventors were also busy in America, seeking to sell Congress a trial semaphore system for only $5,000. Competitive juices now flowing and still convinced of the superiority of his recording telegraph, Morse determined to pursue his project with renewed vigor. First he would have to attend to the stalled request for federal funding by lobbying a lethargic Congress.

Morse now played the Henry card, asking the Princeton professor to endorse the experimental line, and he did: "I have not the least doubt, if proper means be afforded, of the perfect success of the invention." But Henry also predicted that the telegraph might meet with some resistance because of the bad taste lingering from the "chimerical projects" to apply electromagnetism "as a moving power in the arts.[5] Significantly,

Henry expressed his preference for the Morse telegraph over its European counterparts. Scientific authority had again spoken definitively, and this augured well for Morse's enterprise. Of course, conjuring the "proper means" would take much time, money, and inventiveness. An invigorated Morse resumed tinkering and achieved a distance of 33 miles. He also obtained power of attorney from his partners in the essentially moribund company; now he could proceed without having to consult Smith, Vail, and Gale.

On the advice of Representative William Boardman, Morse again publicized his invention, exhibiting it in New York during the summer of 1842. Among the visitors was Henry, who had not viewed Morse's telegraph previously. According to Morse's account of this meeting, Henry called Morse's instrument "the most beautiful and ingenious instrument . . . he had ever seen" and Morse's plan "the only truly practicable plan."[6] A committee from the American Institute, made up of men with technical expertise, reported that Morse's telegraph was well suited for long-distance communication, adding that it was "a most important practical application of high science, brought into successful operation by the exercise of much mechanical skill and ingenuity."[7] Morse was awarded a gold medal by the Institute, and in its annual exposition, a trade show for American inventors and manufacturers, the telegraph was worked all day long and received praise from the press.

In early fall, Samuel Colt was conducting experiments at NYU, developing technology for electrically detonating underwater mines. While giving Colt advice on electrical matters and lending him long wires, Morse tackled one of the problems that would eventually have to be solved for long-distance telegraphy: How could telegraph lines cross rivers? Morse and Colt shared a need to create waterproof cables, and they worked together on this. In a public demonstration at Castle Garden, in the East River, Morse tried to send a message under a mile of water. After a brief transmission, the cable fell silent. It seems that a ship's anchor had hooked the cable, and it broke under the strain. Apparently, well-insulated wire alone did not guarantee underwater transmission. Similar problems would dog submarine cables for decades. Despite this fiasco, Morse's project still enjoyed appreciable support.

Among those who now took an active interest in promoting the telegraph and securing public funds was Representative Charles Ferris of New York City, a member of the Commerce Committee. Ferris suggested that Morse return to Washington and exhibit the latest model. And this he did in December. Alfred Vail, now married, living in Philadelphia, and bereft of funds, was unavailable, and so Morse brought along an NYU colleague, James Fisher. (Leonard Gale had resigned his professorship and moved to New Orleans.) Their successful demonstration, transmitting between committee rooms in the Capitol, even showed that two messages could be sent simultaneously on the same wire. This exhibition, according to Morse's account to brother Sidney, "excites universal admiration."[8]

Morse believed that Congress now would be receptive to his proposal and would quickly pass an appropriations bill. Beyond his telegraph's exemplary performance, he had good reason to be optimistic. After all, nearly 5 years earlier, when he made his first pitch in the Capitol, European telegraphs were just beginning to be reported in the America media. Now that a few systems were up and running in England and Germany, in some cases assisting train travel, the argument for telegraphs in America— where the pace of railroad construction was accelerating—would perhaps fall on sympathetic ears. In a report to Congress, Ferris and the Commerce Committee fashioned a patriotic rationale for backing Morse's experimental line. The inventor deserved support because his telegraph was "calculated to advance the scientific reputation of the country, and to be eminently useful, both to the Government and the people . . . [and thus] he should be furnished with the means of competing with his European rivals."[9] Along with the report, Ferris included Henry's endorsement of the project's scientific soundness. The immediate result was the introduction in the House of a bill that called for a $30,000 grant to Morse for testing "the Practicability of establishing a System of Electro-Magnetic Telegraphs."[10]

Morse stayed in Washington for many weeks, waiting for the bill to come to a vote. This was an especially trying time, for his funds were running out, his clothes were threadbare, and affairs at home were in dire need of attention. Morse drew consolation from his faith that "this delay may be designed by the wise disposer of all events for a trial of my patience."[11] The bill was passed by the House—by fewer than 10 votes—on February 23, 1843.[12] But Morse's ordeal was far from over. The Senate had much business left, yet only 8 days remained in the session to pass an identical bill. Late at night during the Senate's final day, March 3, the telegraph bill was passed unanimously a minute before adjournment, and President Tyler signed it. Morse, with all hope lost and less than a dollar in his pocket, had already gone to bed, having been advised by a senator that there was no chance of passage. Only the next morning did he learn from Annie Ellsworth, the Patent Commissioner's daughter, that he had triumphed. For this success Morse gave generous credit to Providence.[13]

Now, more than 10 years after he had conceived his telegraph on the *Sully*, Morse had the resources to build a full-scale demonstration line that would remove lingering doubts about long-distance telegraphy. He decided to place the underground line between Washington and Baltimore. This 40-mile route was a shrewd choice, in view of Morse's plan to sell the line to the government and the *American* telegraph's increasingly transparent nationalistic function. The decision to bury the line stemmed from concerns raised by others, going back to discussions aboard the *Sully*, about the vulnerability of an aboveground line to sabotage, vandalism, and storms.

Taking charge, Morse summoned his partners in the old telegraph company to Washington. They arrived quickly. For a $1,000 annual salary, Vail would make the hardware; for $1,500, Gale would oversee the manufacture of the lead pipe for containing the wire; Smith received no salary but was expected to help out with legal matters; Morse drew $2,000 a year and reserved for himself the title Superintendent of the Electro-Magnetic Telegraph.[14] In addition, Morse hired James Fisher, the NYU professor who had helped in the recent Washington demonstration, to supervise manufacture of the wire. These assignments indicate that Morse placed a high priority on closely monitoring the quality of materials going into the telegraph; after all, a single defect in the wire could cause an open circuit, and a leaking pipe might create a short. He was also meticulous in keeping the books, for the government required a monthly financial report.

However, in contracting for 160 miles of copper wire and 40 miles of lead pipe, Morse failed to foresee that meeting such large orders on time might strain manufacturing capacity and require his contractors to develop new production techniques. An especially challenging detail was that of putting the insulated wire into the pipe. James Serrell's company was able to produce only 10 miles of pipe, so Benjamin Tatham's company was hired to make the rest. Using a technique invented by Morse and Fisher, Tatham inserted the insulated cable into the pipe as the latter emerged from the shaping tool.

Then there was the matter of how to lay the cable underground. The first estimate for this job—$153 per mile—went over Morse's budget projection, so Smith arranged for the young and impoverished Ezra Cornell to do the job. Moreover, he encouraged Cornell, who had experience as a machinist and a millwright, to construct a trenching machine. Believing he would make a fortune if the machine worked and if telegraphy spread throughout the country, Cornell devised a design that Morse approved. A team of eight mules drew a cart that cut a narrow trench with a plow-like appliance and then inserted the cable. Cornell's workers and machine laid the Serrell cable, beginning in Baltimore at the depot of the Baltimore & Ohio railroad. (Morse had arranged for the telegraph line to be entrenched within the railroad right-of-way to Washington.) However, manufacture of the Tatham cable was delayed, and winter was closing in; Morse shut down operations, which gave him the opportunity to rethink his plans in the face of mounting problems.

On another front, Morse's relations with Francis Smith had deteriorated on account of the lawyer's shady dealings. In one case, Smith planned to inflate the cost to the government of the Tatham contract and keep the $500 difference. In a letter to Sidney, Morse, whose moral code simply did not abide such corruption, despairingly wrote "where I expected to find a friend I find a fiend."[15] Responding to Morse's gentle reproaches, Smith let loose a barrage of insults, even publicly casting doubt on the telegraph itself and on Morse's claim to be its inventor.

_᭶᭶᭶᭶_

Feuding with business associates was the least of Morse's problems. The 10-mile stretch of buried cable leaked, and the Tatham cable proved defective. James Fisher, assigned to test the finished cable, failed to notice that improper manufacture of the lead pipe had damaged the insulation. Morse saw this as a neglect of duty, and so with heavy heart he fired his friend.

Before resuming outdoor work in March 1844, a despondent Morse, who had been advised by Gale and Henry that the underground cable was in danger of failing even if properly made and laid, adopted a new plan. The line would now be strung on poles placed along the tracks. (The lead pipe—more than 20 tons of it—was sold for scrap, but the copper wire was salvaged and reused.) Economy no doubt also motivated this move, for Morse had already spent half of the grant; installing an aboveground cable promised to be much cheaper per mile. Although a surface line would be more vulnerable to vicissitudes of people and nature, it would be easier to track down and repair a break. For this purpose, Morse came up with a tool kit that included an alcohol lamp, solder, wire cutters, matches, and a rope ladder.[16]

Although the cable-laying machine was no longer needed, Morse kept Ezra Cornell on the payroll. It was money well spent, for Cornell ably oversaw the field operations with a crew sometimes exceeding 25 men, including a few Irish Catholics.[17] (In later years, Cornell, in partnership with others, reaped a fortune in the telegraph business; his donations of land and half a million dollars led in 1865 to establishing the university in Ithaca that carries his name. Unlike Yale, Harvard, and other church-dominated colleges at that time, Cornell was nonsectarian and coeducational.)

Cornell's crew began in Washington, emplacing every 200 feet a wooden post that rose 26 feet in the air and carried on a cross-arm two copper wires attached to glass insulators. Progress was swift: in just a few weeks, 7 miles of line had been strung and testing had begun. In fact, as the line bounded to Baltimore, sometimes at the pace of a mile a day, Vail in the field and Morse in Washington were incessantly testing the wires for continuity and honing their skills on the new equipment.

Vail and Morse put the cumbersome port-rule out to pasture, jettisoned the 5,000-word telegraphic dictionary, and replaced the register. In place of the port-rule was a vastly simpler "Morse" key for opening and closing the circuit (figure 12.1, upper). However, operating the key required much practice to ensure that dots, dashes, and spaces (long and short) were of the proper duration and thus the proper length on the register's paper tape. Taking the place of the old register was a new and compact device whose fundamentals would be altered little for decades (figure 12.1, lower). An

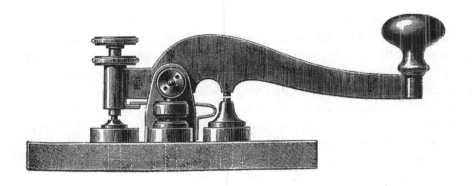

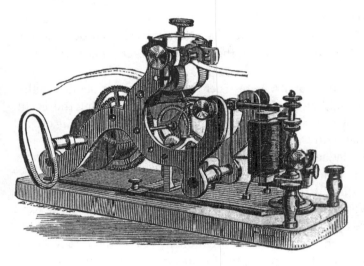

Figure 12.1
Upper: a typical Morse Key. Source: Prescott 1877: 497. Lower: an early Morse-Vail register (not to same scale). Source: Prescott 1860: 75.

electromagnet, operating through a lever and a spring, caused an embossing pen to move upward and downward; the length of time it was pressed upward against the moving paper—whether it indented a dot or dash—was determined by how long the sending key had been depressed. A falling weight or a spring drove the mechanism that advanced the tape, pulling it through two additional rollers.[18]

With the new components, Morse and Vail were soon transmitting dot-dash "Morse code" at dozens of words per minute. In coming years, telegraphers would discover that they could distinguish between dots and dashes on the basis of the electromagnet's clicks, which enabled even faster decoding, and so the register was sometimes augmented or replaced by a simple "sounder."

For weeks Morse and Vail had been teasing the public with demonstrations, including a message announcing the selection, at a convention held in Baltimore, of Henry Clay as the Whig candidate for president. Vail had obtained the news from people on a train stopped at a station about 20 miles from Washington. Sent ahead to Morse, the message beat the train's arrival by more than an hour, a feat that stirred great interest in the capital.

The line was inaugurated on May 24, 1844. There was no suspense about whether it would work, for Morse and Vail had tested it thoroughly before staging this event. Morse honored Annie Ellsworth by asking her to furnish the first message; she offered the biblical passage "What hath God wrought!"[19]

The *Baltimore American* lauded Morse's telegraph as "one of the most remarkable and astonishing triumphs which the science and ingenuity of man ever achieved."[20] No less effusive was the *Utica Daily Gazette*: the telegraph, it said, was "an invention destined probably to revolutionize all our modes and systems of business and to exert an incalculable influence upon the future destiny of the species."[21] Indeed it would.

A few days later, some surprising political information arrived on the wire in Washington because, happily for the telegraphers, the Democratic nominating convention was also being held in Baltimore. Through eight ballots, the delegates were unable to choose between Martin van Buren and Lewis Cass. Vail's periodic dispatches of convention news attracted a swarm of politicians, who gathered around Morse in the Capitol as he read the latest returns. Finally, on the ninth ballot, he announced that the convention had chosen dark-horse candidate James Knox Polk.

The success of the Baltimore-Washington line turned Morse into a national, even international, figure. He was lionized, compared to Benjamin Franklin, and awarded honors and medals. First in painting, again in photography, now in telegraphy, he had achieved personal distinction. The latest success he also attributed to Providence, for God had doubtless helped him to work out the telegraph's devilish details and complete the project under budget.[22]

The demonstration line proved that a Morse telegraph system, with enough repeaters in place, could transmit information to any distance on land. Concerning this aspect of practicality, Morse and Vail had been certain on scientific grounds. But there loomed larger questions of practicality. Would enough people find the telegraph so crucial that they would pay a premium (over mail service) for the privilege of sending messages? Would anticipated consumer demand be sufficient to interest governments or capitalists to invest in new lines? Affirmative answers emerged immediately as people from various walks of life showed up to send messages—some trivial, some of great moment. This robust response to the new medium signaled the presence of a huge latent demand for rapid, if somewhat expensive, communication, which entrepreneurs and capitalists moved quickly to meet.

One of the first messages sent on the Baltimore-Washington line concerned the outbreak of Catholic-Protestant clashes in Philadelphia, which left Catholic churches and homes in ruins and many dozens of people dead or injured. Vail had learned about the violence from people aboard a train arriving in Baltimore. Through Morse he sent the news to Secretary of State John C. Calhoun; in a second message, Philadelphia's mayor sought aid from President Tyler. This event underscored the lesson that France and other countries had learned from their semaphore systems: the telegraph was a potent political technology. Indeed, a U.S. government report claimed that the Morse telegraph demonstrated that the far-flung American republic—reaching, since the Louisiana Purchase in 1803, from the Atlantic to the Pacific—could be governed effectively.[23]

Other early transmissions announced the birth of a new family member carried on a long-distance chess game, and even enabled Samuel Colt to detonate from Baltimore small explosives in a Capitol chamber. The telegraph was also enlisted in a geodesic exercise, that of determining an accurate longitude for Baltimore relative to Washington. In 1839, Morse had suggested to the French scientist François Arago that the telegraph could be used to calculate longitudes, and now it would be tried.[24]

Because electricity was believed to travel almost instantaneously through wires (at "lightning speed"), the telegraph could permit one to learn the exact time at a distant reference place known as a meridian. A meridian is an imaginary arc on the Earth's surface from pole to pole through a particular point. In 1844, the U.S. Capitol defined the prime meridian, the reference longitude in terms of which others in America were determined. By computing local time astronomically, and taking the difference between that time and the meridian time, one could calculate how far west or east—in degrees of earth's curvature—a place was from the meridian. Charles Wilkes, a naval officer with the U.S. Coast Survey, performed the measurements in June 1844 using Morse's telegraph; he found that earlier surveying techniques had slightly misplaced Baltimore.[25] Properly crediting Morse for the idea, Mechanics Magazine regarded this accomplishment as "Among the many wonderful developments of the new telegraph."[26] Indeed, the determination of longitude turned out to be one of the telegraph's earliest and most significant scientific applications. In later years, precise corrections were made for the speed of electricity through copper and iron wires.

Americans' thirst for information of all kinds, including gossip, more than intimated that newspapers could make effective use of telegraph lines, bringing to their readers timely accounts of engrossing dramas such as train wrecks, cliffhanger elections, boiler explosions, catastrophic floods and fires, and riots. Clearly, if one newspaper in a town used the telegraph, others would have to follow or lose readers. The close association of telegraphs and newspapers was cemented in the early days of the Baltimore-Washington line, and soon led to the formation of the first wire services: the Associated Press (in America) and Reuters (in Europe).

Newspapers reported extensively on the many applications of the telegraph, actual and imagined, that were emerging daily. From these descriptions potential capitalists took away lofty visions of quick riches that might be earned by investing in new lines. For investors, the salient element of practicality is a technology's potential to find ample markets. Although market forecasts are notoriously unreliable, faith in telegraphy was not misplaced. Businessmen were eager to commodify information; after all, timely knowledge about conditions elsewhere could translate quickly into savvy decisions and profits. The expense of a telegram was trivial in comparison with the value of the information it might supply. Accordingly, the next several decades witnessed a veritable telegraph craze, with lines proliferating across the country, fueled in part by public stock offerings of telegraph companies.

In the weeks and months that followed the Baltimore-Washington line's success, Morse fielded inquiries from capitalists and entrepreneurs. Some sought to license the patent rights and build their own telegraphs; others wanted Morse to build their projected lines. One Baltimore businessman even negotiated to buy Morse's patent; the inventor was willing to sell, but the deal fell through. Morse also expected that the government would follow the French model and establish a national monopoly, intending to use the telegraph for public good rather than private gain. Indeed, "he hoped the government would buy his patent outright, expand and manage the system on its own, and keep him in place as superintendent."[27] Concretely he urged the government to grant him funds to build a line all the way to New York, but this proposal died from neglect. However, he did receive an additional appropriation of $8,000 to keep the experimental line running for another year. The Washington station was put under the jurisdiction of the Post Office Department and moved to its offices, in a building just down the street from the Patent Office—the "temple of invention"—which had opened just a few years earlier.[28] (Today, on the Seventh Street side of the old postal edifice, one can find a plaque commemorating "the first public telegraph office in the United States.") In the end, federal financing of the Morse telegraph started and stopped with the Baltimore-Washington line.

Meanwhile, Morse and Vail, sometimes working independently, continued to grapple with details. For example, the problem of lightning hitting the line could have been easily foreseen and protected against, but it was not. It took an actual lightning strike that endangered men and damaged equipment to goad Morse into inventing a serviceable lightning arrestor. Then there was the battery problem: the 80-cell Grove battery required constant maintenance at not inconsiderable expense. Together Morse and Vail were able to make modifications so that the line could be worked with just 20 cells.

Morse also tried out a more radical technology to furnish power, calling on Charles Page for assistance. Page was a physician and electrical experimenter whose expertise in electromagnetism was second only to Joseph Henry's in the United States. Indeed, Page's biographer Robert Post points out that he was the first person to invent magneto-electric machines in America (Saxton, recall, worked in London).[29] Page was a prolific inventor of electromagnetic devices, including motors, magnetos, and an early induction coil, which were manufactured and sold by Daniel Davis Jr. of Boston.[30] Page, in consulting for Morse on various telegraph problems, made several suggestions that were incorporated into the design of the hardware.

Encouraged to fashion a magneto that could power the Baltimore-Washington line, Page responded with a large machine of unusual design that cost Morse $99.50.[31] According to Vail's sketchy description, the magneto was a compound machine containing two huge permanent magnets that lay flat on the same plane.[32] The open ends of the huge horseshoe magnets faced each other, and in the space between them revolved, in bearings, two armatures on the same shaft; there was also a built-in commutator. In its overall *mechanical* configuration, Page's magneto was a close descendant of Saxton's machine—times two.[33]

Vail claimed that an electromagnet energized by Page's machine could suspend 1,000 pounds, and it also produced enough power to operate a full-scale telegraph. Indeed, on Christmas Day 1844, messages were sent through the Baltimore-Washington line using Page's magneto. Despite the magneto's technical competence, Morse stuck with batteries. Robert Post suggests the likely reason for this decision: the huge magneto "required the full strength of a man to turn."[34] Page's machine was a technological dead end; the route to high-power compound magnetos began elsewhere.

Although later experiments with magnetos for telegraphy, including Beardsley's innovative machine, were also technically successful, for most of the nineteenth century batteries reigned supreme.[35] After all, a lone telegrapher in a rural station could not easily crank a magneto and handle messages at the same time. But after the proliferation of commercial steam-powered dynamos in the 1870s, Western Union began to experiment with the new generators in its San Francisco telegraph exchange. By this time exchanges in large cities were enormous, handling a welter of converging local and long-distance lines. In these facilities the burning of coal to create steam power could economically replace the zinc and acid being consumed in massive batteries. Thus, in Western Union's New York exchange, a series of dynamos installed in 1880 took the place of more than 19,000 cells weighing 72 tons.[36]

_⁓⁓⁓_

Because the door had been closed to further government support, Morse turned to the private sector to expand telegraphy across the United States and around the world.[37] The mercurial Morse, temperamentally unsuited to becoming an entrepreneur himself, contracted his business dealings out to Amos Kendall, a well-connected Congregationalist who, as Postmaster General, had shown exceptional administrative skills. Kendall organized new telegraph companies, beginning with the Magnetic Telegraph Company (which would extend the Baltimore-Washington line to New York). The worst effects of the Panic of 1837 had abated, and sales of stock in the new companies, mostly to local investors along projected lines, raised enough capital to proceed. The Magnetic Telegraph Company received a controlling interest in each new company (in stock), plus a modest amount of cash.[38] With additions from non-Morse companies, the telegraph network grew so rapidly that by 1851 it was possible to send messages between points as distant as New York and New Orleans (at $2.40 for 10 words).[39]

In 1866, most of the telegraph companies were united by Western Union into a well-integrated national network—the first American industrial monopoly. By the late 1860s, Western Union had 75 percent of the telegraph business.[40] Only in the United States and Canada did telegraph companies remain in private hands after 1868; in all other countries, the companies were nationalized.[41]

Telegraph companies began earning profits almost immediately, for this technology satisfied the cultural imperative of rapid, long-distance communication.[42] In America this imperative had been rather insistent. Americans were in constant motion, following one opportunity or another, moving from the East to the rapidly growing West, from the country to the city, and from city to city (as Morse and Henry had). The result was families dispersed in different communities, even different states, able to keep in touch, if at all, only by mail. Mail deliveries were speeding up in some regions thanks to railroads, but only through the telegraph could one learn almost immediately of a birth or a death in one's family. Many Americans had occasion to turn to this technology from time to time. A poignant example: During the Civil War, while on a trip to New York, Joseph Henry received a telegram that his son William was seriously ill. He rushed back to Washington, and was at his son's bedside in the Smithsonian residence when the young man died.[43]

An even more potent spur to the rapid adoption of the telegraph was interest on the part of the growing business and financial communities, including railroads and the press. An early observer noted that at the telegraph's "very birth, it became the handmaiden of commerce."[44] With the population of the United States increasing dramatically as a result of immigration, and with potential markets expanding in tandem, manufacturers and merchants salivated at the prospect of keeping in touch with distant suppliers and representatives so that they could match supply to demand and consummate deals in a day. And if one business adopted the telegraph, its competitors would have to follow suit, in a pattern of contagious adoption.

In view of these baseline expectations of demand for telegraph service which were rapidly realized by the first lines, investors could plunge with confidence that a new line would not sit idle. Not surprisingly, "merchants, small goods producers, and bankers who came to rely on the telegraph furnished much of the necessary funds for telegraph industry investment."[45] Indeed, the telegraph's effects on the American economy were wide-reaching and profound. It lowered transaction costs, made possible the first national commodity markets, created demand for futures markets, and stimulated the trading of stocks and bonds on Wall Street.[46] The economist Richard Du Boff maintains that the telegraph played a pivotal role in transforming business operations and, along with the railroad, contributed appreciably to economic growth and the concentration of economic power in the late nineteenth century.[47]

The telegraph was clearly a good fit for American capitalism, and that is why it was adopted so rapidly and thoroughly—at an average cost of about $150 per mile.[48] By 1851, most of the large cities east of the Mississippi were connected by wire. By 1855, there were at least 32,000 miles of telegraph in North America, outdistancing the 21,000 miles of railroads.[49] By the early 1870s, the United States had 180,000 miles of telegraph lines and about 6,000 stations.[50] Western Union alone transmitted 40 million messages in 1869, plus newspaper copy, and earned profits of more than $2.5 million on $10 million in receipts.[51] Not surprisingly, telegraph usage was higher in the United States than in other countries. Scientific American in 1867 reported that in the United States one message—at an average cost of 57 cents—was sent annually for every 2.5 people, whereas the ratio was 1:18 in France, 1:9 in Prussia, and 1:5 in Great Britain.[52]

The promise of profit was so great that the telegraph business lured inventors and entrepreneurs seeking to evade the Morse patent. Among the new telegraph designs was one created by a man with the improbable name Royal House; it employed a piano-like keyboard on the sender and a daisy-wheel-like device that printed letters. Both were exceedingly complex mechanically; indeed, Morse was fond of pointing out, with more than slight injustice, that they were simply *his* instruments "made complicated."[53] Predictably, some entrepreneurs were attracted to alternative telegraph systems and constructed lines that used them. The most persistent and dangerous competitor was Henry O'Reilly, who at first employed the House system and built a line from Philadelphia to the Mississippi River east of St. Louis.

Reluctantly, the Morse interests became locked in messy and protracted court battles with O'Reilly. The Eastern press, highly dependent on the telegraph for fresh news, took great interest in these affairs; after all, competition among telegraph companies might result in lower costs for gathering news. And so it was that Morse, once the darling of the press, came often to be vilified for trying to monopolize all telegraphic communication.

In the course of these trials, which came to include 15 U.S. Supreme Court cases, the aging inventor learned that patents were not so much a protector of intellectual property as a license to litigate. This unexpected turn of events caused Morse much anguish, mitigated by a new wife 30 years his junior and a lovely Tuscan-style villa on the Hudson near Poughkeepsie. The telegraph had at last made Morse prosperous, but lawyers' fees ate deeply into his fortune and preparing for trials into his time.

Perhaps Morse's greatest vexation in legal matters was Joseph Henry's testimony.[54] Their estrangement dated from the 1845 publication of Alfred Vail's book *The American Electro Magnetic Telegraph*. Morse took no part in writing the book and even discouraged Vail from publishing it, fearing that detailed disclosures might jeopardize patent applications in Europe. But Vail proceeded anyway. Once the book came out, Henry learned that his new principles of electromagnetism, which Gale had passed on to Morse, were unmentioned, despite their crucial role in creating a telegraph that could transmit farther than 40 feet. Morse tried to placate Henry with disclaimers, but he could not be mollified. Henry was further incensed when an 1847 edition of Vail's book did not remedy the omission.[55] Even so, in the court cases that would drive them irrevocably apart, Henry was the reluctant witness, forced to testify by subpoena.

Henry did respect Morse for bringing the telegraph to fruition, and Morse respected Henry for his scientific acumen, but shortcomings of the patent system drew the two men into adversarial roles that neither of them sought or relished. And it did not help matters that both the scientist and the inventor were prideful, pious men who asserted their claims to earthly immortality with a vigor and righteousness befitting a Calvinist preacher.

O'Reilly vs. Morse went to the Supreme Court late in 1852 and was not decided until February of 1854—the year in which the original telegraph patent was set to expire. To an outsider, Henry's deposition hardly appeared inflammatory, but it did manage to stir Morse's passions in statements such as these: "I am not aware that Mr. Morse ever made a single original discovery, in electricity, magnetism, or electro-magnetism, applicable to the invention of the telegraph. I have always considered his merit to consist in combining and applying the discoveries of others in the invention of a particular instrument and process for telegraphic purposes. I have no means of determining how far this invention is original with himself, or how much is due to those associated with him."[56] The last sentence was tantamount to doubting whether Morse had even made an invention. Certainly Henry knew that the early port-rule and recording receiver were unprecedented, and that their conception was entirely Morse's. But all Henry could see in Morse's hardware was the materialization of his own principles; for him, the sophistication of the hardware and the telegraph's emergent performance characteristics were of no consequence. Never before had Henry come so close to denying Morse credit that was unquestionably due him.

While the justices were considering the case, Morse penned a blistering 90-page answer to Henry's testimony. In a marvelous circumlocution, Morse accused Henry of lying about dates and events (Henry was "*not in ignorance of facts which make his statements incorrect*"), and took it as his "duty to the cause of Historic truth . . . to expose as I shall be able to do, the utter *non-reliability* of Prof. Henry's testimony."[57] Morse did show that Henry got a few historical facts wrong and that Henry's own publications had built on prior work that he did not in every instance explicitly acknowledge. However, Morse went too far by denying that Gale had conveyed to him any useful information from Henry's electromagnetic researches.[58] In his accusations and intimations wrapped in heated rhetoric, Morse completed the alienation of Henry and, in the process, rendered himself a less sympathetic figure to later writers on electrical history.

Henry's deposition did little more than demonstrate that the Morse telegraph drew upon earlier inventions and scientific principles. This was hardly damaging to Morse's case, since the justices understood that all inventions arose in this manner. And so a year later the Supreme Court affirmed all but one claim in Morse's patent. O'Reilly was enjoined from continuing his telegraph enterprises using infringing components. The rejected claim—for using electromagnetism to register intelligence at a distance— was overly broad, for it would have prevented others from making improvements on the Morse telegraph, an unjustifiable check on progress. Soon Morse received another gift from the government: a 7-year extension on his patent, which affirmed that he had not been fairly compensated during its original 14-year term. Telegraphy would remain ensnared in legal warfare for the rest of the nineteenth century, establishing for electrical technologies an ugly precedent.

Henry did not deign to reply to Morse's extended diatribe. As Secretary of the Smithsonian Institution (a position he had assumed in 1846), Henry had other tools at his disposal for smiting the ingrate inventor. Henry asked the Smithsonian Regents to investigate the matter, and they appointed a distinguished committee that included the president of Harvard College. The committee's published report embellished Henry's contributions and diminished Morse's, firmly denouncing the inventor for his unseemly attack on the Secretary.[59] Almost to his dying day, Morse continued to publish tedious tracts defending his position as *the* inventor of the electromagnetic telegraph.

None of this esoteric squabbling had much of an effect on Morse's public reputation (beyond Smithsonian circles). The distinguished inventor received numerous honors from foreign heads of state, including a Danish knighthood and the French Légion d'Honneur. Perhaps best of all, a consortium of continental European nations, led by France, in 1860 granted Morse an indemnity of $80,000 for having employed his telegraph technology without benefit of patent licenses.[60] This was a pittance, to be sure, but the acknowledgment of Morse's rights as *the* inventor satisfied his sense of justice. In fact, Morse did not need the money; his stock in telegraph companies, including Western Union, had made him very wealthy, continuing to yield dividends long after his patents ran out.

Morse at last achieved in abundant measure the personal distinction he had so earnestly craved. He was, after all, the driving force behind commercializing the technology that, through the favorable judgments of many players, achieved practicality in every respect. Before long, land lines and submarine cables extending hundreds and then thousands of miles would enmesh the nations of every continent in a worldwide communication web.[61] Most of these telegraphs would use Morse-derived technology, for, according to Silverman's assessment, Morse had "created a telegraph system that against many competitors repeatedly proved itself to be the cheapest, the most rugged, the most reliable, and the simplest to operate."[62] Henry and Morse probably agreed on these performance advantages.

_⁀⁀⁀_

Beyond bringing families closer together and profoundly altering business practices, news gathering, and diplomacy, far-flung telegraph networks had consequences for later technologies—electrical and otherwise—in a rapidly industrializing America.[63] As companies formed to build telegraphs, demand surged for poles, insulators, wire and cables of many kinds, keys, registers, paper tape, batteries, and all other components.[64] This demand stimulated the growth of older manufacturing firms and invited the entry of new ones. The result was the establishment of an infrastructure for producing electrical components. Given this unrelenting demand, manufacturers and entrepreneurs easily obtained capital for scaling up operations and founding factories. At the same time, opportunities arose for specialty manufacturers that could produce in quantity a limited range of parts, such as electromagnets whose coils were wound by machines.[65] As the dynamo, the telephone, and other new electrical technologies were commercialized in later decades, firms making telegraph equipment responded with little difficulty. Indeed, a year after Bell's invention in 1876, the first full-scale telephone system went into service in Massachusetts, and by 1880 there were nearly 50,000 telephones in use.[66]

The rapidity with which the telephone was commercialized clearly owed much to telegraph technology and its manufacturing infrastructure, but there was also the contribution of skilled labor. The telegraph recruited legions of young men seeking adventure, prestige, and economic opportunities, eager to learn Morse code and the inner workings of the telegraph office.[67] Perhaps introduced to the subject by the many textbooks on telegraphy that followed Vail's 1845 treatise, telegraphers, many of them itinerant, became familiar with principles of electricity and learned operations such as soldering, maintaining batteries, troubleshooting circuits, and repairing mechanical parts; they also became adept at improvising when routines failed. Men who mastered the telegrapher's trade—Thomas Edison among them—would be in the vanguard of electrical invention and would constitute a skilled labor force that would be tapped for new ventures such as the telephone and electric light and power systems.[68]

As telegraph networks grew in size, in numbers of employees, in messages sent annually, and in complexity of tasks, no longer were loose partnerships able to meet the incessant demands on management. Clearly, new forms of organization were needed. Western Union and the railroads invented the first large private-sector bureaucracies. Among other features, these corporations adopted rigorous record keeping and cost accounting, rigidly defined operational divisions, specialized jobs (including full-time managers), administrative hierarchies, and the separation of ownership (stockholders) from management (employees). Later in the nineteenth century, as other businesses based on new technologies grew in scale and in scope, they turned to Western Union and the railroads for models of suitable organizations. The result was the proliferation of the modern bureaucratic corporation.[69]

The telegraph also affected the worldviews of nineteenth-century Americans, rich and poor, Catholic and Protestant, East and West. Here was an electrical technology that could—as it was so often put—annihilate time and space. No longer tethered to a person traveling by horse, ship, or train, a message could reach its destination without a messenger. That information could move through a stationary medium, divorced from any conventional conveyance, was a mysterious feat purportedly accomplished by an invisible fluid coursing silently at lightning speed through wires across forests and prairies, through fields and pastures, from building to building and eventually beneath the vast seas. Contemplating this extraordinary technology, people in America and in other countries could not help but be infected by an enthusiasm to put electricity to work in other ways, for the horizons of new applications surely seemed endless. Telegraph technology—the Morse-Vail register in particular—inspired an explosion of inventions, many of which incorporated electromagnets and produced mechanical effects at a distance.

Moreover, as the first successful capital-intensive electrical technology, the telegraph fostered a tempered enthusiasm for other ambitious electrical ventures.[70] The rapid commercial triumph of the telegraph, broadcast far and wide by the press, was an emphatic cultural experience that awakened moneyed men to the possibility that bringing a new electrical technology to market might yield fantastic profits, despite the need to first traverse a sizable developmental distance, possibly at great and unpredictable expense. Now, as never before, investors were willing to take seriously proposals for commercializing other capital-intensive electrical technologies, such as the telephone and electric light and power systems. Thus, American electrical inventors after mid-century had access, in principle, to sources of capital in addition to family, friends, and governments. Yet, because investors made funding decisions on a case-by-case basis, a technology denounced as impractical by scientific authorities like electric motors could languish for decades.

When Edison and others began work on electric light and power systems in the 1870s, they could draw upon the organizational, financial, technological, and human resources begot by the telegraph. But there was more. The telegraph spawned the formal science of electrical measurement, which became the foundation of electrical engineering as it developed into an organized profession later in the century under the spur of electric light and power.[71] Such contributions consisted not only of standardized units of measurement, including the now-familiar volt (for electromotive force or tension, formerly intensity), ohm (for resistance), and ampere (for current, formerly quantity), but also of sophisticated apparatus for determining their values in the field.[72] Insights and apparatus came from well-known academic scientists, mostly working in Europe (including Wheatstone and William Thomson), who engaged telegraph-inspired problems, as well as from telegraphers themselves.[73] Instrument makers successfully commercialized devices such as the Wheatstone bridge for measuring resistance and Thomson's mirror galvanometer for detecting tiny currents.

Accurate information on electrical quantities became essential for creating and operating light and power systems. Had measurement science and apparatus not been available already, Edison and others would have been obliged to invent them in the course of their projects. Clearly, commercialization of the telegraph created a plethora of human, intellectual, and material resources that Edison and other builders of electrical systems could readily exploit.

13 Magnetic Power Derailed

During the 1830s and the early 1840s, inventors in Europe had, like Thomas Davenport and Ransom Cook, built electric motors and exhibited them driving machines. William Sturgeon, for one, used his motors to pump water, saw wood, and pull a wagon.[1] Some projects were even more ambitious. With generous support from the Russian government, Moritz Jacobi assembled a large motor, installed it in a 28-foot boat, and with a dozen passengers cruised up and down the Neva River at about 3 miles per hour.[2] In Scotland, Robert Davidson constructed an electric coach (named *Galvani*); it underwent some testing on British rails.[3]

Like the electromagnet, the electric motor inspired a spurt of inventiveness.[4] Its most general performance characteristic—the ability to produce continuous rotary or reciprocating motion—invited inventors to envision the substitution of galvanic power for steam, animals, human exertion, and other prime movers in countless activities. This idea was so captivating that Edward Palmer, an instrument maker in England, "had in his window a very great variety of rotating engines, with attached models of machinery which are kept in motion by them."[5]

But the electric motor was an unconventional prime mover, for its power depended on electricity generated by the consumption of zinc and acid in batteries. An appreciation for the implications of this disquieting fact set the agenda for discussions of the practicality of electric power between 1840 and about 1875. Scientific authorities constructed quantitative arguments, based on the relative costs of burning coal and consuming zinc, that buttressed Henry's early forecast that electric power would be prohibitively expensive. Manufacturers and investors attended closely to these arguments and concluded that the electric motor could not become a practical power unless a vastly cheaper source of electricity appeared.

Nonetheless, people continued to conceive new and varied motor designs, for the possibilities are endless. Sometimes inventors claimed near-miraculous efficiencies for their motors or called attention to performance characteristics such as safety of operation, that supposedly would—or should—trump the greater costs of electric power in particular applications. Occasionally, an electric motor was put to work in a conspicuous display, usually to advertise an inventor's technical virtuosity or an instrument maker's handcrafted creations, but no one was manufacturing a motor that could be regarded as a competent industrial product. Before about 1880, the electric motor had not reached anything like the kind of "closure and stabilization" in design that, according to Wiebe Bijker, marks a maturing technology.[6] Like Henry's teeter-totter, electromagnetic engines remained "philosophical toys." This flirtation with electric motors furnishes evidence that a near-uniform consensus of scientific authorities can sometimes dissuade moneyed men from commercializing a new technology. In this case, the advocates of electric motors—inventors and potential consumers—were powerless to affect the outcome.

Although scientific authorities, manufacturers, and investors had proclaimed electric motors impractical, one well-connected American inventor, following in Jacobi's footsteps, persuaded his government to invest in a grandiose project that could not have garnered support from private enterprise. Jacobi's counterpart was Charles Grafton Page (1812–1868), who secured with relative ease a large federal grant to build a battery-powered locomotive. Not surprisingly, this event raised questions about federal technology policy.

_⌒⌒⌒_

Charles Page, a physician and an electrical experimenter when he fashioned the large magneto for Morse's Baltimore-Washington line, was in the late 1840s an examiner in the U.S. Patent Office. Page's biographer Robert Post notes that when Joseph Henry arrived in Washington to assume the duties of Smithsonian Secretary, it was Page who helped him establish a social network of influential people. Together, Page and Henry worked to heighten respect for scientific expertise in the federal government. Page also supported Henry's efforts to shape the Smithsonian into a world-class research institution, not merely a repository for exotic things as some members of Congress preferred. Although both men invented electromagnetic technologies, Henry was far more familiar with physical theory than Page.[7]

Before his arrival in Washington, Page had collaborated with the Boston instrument maker Daniel Davis Jr., making varied apparatus for exhibiting electromagnetism. One especially fascinating device was Page's "compound magnet and electrotome," an early version of the induction coil whose technological descendants would find many applications (figure 13.1). In modern terms, the induction coil is a transformer that can step up the tension by a factor of hundreds or thousands or more. That is, when pulsating current of low intensity is fed into the primary coil, the secondary coil—with many more turns—generates a long spark, the kind of effect normally produced by an electrostatic generator.[8] Page published his invention—a distant echo of Faraday's

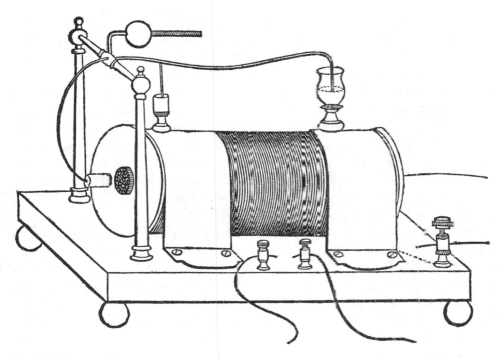

Figure 13.1

Page's "compound magnet and electrotome" of 1838. Adapted from figure 2 of Page 1867.

iron ring and two coils—in *Silliman's Journal* in 1839, and it was reprinted almost immediately in *Sturgeon's Annals*.[9]

The core of Page's apparatus was an iron tube packed tightly with iron wires running lengthwise. The tube was the foundation for winding two coils of insulated copper wire (the compound magnet), first the primary and then the secondary. Had he stopped there, Page could have operated the apparatus by manually making and breaking the circuit from the battery to the primary coil. But he went further, for the electrotome repeatedly interrupted the primary current *automatically*.[10]

The electrotome worked as follows: Included in the primary circuit was a small iron cylinder attached to one end of a thick copper wire, the opposite end of which swept upwards, through a transverse shaft, and then dipped into a cup of mercury. The cylinder was placed very near the coil's iron core. When the primary circuit was closed, the core attracted the cylinder, which caused the delicately balanced copper wire, acting as a lever, to lift its other end out of the mercury. The circuit now broken, the copper wire fell back into the mercury, which again closed the circuit. The result was an oscillating motion that repeatedly interrupted the flow of current through the primary, which in turn induced in the secondary pulses of high tension.

Page's compound magnet and electrotome, brought to market by Davis in 1838, was offered in instrument catalogs for decades. Initially its price was $8; in 1848 it was $10–$12; in 1857 it was $20.[12] This invention was a quintessential demonstration device for natural philosophers and public lecturers. Surprising and pleasing in its automatic operation, the apparatus buttressed the belief that means could be found to convert any kind of electricity into any other. Sometimes, however, the apparatus was illustrated with a pair of graspable conductors connected to the secondary, indicating that it could deliver shocks in place of the magnetos, electrostatic machines, and Leyden jars employed in electromedicine. Commercialized by dozens of instrument makers and entrepreneurs, "medical batteries" of the middle and late nineteenth century often incorporated small induction coils. These products were also judged practical by more than a smattering of physicians and patients.[13]

In the 1850s and the 1860s, people in the United States, France, and England contributed to the design of induction coils that generated impressively long sparks, potentially able to replace the largest electrostatic machines in experiments and lectures.[14] Beyond the straightforward technique of adding turns to the secondary coil, experimenters discovered that the coils' wires and layers had to be heavily insulated to prevent internal arcing. They also learned that adding a capacitor (then called a "condenser"), often made of alternating layers of waxed paper and metal foil, to the primary circuit, in parallel with any sort of circuit breaker, appreciably augmented the tension.

Although electrostatic machines could generate long sparks, the amount of current was negligible. In contrast, the induction coil could, in principle, produce high tension and nontrivial current. In 1869, a man named A. Apps built a huge induction coil for the Royal Polytechnic Institution in London, a venue for popular science demonstrations (figure 13.2). Nine feet long and 2 feet in diameter, it had 3 miles of wire in the primary and 150 miles in the secondary, and made extensive use of ebonite (a hard rubber invented by Charles Goodyear) for insulating the coils from each other and for enclosing the entire apparatus. Powered by a 40-cell battery, the Apps coil yielded a 29-inch flaming arc and caused asbestos—an excellent thermal insulator—to turn red-hot.[15] No electrostatic generator could match these current-dependent effects. However, weighing nearly a ton and encumbered by many accessories, the Apps coil was not the acme of elegant design, and reportedly failed within a few years.

Perhaps the most skilled builder of large induction coils was the Boston instrument maker Edward Ritchie. In 1871 he made one for Henry Morton, president of the Stevens Institute of Technology in Hoboken, New Jersey.[16] This coil measured 40 inches long and 19 inches tall; with condenser it weighed only 335 pounds. The primary was composed of 200 feet of wire and the secondary of 234,100 feet. These windings were well insulated throughout. Used very conservatively with only a three-cell battery, Ritchie's induction coil generated a 21-inch spark and shattered a block

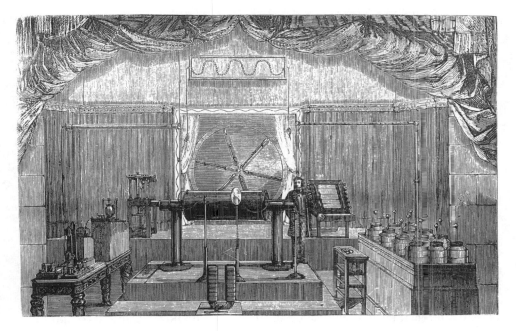

Figure 13.2
Apps's huge induction coil at the Polytechnic Institute. Adapted from Foutledge 1879: 385.

of glass 3 inches thick. Predictably, the antics of behemoth induction coils became a staple in public exhibitions of electricity, dazzling audiences on both sides of the Atlantic.

After 1850, instrument makers offered induction coils in countless varieties. Most were smaller than the Apps and Ritchie mega-coils and were adopted for specialized applications. In addition to being incorporated into medical devices and used in experiments and lectures, induction coils were sometimes employed for detonating explosives and lighting gas lamps. In the early 1860s, an induction coil supplied spark for the first 500 Lenoir gas engines.[17] Through much of the twentieth century, induction coils were used to fire the spark plugs of gasoline automobile engines.

Around 1860, John Gassiot reported that he had produced a "brilliant white light" by passing an induction coil's current through a spiral glass tube filled with carbonic acid gas (carbon dioxide). A gas-filled or partially evacuated tube with an electrode at each end was known as a Geissler tube, after the German physicist who had invented it. Sold in many shapes and sizes, these artifacts are still made today. Although no one commercialized Geissler tubes for lighting homes and businesses, this illuminating technology did find a few applications. After all, putting together a light required only a battery, an induction coil, and a Geissler tube containing a suitable gas, all of which were available from instrument makers. The most conspicuous applications were portable lights sometimes used by miners and divers.[18] Small Geissler tubes also made their way into medical instruments that could illuminate body cavities.[19] Today the Geissler tube thrives as the "neon" lights so ubiquitous in outdoor advertising, but only the reddish-orange light contains neon gas; other gases emit different colors.

Because of the induction coil's unique performance characteristics—a reliable source of high tension and appreciable current—chemists and physicists delighted in trying them out. At first induction coils merely replaced electrostatic machines, but soon they entered new experiments. Of particular note was the use of induction coils in optical emission spectroscopy. In this technique, a sample of material is raised to a high temperature and emits light spectra characteristic of its chemical constituents. Gustav Kirchhoff and Robert Bunsen, who developed emission spectroscopy in 1860, at first used Bunsen gas burners to supply heat, but soon other investigators were using the hot arc of an induction coil.[20] By characterizing the spectrum of each element, investigators established standards that made it possible to identify the elements present (and their relative abundance) in a sample of unknown chemical composition. For many analytical chemists, a spectroscopy apparatus with induction coil was indispensable.

Other experiments with induction coils revealed startling effects that gave rise to new physics as well as new technologies.[21] The Crookes tube, or simply "vacuum tube," was a variety of Geissler tube having a high vacuum that sometimes contained additional electrodes. Crookes tubes enabled the discovery of x-rays and cathode rays, which fostered new research questions and, in the early twentieth century, became the mainstay of electronic technologies.

As applications of the induction coil grew in number and importance, Napoléon III awarded the Paris instrument maker Heinrich Rühmkorff the prestigious 50,000-franc Volta Prize for his application of voltaic electricity in the "Rühmkorff coil." This 1864 award smacked of blatant nationalism, oblivious to the fact that the induction coil was yet another invention without an inventor. The induction coil had deep roots in the researches of Faraday and Henry and in Page's "compound magnet and electrotome," and it also embodied modifications introduced by other experimenters and instrument makers in several countries. Not until Rühmkorff dissected one of Ritchie's models was he able to improve the insulation and windings of his own coils and achieve high tension.[22] Clearly, by bestowing the Volta Prize on Rühmkorff, the French government demonstrated its power to create its own version of technological history.

Because the induction coil had contributed much to science, its authorship became a matter of personal and national contention.[23] Page was not pleased by the attention and credit that flowed to Rühmkorff, for in his view the induction coil had been largely an American invention, *his* invention. In a scholarly book published in 1867, Page argued his case at length on the basis of an exhaustive historical survey. His aim was to secure a patent and bring to America just credit. However, as an employee of the Patent Office, Page could not receive a patent without an act of Congress. Surprisingly, the requisite law was enacted—a gracious gesture to a dying man that revealed yet another governmental power to fashion technological history. The patent was awarded in 1868, not for a compound magnet with electrotome, but for an "induction coil."[24] Although a few anglophone histories recognize Charles Page's seminal contribution to the induction coil, he is remembered more often for his electric locomotive.

Recall that Page agreed with Henry that Davenport's enterprise had been debased by "mercenary speculation." However, Page believed that, because no one had established an absolute limit to how much magnetism a given battery could produce, the door remained open for further experiments. Perhaps, Page suggested, new kinds of electromagnets might make galvanic power less expensive.[25] On this issue Page and Henry would never agree.

Long before beginning the locomotive project, Page had made reciprocating electric motors, which he had called "electro axial engines." They were solenoids with one or more stationary coils and movable iron cores. In his view, it was important to learn *through experiment* how far such motors could be scaled up and operated economically.[26] The potential economy of a reciprocating design was, however, an optimism few other experimenters expressed. Moritz Jacobi, for one, had argued years earlier that a reciprocating motor was intrinsically inefficient because momentum was lost at the end of every stroke.[27] In the 1840s, Page—undeterred by these arguments or unfamiliar with them, and obsessed with the belief that they would become more economical, perhaps up to some determinable limit—built reciprocating motors of ever greater power.[28]

Although the economy of a battery-motor system is affected by many factors, Page was most concerned with the efficiency of converting magnetic force into mechanical force. Surely this sort of efficiency varied greatly among the various motor designs of the 1840s and later, but there was no reliable way to measure these differences and no generally accepted first principles from which to proceed. Instead, experimenters holding differing theoretical views made motors that differed dramatically in design and performance. Indeed, design principles for electric motors would remain fluid for decades. But far more relevant to motor economy was the amount of magnetic force obtained from consuming a given amount of zinc in a battery.

Page and others were attentive to this factor, but they believed that improved electromagnets would continue to yield greater dividends. Sturgeon, too, held this belief, even offering a token prize—a fancy bound volume of his *Annals*—to the person who could make the most powerful electromagnet in relation to the weight of an iron core exceeding 10 pounds. This ratio, however, was a measure of clever design and the conservation of copper, not of the economical use of current. Although new electromagnet designs were reported in *Sturgeon's Annals* (and elsewhere), these creations could not be compared on an absolute scale of energy efficiency because no such scale yet existed. These limitations did not prevent experimenters from proffering numerous empirical "laws" of electromagnetism ostensibly applicable to motor design. These formulations, however, were special cases not generalizable beyond the conditions of a given experiment, for too many variables had been left uncontrolled.[29]

In the meantime, one motor experimenter helped to establish a scientific principle—the conservation of energy—so general and so fundamental that its author, James Prescott Joule, was elevated in later decades to the pantheon of immortal physicists just one or two tiers below Newton, Maxwell, and Einstein. As has been noted in earlier chapters, the unity of forces was a notion gaining currency during the early nineteenth century. This doctrine implied that any one force could be converted into any other. Joule's major contribution was a series of ingenious experiments that yielded precise measurements of the *quantitative* relationships between specific forces (that is, forms of energy)—particularly the mechanical equivalent of heat. Joule himself gradually worked out the dire economic implications of the energy principle for replacing steam engines with battery-powered motors.

Joule (1818–1889) grew up in Manchester, England. Among engineers in that industrial city there was an insistent interest in quantifying "processes of conversion, particularly that in which fuel was converted to mechanical effect by a machine such as a steam-engine."[30] It was precisely in this industrial context that James Watt had defined one horsepower as the work needed to raise a weight of 33,000 pounds one foot in one minute. As a young man, Joule had been home-schooled, tutored for a time by chemist John Dalton, a founder of atomic theory. At age 19 he became a protégé of William Sturgeon, acquiring from him an enthusiasm for the possibilities of electric power.

In 1838, in his first published account of an electromagnetic engine, Joule suggested that "wheels or paddles may be affixed [to its axle] so as to answer to either locomotive or sailing purposes."[31] After further experiments, he presented a quantitative law: "The attractive force of the electro-magnet is directly as the square of the electric force to which its iron is exposed."[32] Reasoning from his law, he argued that the cost of electricity increases linearly, but the electromagnetic power increases by squares. "I can scarcely doubt," he wrote in 1839, "that electro-magnetism will eventually be substituted for steam in propelling machinery."[33] The reasoning seemed compelling, but the law was erroneous; the young Joule was not yet a sophisticated experimenter.[34]

Joule made further attempts to discern the laws of electromagnetic force, and in 1840 he suggested three ways to obtain power economically: increase the thickness of the electromagnet's wire, use a battery of higher tension, and optimize the arrangement of electromagnets. Working out such design principles was, Joule believed, a first step in developing electric power: "I have neither propelled vessels, carriages, nor printing presses. My object has been, first to discover correct principles, and then to suggest their practical development."[35] This was a rather more measured assessment than his earlier claims, for now he believed that actual applications were somewhat premature in view of the dearth of relevant design principles.

After still more experiments, Joule concluded that as the weight of an electromagnet increases, its lifting power as a ratio to that weight *decreases*. Thus, his smallest electromagnet, 9 grains, carried "2834 times its own weight," whereas one of 174 pounds held only 2,090 pounds—just 16 times its weight.[36] This was not an encouraging result. Moreover, for electromagnets large and small, the fundamental question of economy remained: How could the efficiency of converting galvanic electricity to magnetism be quantified? Joule now addressed this question, taking a cue from Faraday's studies in electrochemistry and inventing his own galvanometer.[37] This line of research led directly, in 1841, to Joule's momentous conclusion on the conversion of chemical into mechanical force (i.e., energy) in a battery-electromagnet system: ". . . every pound of zinc consumed in Grove's battery produced a mechanical force (friction included), equal to raise a weight of 331,400 lbs. to the height of one foot. . . . The duty of the best Cornish steam-engines is about 1,500,000 lbs. raised to the height of one foot by the combustion of each pound of coal. . . . I must confess that I almost despair of the success of electro-magnetic attractions as an economical source of power. . . . the expense of the zinc and exciting fluids of the battery is so great, when compared with the price of coal, as to prevent this class of magnetic engine from being useful for any but very peculiar purposes."[38] In just a few years, Joule had gone from enthusiast to pessimist on the use of electricity for propelling machinery. He had done so on the basis of quantitative analyses that seemed difficult to refute. In other experiments of unsurpassed elegance, Joule would add to the weight of evidence supporting the energy principle—a pillar of modern science. But the first fruit of his labors on this principle was quantifying the exorbitant costs of electric power in comparison with steam.

Joule's findings met a mixed reception. Many scientists and engineers understood immediately that the mechanical work that could be done by electromagnetism was determined by the amount of zinc consumed in the battery. But Charles Page and other motor enthusiasts still believed that better motor designs could squeeze ever more work out of electromagnetism, and that eventually it would replace steam power.

‿⌒⌒⌒‿

In 1844, at a meeting of the Association of American Geologists and Naturalists, Page exhibited a large "axial reciprocating engine."[39] It consisted of an iron rod that moved freely within two adjacent coils, each 4 inches long and 3 inches in diameter. When the coils were alternately energized, the rod moved vigorously back and forth. The iron rod was connected to a crank that turned a flywheel, and so produced rotary motion for driving machines such as saws and planes.[40] Several of these motors did find singular applications. In one instance, a printer in Washington used a Napier press outfitted with a Page motor to run off 1,200 impressions per hour.[41]

In the article reporting his new motor, Page also described its apparent offspring, a "magnetic gun." The gun's barrel had four coils arranged in a row such that an iron bar could pass freely through their hollow centers. The bar, carrying a wire connected to the battery, energized and de-energized the coils in succession as it passed through them. In this fashion, the bar was "projected to the distance of forty or fifty feet."[42] The magnetic gun embodied a design principle that would reappear in Page's largest motors.[43]

Also in the mid 1840s, Paul-Gustav Froment attempted to establish, at least in a very limited context, the practicality of electric power along lines first explored by Thomas Davenport. Froment was a French instrument maker who had devised rotary motors (figure 13.3). Not content to sell them merely as display devices, he coupled them to tools in his shop. Because his drills and lathes and so forth would have been used intermittently, the costs for battery maintenance might have been acceptable. In 1857 Froment received from Napoléon III the Volta Prize for his innovations.[44] Davenport, now dead, was spared the indignity of learning about the Frenchman's good fortune.

By 1850, Page wanted to demonstrate that his motors could replace railroad steam engines. However, building a full-scale motor and a locomotive to be powered by it was clearly beyond his means. He would need money—big money—for materials and for skilled labor to design, cast, machine, and assemble the many parts. No doubt aware that manufacturers and investors would decline to finance this project, Page petitioned the Senate to support "an investigation of a mode discovered by him of applying electro-magnetic power to purposes of navigation and locomotion." This could, he suggested, result in "a general substitute for the dangerous agency of steam."[45]

Having witnessed the excitement aroused by Morse's telegraph scarcely 5 years earlier, Page may have believed that Congress might fund another ambitious develop-

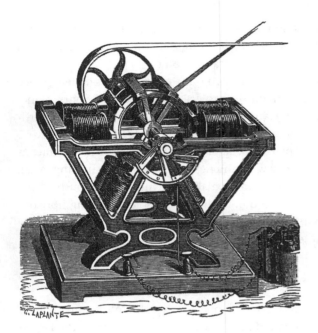

Figure 13.3
A Froment motor of 1845. Source: Du Moncel et al. 1883: 63.

ment project on the frontiers of electrical technology. Page was an international authority on electromagnetic devices, and his expertise could hardly be disputed. Moreover, Page could exploit his social network in Washington to lobby on the project's behalf. And he had a powerful champion in Senator Thomas Hart Benton, chairman of the Committee on Military Affairs. The senator's knowledge of scientific matters was limited, but he was aware of Page's experiments and "inclined to consider them of importance."[46] To strengthen his case, Page exhibited a motor before Benton's committee, bloviating about the electric power's potential to do real work.

Page's assessment of congressional inclinations was accurate, and on March 3, 1849, with no strong opposition, Congress approved legislation to grant Page $20,000 for investigating whether electric motors could be made sufficiently powerful to propel ships and trains. Page's advocacy of his own project was the only endorsement by a man of science that Congress needed. Apparently, Joseph Henry was not consulted.

Page set up shop in the Navy yard in southeast Washington. With access to people in the Navy's departments of engineering, blacksmithing, and plumbing, he built ever larger motors. He also hired a machinist, laborers, assistants, and a chief engineer (Ari Davis, brother of the instrument maker Daniel).[47] However, Page spent most of the grant on tools and supplies. Compulsive about ensuring that the materials going into the coils and mechanical parts were of high quality, Page used strength-testing equipment to assess samples from various vendors. He also tested the motors' output in horsepower. That Page had organized his project around first-rate men and materiel seemingly raised the likelihood of a good technical outcome.

In his progress report to the Senate, Page pointed out that, as a full-time employee of the Patent Office, he was able to work on the project only during off hours. Nonetheless, he had built several motors. The largest had two huge solenoids oriented vertically; after much tweaking and enlarging of the battery, it yielded an impressive 2 horsepower. To exhibit this motor's "practical character," by which Page merely meant sheer power, he geared it to "a circular saw ten inches in diameter, the turning lathe, and the grindstone of the workshop" and "worked all of them simultaneously."[48]

Re-using parts, Page configured a new motor with a single solenoid lying horizontally. Exhibited at the Smithsonian Institution, the 4-horsepower motor reportedly won Henry's commendation as a technical accomplishment.[49]

Even Page acknowledged that the cost of operating his motor exceeded that of operating an inexpensive steam engine, but he asserted (without supporting evidence) that "the expense was found to be less than the most expensive steam-engines." Yet he also mentioned recent computations by European "experimenters and men of science" showing that electric power would be vastly more expensive than the largest and most efficient steam engines. Making no attempt to reconcile these contradictory claims, he asserted that the costliness of electric power would be "no obstacle to its introduction, considering its immense advantages in other respects."[50] Perhaps seeking to save face or merely grasping at straws, Page maintained that only scaled-up experiments, not "calculation or process of reasoning," could determine whether engines on the order of 100 horsepower might turn out to be very efficient.

Funds for Page's project began to dwindle, owing to expenses such as $2,617 for the "platina plate, wire, and foil" needed to make the Grove batteries.[51] To help Page sustain the project, Benton appealed to the Senate for a supplemental appropriation of $40,000. The major objection raised in debate was "Where would it stop?" Would every starry-eyed inventor with a sketch in hand petition Congress for a grant? And surely, added Senator Jefferson Davis, such appropriations were not sanctioned by the Constitution. The opposition carried the day, and Page received no more federal money.[52]

An issue not raised then—one that inheres in all complex projects, even today—was whether the inventor could traverse a great developmental distance with the requested funds. Morse had been lucky, or very skilled, in forecasting the cost to build the Baltimore-Washington telegraph line. In contrast, Page had badly underestimated his financial needs. How could members of Congress, most of whom lacked scientific or engineering expertise, determine if a petitioner was the next Morse or the next Page?

In an 1853 editorial, *Scientific American* weighed in with an accounting of federal funds spent on inventions.[53] The pessimistic conclusion was that of nine projects supported—for a combined outlay of $110,000, including Colt's submarine battery—only Morse's telegraph was successful. In raising issues of federal policy, *Scientific American* staked out a middle course: an inventor was entitled to petition the government for funds to test a promising invention unless he had adequate means or could get private funds. The government also purchased patents, and the editorial raised no objection to this practice, except to observe that some inventions acquired in this way had failed to work.

To my knowledge, during the remainder of the nineteenth century no inventor received a direct congressional grant to develop an electrical technology *per se*. However, the government did let contracts to buy electrical equipment in the marketplace. (Among the items purchased under such contracts were Beardslee's magnetos for Civil War telegraphs and electric lighting systems for federal facilities like lighthouses.) Sometimes, however, an existing technology had to be modified for a challenging application, such as the electric gas-lighters that Samuel Gardiner Jr. installed in the new Capitol dome and the mine detonators developed for the Navy. Also, the federal government subsidized the Atlantic cable by agreeing to purchase services and supplying Navy ships and crews. Thus, the development of electrical technologies was not infrequently subsidized through projects tucked into various agencies' budgets.

__/\\/\\/_

Embarrassed by his failure to fashion a working locomotive, the beleaguered Page began pouring his own money into the project, and then borrowed from friends. Months went by but still there was no vehicle. Finally, on April 29, 1851, Page was ready to reveal his handiwork by traveling the rails to Baltimore and back on the same route as Morse's first line.

The locomotive, resembling a strange omnibus or coach (figure 13.6), weighed 21,000 pounds and carried seven passengers on its maiden run. Two large axial engines, estimated by Page to be at least 12 horsepower each, provided power independently to two driving wheels. The solenoids had an unusual design that descended, I believe, from Page's magnetic gun. Each consisted of a series of adjacent coils surrounding a heavy iron rod, 5 inches in diameter and 4 feet long, that moved back and forth. A sliding switch activated the coils, one after another, at just the right instant to add impetus to the moving mass of iron; the sequence was reversed on the return stroke. To ensure that the coils were activated on time, Page had coupled the sliding switch to the iron rod.[54] The rods were connected to the driving wheels.

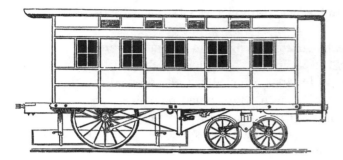

Figure 13.4
Page's electric locomotive. Source: *Greenough's American Polytechnic Journal* 4 (1854): 257.

To motivate the motors, Page constructed his own battery of 100 Grove cells, each consisting of an inner and an outer jar.[55] In the outer jar was a zinc electrode and dilute sulfuric acid; the inner jar held a platinum electrode and concentrated nitric acid. Page learned on the way to Baltimore that his Grove cells, suspended in a trough below the carriage, had a weakness: the inner jar, made of low-fired pottery, was fragile. Even before the journey began, jars in two cells broke, reducing their effectiveness as the acids mixed.[56] More than a dozen had already given way as the locomotive limped toward Bladensburg, barely past the District of Columbia boundary. Had the battery sat stationary, the cells might have given little trouble, but they had to endure the engine's vigorous vibrations. Passengers also might have experienced discomfort from the locomotive's "oscillating motion."[57] Out of economic necessity, Page had designed the locomotive's wooden carriage and running gear himself, and the amateurishness of his efforts was evident.[58]

Declining battery power wasn't the only problem. The insulation in the motors also began to break down. As a result, several coils had to be disconnected. With little hope of reaching Baltimore, Page put the engines in reverse at Bladensburg—35 miles short of his destination—and returned to Washington. In all, the trip was interrupted seven times as the inventor attended to failing cells and arcing coils.[59]

In a progress report to the Secretary of the Navy, Page put the best face on the "magnetic locomotive's" test run, which most others regarded as an abject failure. Obviously, no one could gainsay Page's claim to having made an axial engine as powerful as a steam engine; surely larger ones were also possible. Moreover, the locomotive had reached a maximum speed of 19 miles per hour, a respectable achievement for its time.[60] However, beyond technical competence, potential manufacturers and consumers would be interested in economy of operation. On this issue, Page merely asserted, without support, that "the larger my engines the greater the economy of power."[61]

Although insisting "that the present condition of the experiment is more encouraging than ever," Page also acknowledged that a rotary engine, less prone to vibrate strongly, might be better for this application.[62] Since he had not worked on large rotary engines, this was a devastating concession; perhaps all his toil on axial engines had been in vain. Page, $6,000 in debt, abandoned the project.

Page's axial engines were the most powerful electric motors of their time, but they had no technological legacy. When "linear motors"—essentially axial engines—were reinvented in the late twentieth century for various specialty applications, including roller coasters, Page's pioneering work had long been forgotten. But the story of his electric locomotive's supposed debacle lived on.

_____~~~~~_____

During the 1850s, debates raged over the practicality of the electromagnetic engine as a motive power in real-world applications. On the one side were scientists and engineers, in firm possession of the energy principle, who trumpeted the diseconomy of electric motors in relation to steam engines. Surprisingly, in 1850, while Page was still assembling his locomotive, a stark assessment emanated from his boss, Thomas Ewbank, the newly appointed Commissioner of Patents. Apparently drawing on Joule's work, Ewbank wrote: "At the present cost of metallic fuel, electro-magnetism cannot become *commercially* valuable, nor in any of the ordinary applications of steam can it come into competition with that agent."[63]

The most outspoken critic of electric power was Robert Hunt, a British professor of physics. In his 1851 introductory textbook, Hunt laid out what he believed were the most important economic parameters: "A grain of coal burnt in the boiler of a Cornish [steam] engine, lifted 143 lbs. one foot high. A grain of zinc consumed in a battery to move an electro-magnetic engine, lifted but 80 lbs. one foot high." Zinc, Hunt noted, was 24 times more expensive than coal per unit weight. Thus, he concluded, for an equal amount of work, the operating expense of "an electro-magnetic engine . . will be more than fifty times greater."[64]

Hunt lectured on electric power's liabilities at a meeting of the Institution of Civil Engineers in 1857. In discussions lasting the entire evening, no one dissented from his gloomy assessment.[65] The consensus of the engineers was definitive: "There could be no doubt, from what had been said, that the application of voltaic electricity, in whatever shape it might be developed, was entirely out of the question, commercially speaking." In view of the present cost of electricity, "electro-magnetism must be confined to special purposes, where the danger of steam and creation of vapour were sought to be avoided, or where economy of space was a consideration. But it was far removed from general application as a motive power."[66] Although that same discussion also showed that electromagnetic power was 50 percent cheaper than "manual power," no one waxed eloquent about using motors to relieve or replace human labor.[67]

Three years later, Hunt noted in another paper that zinc was now selling for £35 per ton, yet coal was less than £1 per ton. In an arresting conclusion, perhaps not altogether tongue-in-cheek, he observed that "it would be far more economical to burn zinc under a boiler, and to use it for generating steam power, than to consume zinc in a voltaic battery for generating electro-magnetic power."[68] For his studies of electric power, Hunt received the prestigious Telford Medal from the Institution of Civil Engineers.

Hunt's findings were given great credence, quoted often in technical and popular writings, extended by other scientists and engineers, and absorbed by potential motor investors and manufacturers. The famous chemist Justus von Liebig chimed in, insisting that people who advocated electricity as an inexpensive power were deluding themselves, for they had apparently not done the calculations.[69]

However, there were voices on the other side, including some inventors and a few men of science, who offered creative remedies to make electromagnetic power more economical.[70] J. F. Mascher of Philadelphia suggested that if nitric acid were to be used as the electrolyte and if silver were to be used for the zinc electrode, the battery's operation would produce a valuable by-product: silver nitrate, an expensive salt used in photography. In experiments with his own motor, Mascher calculated the consumption of silver, nitric acid, and the yield of silver nitrate. After converting these quantities to current prices, Mascher concluded that his system could turn a profit so long as the expenses were low for "collecting and casting the nitrate into sticks ready for market."[71]

In England, Dr. Joseph Watson made a similar proposal. He would use a lead electrode instead of zinc, and an electrolyte of potassium bichromate. The salable byproduct would be lead chromate, "a beautiful yellow pigment." According to the brief report in *Scientific American*, Watson had formed an "Electric Power and Color Company," and it was soon to build an establishment "to carry out the project."[72] This apparently did not happen, for Watson's scheme assumed that his batteries, along with electric lights and motors, would be commercialized and adopted in sufficient numbers to furnish an economically viable supply of pigment.

The electrometallurgist James Napier, writing in *The Engineer*, dismissed these suggestions for making electric power more economical: "It is true that efforts have been made to make the salts or products of the battery available for other purposes, so that by their sale the cost of production is lessened; but as yet these efforts are not practicable" (because of the cost and difficulty of storing and processing the salable by-product).[73]

Over the years, *Scientific American*, owned by a patent agency, promoted invention and patenting by encouraging inventors while also acknowledging the economic burden of electric power. For example, an 1851 editorial invited inventors to seek an "important discovery in electro-chemistry" that might reduce the cost of galvanism—a seemingly disingenuous suggestion. Inventors could, the editorial continued, "labor with hopes of ultimate success."[74] And *Scientific American* had published a glowing account of Page's large axial engine, even suggesting that a compact electric motor might one day be used for "aerial navigation."[75] This sympathetic treatment was understandable; after all, Charles Page was a patent examiner, and it behooved *Scientific American's* proprietors to remain in his good graces, although—as subsequent editorials indicated—they held no illusions that galvanic power could replace steam engines.[76]

In the period 1850–1880, *Scientific American* also described new electric motors, pointing out interesting design features. The magazine also publicized applications of one-of-a-kind motors, such as driving a sewing machine, and suggested new applications, including opening and closing inaccessible vents or valves. In 1874 the magazine was calling for easy-to-maintain batteries so that electric motors could be used in domestic and commercial settings.[77]

The arguments of scientific authorities notwithstanding, inventors could advertise their ingenuity by crafting and patenting original motor designs. Sometimes, however, inventors also emphasized the performance characteristics that might make electric motors desirable, at least in applications where operating cost was not terribly important. Electric motors presented no danger of explosion, consumed no fuel while idle, could be started and stopped instantly, took up little space, were simple in construction, were quiet, generated little heat, could be used by one person with little training, could be made to produce low and even fractional horsepower (the range in which steam engines are least efficient), and might be readily coupled to individual machines in homes, workshops, and factories.[78] Indeed, nearly every issue of *Scientific American* featured new machines that, in principle, might be driven by electric motors, from cider presses to washing machines. Recall that prospective customers writing to Davenport and Cook had also expressed a need for low-power motors. That a large latent demand existed for small motive powers was underscored by the stream of inventions that, in later decades, would become a river, eventuating in varied internal and external combustion engines, some of which enjoyed commercial success.[79]

However, not until the late 1870s did a wide variety of small and large industrial-quality electric motors, reflecting a significant investment of mechanical as well as electromagnetic expertise, become available in the marketplace.[80] Before that time, motors sold in instrument shops were, with few exceptions (e.g., Froment's), too unrefined and flimsy for serious work, for they lacked rigid frames, lubrication, and even pulleys for taking off power. Clearly, a considerable developmental distance had yet to be spanned. As a result, potential consumers had no good options for replacing other motive powers with electromagnetic engines. How can one explain this state of affairs?

In considering whether to commercialize a technology spawned by natural philosophy, manufacturers and investors often sought advice from scientific authorities. From Henry to Hunt, men of science spoke almost in unison about the electric motor's exorbitant operating costs, despite the occasional throwaway line about "special applications." These opinions mattered to businessmen and affected their investment decisions. The philanthropist Peter Cooper, who did much to encourage invention and technical education, was often beseeched by inventors to support their work on motors. For advice on this matter he relied on Dr. P. H. Vander Weyde, who on the basis of experiments from 1843 to 1848 had concluded that battery-powered motors were impractical. Accordingly, he advised Cooper to resist the requests, and claimed that "Mr. Cooper never spent a single dollar on account of electro-motors, except on such small specimens as were required for class instruction."[81]

Mature and rugged electric motors did not make it to market because manufacturers and moneyed men, relying on the practicality assessments of scientific authorities, declined to commercialize them. After all, countless inventions were in need of financial backing, and many seemed much more likely than electric motors to turn a profit. Unlike Edison in later years, motor inventors were apparently not in a financial position to develop and bring to market their own creations. And no instrument maker, not even Daniel Davis, commercialized Page's large axial engines.

In 1853, as president of the short-lived Metropolitan Mechanics' Institute of Washington, Joseph Henry delivered an address summarizing his long-held views on the contributions of scientists to the development of inventions. The scientist, Henry insisted, is the person most qualified to judge technical feasibility: "There can be no reality in science if at this late day it cannot predict that certain proposed inventions are impossible, as well as declare that others are in accordance with established principles."[82] In expressing an honest opinion about a proposed invention, the scientist rendered a valuable public service by sometimes identifying and exposing frauds, saving investors from financial ruin and embarrassment. In Henry's construction of this role, the scientist merely determined whether an invention was permitted by known principles. However, the case of the electric motor shows that scientists—even Henry—could extend their power and influence over technological development by also offering opinions, for example, on the relative costs of electric and steam power *and* the likely response of consumers to expensive electricity. Henry's contention that electric power was "in opposition to the best-established truths of science" irretrievably conflated science with issues of economy and consumer behavior.[83] Blurring and extending the boundaries of scientific authority in this way gave Henry and other men of science a sweeping mandate to pontificate.

In comparing electric and steam power, Henry, Hunt, and others focused on the expense of fuel (zinc vs. coal) in producing equal quantities of mechanical power. Left out of these calculations were the costs of buying and installing the power systems, acquiring the necessary space, hiring labor for operation and maintenance, and depreciation. And at no time were the risks to human life factored in. Moreover, the estimates were context-free, unconstrained by the local factors that a proprietor would take into account in a specific installation, such as the amount of power needed or the anticipated frequency and duration of use. Also ignored was the larger issue of potential competition with non-steam power sources, such as human labor and (in later years) gas engines. With the benefit of hindsight, these omissions seem especially glaring in analyses that were purportedly scientific. Indeed, in asserting the impracticality of electric power on economic grounds, scientific authorities were not doing science *per se*; rather, they were engaging in a crude kind of cost accounting. Nonetheless, these analyses would have appeared to be scientific because they were quantitative and alluded to principles of science—*and because they were put forward by scientific authorities*.[84]

The conclusions of these scientific men no doubt seemed plausible to manufacturers and investors because all parties shared certain cultural beliefs about the parsimonious behavior of consumers, especially profit-making firms. It was assumed, tacitly to be sure, that a businessman considering motive powers would be swayed by the narrowest of economic considerations. However correct that assumption might have been in many cases, we know that an expensive new technology is often adopted at first because it has utilitarian and/or symbolic performance characteristics that, in important activities, outweigh economic disadvantages. (Think here of medieval cathedrals, missile defense, and carbon-fiber bicycle frames.) No laws of physical science prevent people from acquiring costly new technologies.[85] Moreover, it would appear that forecasting (or divining) consumer behavior is a matter for social and behavioral science, not chemistry or physics.

In the middle of the nineteenth century, when cost accounting was still in its infancy and formal marketing research did not yet exist, men of science expanded their social power in an understandable way, colonizing intellectual territories as yet unoccupied by other specialists. And, having little other recourse, manufacturers and investors placed faith in their judgments. As a result, scientific authorities sometimes affected, positively or negatively, the commercialization of "scientific inventions" such as the electric motor.

Motor inventors tried to refute the scientists' arguments, but they were interested parties, their motives were suspect, and they lacked appropriate authority and social power. In controversies over *apparent* scientific issues, a Robert Hunt, a Joseph Henry, or a Justus von Liebig easily prevailed over inventors, at least in the short run. More-over, naysayers could embellish their arguments by pointing to conspicuous failures, such as Page's locomotive.

Inventors, I suggest, were not entirely blameless. They had committed a serious strategic error. If debates about electric power had been only about small motors driving machines intermittently to replace human exertion, perhaps scientific authori-ties might have framed their analyses differently, arrived at rosier economic conclu-sions, and judged electric power practical for consumers much earlier. Instead, inventors challenged the steam engine, invoking the boiler explosions that routinely killed many people at a time.

Why did they challenge steam power? Perhaps inventors believed that the danger of steam boiler explosions, apparent to all from lavish media coverage of disasters, would predispose businessmen to finance development of their safer alternative. Although other entrenched motive powers, such as wind, water, and animals, had some undesirable performance characteristics, none was as threatening to human life as steam, and so inventors could trade on its emotional salience as a technology that killed. It would seem, however, that targeting the demon of steam was a strategy fatally flawed because it set the agenda for the economic analyses performed by scientific authorities.

In the late 1870s, steam-powered dynamos began to produce electricity that was vastly cheaper than that produced by batteries, and so scientific authorities could at last endorse electric power. As a result, a host of firms developed and brought to market an endless variety of sturdy, powerful motors capable of continuous duty. According to the economic historian Malcolm MacLaren, "by 1887 there were fifteen well-known manufacturers of small motors in the [United States] who had produced over 10,000 motors of fifteen horsepower and below."[86] As late as 1910, however, only 10 percent of American homes had electricity supplied by central stations. Even so, the remaining 90 percent of homeowners desiring motorized appliances could buy items such as electric fans, sewing machines, and phonographs, and dentists could purchase electric tools for drilling, shaping, and plugging.[87] Surprisingly, well into the twentieth century the motors in all these products were powered by batteries.

The hundreds of one-off motors devised before 1875 had few if any technological offspring. In fact, the immediate ancestor of later electric motors was not a motor at all but a generator, and many of the earliest electric motors successfully commercial-ized were installed in trolleys and streetcars, where they took the place of horses and steam engines. As I will show in later chapters, these two curiosities are related.

14 Humbug!

The advent of large and complex technological systems, including the telegraph, the railroad, the steamboat, mechanized farming, steam-powered factories as well as public water, gas, and sanitation systems, presented inventors with an endless array of new technical problems to work on. And work on them they did. The outpouring of inventions increased phenomenally, spurring the creation of new institutions and the remodeling of old ones.

By the start of the nineteenth century, most countries, hoping to encourage invention, had enacted patent laws and established offices to administer them. Early legislation, such as the U.S. patent law of 1791, merely provided for a listing of inventions, with little or no scrutiny of their merits. As patent institutions became burdened with applications, the laws were revised to make the examination process more rigorous. The staff of the U.S. Patent Office came to include experts in particular domains of science and technology, and their numbers grew. In 1837, the year after the enactment of the new U.S. patent law, there were two examiners; in 1861 there were twelve.[1]

As the patenting process became more formal, inventors lacking a legalistic turn of mind found themselves disadvantaged. Clearly, they would need professional help in crafting applications and in negotiating with patent examiners. This need led to the emergence of a new occupation: patent agent. The patent agent became a significant player—often a gatekeeper—who was in a position to judge an invention's patentability.

In the United States, the business of patent agencies expanded at an astounding pace, keeping up with the exponential increase in submissions to the Patent Office. In 1844 the Patent Office processed a little more than 1,000 applications; in 1872 it processed more than 18,000.[2] By that time, most applications were submitted by patent agencies on behalf of inventors. In the United States and in England, patent agencies came to own invention-oriented periodicals that promoted their primary business of collecting fees from inventors: *Scientific American* was founded in 1845, *Mechanic's Magazine* in 1823. Besides serving the inventor community, these periodicals carried out important functions during the golden age of the independent inventor.

To build and retain their credibility with patent examiners, an agency's experts had to vet inventors' submissions. Some inventions were imperfect but promising, a few were apparently the product of charlatans, and far too many showed an unacceptable ignorance of previous work and established knowledge. Over the years, respectable patent agencies, growing in power and influence, became more selective, declining to submit inventions (such as perpetual-motion machines) that might embarrass them and compromise their efforts on behalf of strong cases. In the early 1850s, *Scientific American* put it bluntly: "No case is . . . permitted to leave the office until it has passed the ordeal of our criticism. This is perhaps one of the principal reasons of our great success in obtaining Letters Patents for new inventions."[3] At the same time, however, these agencies pushed for a more liberal interpretation of patent laws to increase the likelihood of an application's success.

Patent agencies such as Munn & Co., owners of *Scientific American*, had now become players in technological development. They judged an invention on one criterion: whether the invention, if properly specified in an application, was apt to be approved by a patent examiner. That is, did the invention meet the ambiguous standards of originality and utility specified in the patent law of 1836? Patent agents also had to consider who would be judging an application, for the examiners varied in special expertise, in diligence, in their interpretations of patent law, and in willingness to respond to pressures that might come from *Scientific American*, patent agencies, the inventor community, Congress, and the Commissioner of Patents (a political appointee). As Robert Post observed, "Passing judgment on patent applications is not an objective process."[4] We may suppose that, all else constant, an application properly shaped by a reputable patent agency had a greater chance of approval than one crafted by an inventor. But the awarding of a patent was no guarantee that an invention would work or was likely to be useful; sometimes it did not even guarantee originality.

Daily confrontations with the raw products of creativity gave *Scientific American* a strong incentive to educate inventors—indeed, to inculcate a doctrine of "scientific invention."[5] Cultivating a community of enlightened inventors would obviously benefit the patent agency, and it might elevate the status of American inventors at home and abroad. In the second half of the nineteenth century, *Scientific American* refined and reiterated this doctrine in dozens of editorials while also acknowledging the essential contribution of the inventor's ingenuity and perseverance.

The tenets of scientific invention, which were simple and few, helped lay the foundation for a strong patent application. First, the inventor had to be familiar with earlier improvements in relevant areas of invention. Second, the inventor (especially the inventor of a mechanical device) had to be acquainted with mathematics and the "first principles of physics," so that machines could be properly designed. And third, the inventor had to possess an appreciation for the "first principles" underlying modern industries, such as the "peculiarities" (i.e., performance characteristics) of different fibers in relation to manufacturing processes. These tenets implied that an inventor had to be completely forthcoming in describing an invention and how it worked; there could be no secret parts or processes.[6] Although these tenets had a theoretical tilt, *Scientific American* tried to strike a balance between theory and practice: ". . . while we are ready to admit that theory alone cannot subserve the purposes of the inventor or the mechanic, we maintain that practice alone will not answer. The truly great inventor gets as much of both as he can."[7] This was a position that Joseph Henry also espoused: "We have practical men in great numbers without theory and theoretic men without practice. Now it is evidently the union of these two in the same individual from whom we must expect the greatest and most successful efforts."[8]

To help the inventor acquire broad familiarity with first principles, *Scientific American* defined for itself a new role in journalism. It published brief biographies of inventors that drew object lessons; multi-issue articles authored by eminent engineers and scientists (including Faraday); detailed, image-filled articles on the history of particular technologies; descriptions of contemporary commercial and industrial processes; abstracts of recent American patents; brief accounts of foreign inventions; and articles about selected inventions (almost always advertisements for its clients).[9] *Scientific American* also "exposed the cant and rant of . . . pretenders to scientific knowledge," sometimes engaging in lengthy controversies that brought to light fraud and deception posing as invention.[10] The expectation was that reading each weekly issue would prepare a person to pursue scientific invention to a successful conclusion: a strong patent that might lead to riches.

While *Scientific American* raised the bar for inventors by hammering home the tenets of scientific invention, it also emphasized that fields ripe for invention continued to expand. The magazine frequently pointed to problems, large and small, that "geniuses" and "lesser lights" might tackle.[11] For example, it reported in 1852 that a prize had been offered in France for rendering the battery "applicable, with economy, to industry, as a source of heat—to lighting, chemistry, mechanics, or medical practice."[12] Making galvanism more economical was a longstanding cultural imperative that impelled inventive activities. More typically, *Scientific American* called for train seats that become "sleeping couches" at night, for steam-driven plows, and for a substitute for ivory.[13] There were also needs for machines to husk corn, to cut down trees, and to milk cows.[14] Perhaps taking their cues from *Scientific American*, inventors brought all these to fruition before the end of the century.

Although the doctrine of scientific invention furnished patent agents with an *apparently* objective basis for weeding out weak cases and scams, an invention could venture into regions of science where squabbling authorities made clarity elusive. With electrical inventions, especially, there was sometimes room for dispute about what might be technically feasible. This chapter focuses on the American inventor Henry Paine, creator of the "Hydro Electric Light," the benzole light, and a seemingly miraculous electromagnetic motor. Controversies surrounding Paine's inventions played out in the pages of *Scientific American*, with heated debates taking place between Paine, his supporters, and scientific men. Sometimes lasting years, these exchanges help us to see the difficulties of enforcing the doctrine of scientific invention.

_⌒⌒⌒_

In 1847, J. W. Orton of Oxford, New York, invented an unusual system of lighthouse illumination that employed neither carbon arc nor battery but was still electrical. Atop a lighthouse Orton proposed to set a windmill that would transmit power through gears and shafts to two Saxton-style magnetos. The currents would be combined and conveyed to vessels, where it would be used to decompose water. Burning the hydrogen would generate heat, which would be concentrated on a piece of lime to produce brilliant limelight.[15] There is no evidence that this system was ever built, much less installed in a lighthouse. Even in its time, I suspect, this invention would have been regarded as fanciful, yet it was permissible by accepted scientific principles. Indeed, Orton's mechanical-magnetic-electrical-thermal-chemical light materialized the belief, given scientific grounding in Joule's conservation-of-energy principle, that a clever inventor with the right gizmo could convert any form of energy into any another. Since science ruled out perpetual motion, the energy principle was the next best thing, for it intimated that the horizons of invention were limitless—so limitless, in fact, that they invited unscrupulous people to contrive electrical devices of bewildering complexity. Purporting to accomplish astounding feats of energy transformation, these technologies functioned mainly to lure naive investors and as a vehicle to satisfy the inventor's desire for celebrity. With complex electrical apparatus, a charlatan could easily obscure what a device was doing—or not doing.

Orton's invention disappeared without a trace, but others quickly took its place in the incessant quest to create inexpensive lighting technologies that might compete with coal gas. Like Orton, many inventors claimed to be able to convert, in several steps, one form of energy (usually mechanical) into light at little cost. Henry Paine of Worcester, Massachusetts, was the most notorious of these inventors, attracting attention in newspapers, magazines, and scientific journals.

In 1848, several weeks before Christmas, Paine circulated an immodest circular in which he claimed to have "produced a light equal in intensity to that of four thousand gas burners of the largest 'bat's wing' pattern, with an apparatus occupying four square feet of room, at a cost of One Mill per hour, the current of electricity being evolved by the action of machinery."[16] Such an invention raised the tantalizing possibility of loosening the grip of the gas companies. But Paine did not disclose how his invention worked, saying that he feared others would soon 'reap its benefits" by defying his patents (though as yet he had none). However, he supplied enough clues to support the inference that his invention, like Orton's, used a magneto to decompose water, burned the products, and somehow generated light.

In publishing Paine's circular, *Scientific American* also discounted his claim: 'It will indeed be something new to the Scientific World, when *mechanical electricity* will be exhibited decomposing water and sustaining for ten hours 4000 lights . . . for one cent."[17] In an indignant response, Paine announced that his light would be exhibited soon in Worcester.[18] Some months later, *Scientific American* printed a testimonial by "G.C.T.," who witnessed Paine's light issuing from a parabolic reflector in the cupola of Worcester's Exchange. It was so bright that a newspaper could be read a mile away. In addition, G.C.T. asserted, Paine had demonstrated the technology's fitness for city and lighthouse use by installing on a tower a large system driven by a 69-pound weight that fell 16 feet; it had to be wound up only four times a day. G.C.T added that Paine's parlor and laboratory were lighted by gas coming from a small apparatus powered only by "a common brass eight day clock wound up every morning."[19] Asserting that "Mr. Paine has fully accomplished all he promised in his circular," G.C.T. wondered why the press had ignored these feats. According to Paine, whom G.C.T. queried on this very point, editors who had witnessed the light declined to report it because Paine had refused to explain how it worked. G.C.T. concluded with the remarkable statement that "no mercenary motive prevents [Paine's] making his discovery public, for he has sold a portion of it for a sum that puts him beyond pecuniary embarrassment." This claim appears calculated to attract ordinary people looking for investments that capitalists had already endorsed.

In a brief comment appended to G.C.T.'s letter, *Scientific American* insisted that an inventor whose project had been challenged on the basis of scientific argument would have to refute that argument before the invention could be taken seriously. Scientific men had shown that using galvanism to produce light, by carbon arc or decomposing water, was more expensive than gas. If Paine's method did produce a vastly cheaper light, then he was obligated to make its operation transparent. One letter writer invited Paine to defuse the growing skepticism by bringing his light immediately to New York, where "the inhabitants . . . are aroused against the present gas companies, and would at once patronize any other cheaper mode of illumination."[20]

In defending his refusal to reveal the invention's innards, Paine asserted that patent laws failed to protect inventors from "piratical infringements or ruinous law suits."[21] However true that statement might have been (and still is), Paine was not playing by the genteel rules of scientific invention. One writer put it this way: "So far as the opinion of men of science is concerned, they cannot be satisfied with the mere exhibition of the Hydro Electric light—that is nothing to them; it is the new manner of producing it. Until this is done by Mr. Paine, in a public lecture, or description, the reported discovery will be viewed as something suspicious."[22]

⁓⁓⁓

In responding to a persistent critic whom he accused of shilling for gas companies, Paine maintained that scientific facts were not fixed forever. Natural philosophers had believed that there were only four elements and that the earth was immovable—"facts" that were eventually shown to be in error. Paine claimed that current scientific facts about water were also wrong, and that he would replace them with his own discoveries. For example, "water is a simple substance—and oxygen water [is] *held in solution* by positive Electricity, and hydrogen by negative."[23] Paine also took credit for discovering a "*new* principle" of electricity: its "ponderability, materiality, and obedience to the laws of gravitation." Paine asserted that the new principle enabled him to devise an apparatus for decomposing water far more effective than the usual galvanic method.[24]

Paine's radical claims might have resonated strongly among inventors who espoused unconventional ideas and felt oppressed by theoretical authorities. Moreover, his seemingly outrageous statements were not without benefits to the inventor-*cum*-subversive, for they could boost his persona in the print media, raise his status among kindred spirits in the inventor community, and attract gullible investors. I suspect that he was fully aware of the possibilities of the rebel role and played it to perfection.

No dispute about electrical science and technology so widely publicized could long escape the notice of Joseph Henry. Correspondents sometimes sent him newspaper clippings on Paine's discovery and asked for his opinion of its practicability. Henry was now the major spokesman for professional science in America, and speak he did.[25] In his 1850 presidential address to the American Association for the Advancement of Science, he warned against unscrupulous people who would undermine the search for truth. In discussing "humbugging," Henry almost certainly had Paine in mind: "This species of deception frequently begins in folly and ends in fraud. The author of it generally imagines at first that he has discovered some very important principle in nature which is directly applicable to useful purposes in the arts." After convincing friends and neighbors, the humbug seeks and receives attention in newspapers, and this makes it more difficult to admit error. He then extends the deception, "defrauding all who may have become his dupes."[26] Henry did not go out of his way to denounce Paine publicly, but his negative assessments of the Hydro Electric Light were quoted by others.[27] Confiding in the anthropologist Henry Schoolcraft, Henry acknowledged that "if I attempt to run a tilt against all the humbugs of the day I shall have more than my hands full."[28] But to anyone who asked, Henry did not hesitate to declare Paine's claim impermissible by the laws of physics.

According to another critic, Paine's claim also violated the laws of chemistry.[29] Paine had insisted that his white light burned only pure hydrogen extracted from water, but chemists knew that hydrogen alone gives off only a faint bluish flame. To burn brighter and whiter, hydrogen must be combined with carbon. That Paine's light was no exception was demonstrated in mid 1850 by a committee of chemists and engineers appointed by Worcester's gas companies. Although Paine did not show up at the appointed time and place, his representative exhibited the light at his home. However, the committee was not permitted to see the magneto that was supposedly decomposing the water.

The light was bright, as advertised, but the illumination gas itself—before combustion—had "the odor of oil resin gas." Apparently forced by the critics to account for the added carbon, Paine, in his preparations for the demonstration, had brought into the open a bottle of turpentine through which the hydrogen gas was allegedly passed. However, the plumbing arrangements were complicated, and the committee demanded clarifying experiments. The hydrogen tube was disconnected and lit; it produced but a feeble light. However, when it was dipped directly into the turpentine, the gas that emerged yielded a brighter light, but not as bright as that in the original set-up. This result left everyone confused. The committee was prevented from performing more experiments, and so the report could only imply that a source of carbon *beyond* the turpentine had carbureted the hydrogen.[30]

In ruminating over the committee report, *Scientific American* insisted that turpentine alone could not turn hydrogen into a good illuminating gas. The editorial gloated that the additional carbon—whatever the source—entailed an expense, and so Paine's light could not be cheaper than gas; "this discovery is a downright error."[31] Paine answered with a verbose and belligerent barrage that boiled down to one point: "practice is better than theory." But he added that the invention was valid because "men of high standing certify that I am doing all that I claim to do, and that they have seen it done." His financial backers were satisfied, Paine insisted, and he didn't have to convince anyone else.[32]

_⁓⁓⁓_

Suddenly the dam of scientific certainty in matters chemical began to breach. George Mathiot, an electrometallurgist employed by the U.S. Coast Survey, reported that hydrogen bubbled through turpentine did produce a much brighter light than hydrogen alone; mysteriously, the consumption of turpentine was negligible. Hydrogen treated in this manner was, Mathiot asserted, a fit illuminant. It remained, he suggested, for the "scientific world" to explain how such a small amount of turpentine could achieve this unexpected effect. Mathiot's letter in *Scientific American* was followed by an editorial comment.[33] Here, wrote the editor, was testimony from a man who "possesses a vast amount of practical scientific knowledge" and whose experiments were witnessed by "scientific gentlemen attached to the survey." Thus, Mathiot's testimony could not be discounted as those of Paine's friends and backers could; perhaps, after all, Paine had made an original chemical discovery. Nonetheless, the editorial returned the burden of proof to Paine, insisting again that he reveal the operation of the entire apparatus. And again he refused to divulge its secrets.[34]

Acknowledging that hydrogen and turpentine vapors could somehow produce a white light, *Scientific American* shifted the debate toward the calculation of costs, reiterating that coal gas would still be cheaper.[35] Clearly, beyond the issues of scientific validity that Paine's "discovery" raised, the practicality of the hydrogen light for consumers would ostensibly hinge on its cost relative to that of coal gas.

Seeking to widen the breach among men of science, Paine invited E. Wright, editor of the *Boston Chronotype* and a retired professor of chemistry, to view his invention in action. Paine exhibited the electrical components for the first time, but did not reveal the special electrode that supposedly produced only hydrogen. In print Wright described the apparatus and praised its maker, who had "extorted from nature the secret of the artificial production of light at a nominal cost." This discovery, Wright continued, "enables [Paine] to command a new force of nature, which is soon to supersede most of the forces now employed—something which is destined to work a revolution both in science and art."[36] Other Massachusetts papers, including the *Boston Transcript*, chimed in with earnest testimonials in support of the local hero who, in daring to challenge the experts, had emerged victorious.[37]

Scientific American did not question Wright's credentials or powers of observation, but conceded that resolving the scientific issues would require more experiments. By this time, Henry, seeking to establish criteria for exercising scientific authority, had argued that men of science were specialists. Thus, chemists should pronounce only on matters of chemistry, geologists on geology, and so on. It was specialized expertise—knowledge and experience—that conferred upon a person the authority to offer legitimate scientific opinions. Not only was it important to distinguish scientific men from mechanics and inventors; it was also necessary to recognize distinctions within the scientific community and to comprehend their implications for resolving controversies.[38]

Relevant expertise was an issue that *Scientific American* could have raised in relation to Wright's opinion of Paine's invention, but didn't. Clearly, Wright's description of the electrical components—two permanent horseshoe magnets, between whose opposed poles revolved several coils—betrayed the fact that he had little acquaintance with electromagnetism. Not only did Wright refer to these components as a "sort of electro-magnetic condenser," perhaps parroting Paine; he also insisted that the coils revolved easily, hindered only by slight mechanical resistance. By this time, however, it was well known that self-induction created a counter-electromagnetism in the coils that resisted their rotation. Simply put, one had to perform work in rotating the armature.

Joseph Dixon, "famous for his crucibles, manufacture of American steel, and practical chemical knowledge," furnished a possible explanation for why Paine's coils rotated so easily.[39] Dixon believed Paine's claims to be "utter nonsense" and bet $5,000 that Paine "cannot resolve water entirely into hydrogen, nor decompose it with only one pole of a magnetic connection." In a lecture on light, Dixon showed "how easily *wise* people might be deceived with perpetual motions, and new gas lights." He had assembled an apparatus such as Paine had reportedly employed, complete with magnets and coils, secret electrode, turpentine, and the rest. Like Paine, Dixon produced a fine light, burning the gas that was supposed to be turpentine-tainted hydrogen. Then, with a dramatic flourish, he revealed that the secret electrode was nothing but a battery that itself decomposed the water. Not only were the magnets and coils superfluous; the latter were made of wood and thus could rotate easily.

In a curt reply, Paine insisted that his "electrodes were taken to pieces in the presence of Dr. Channing, of Boston, Dr. Doremus and President Young, of the Manhattan Gas Co.," all of whom could identify a battery. (Curiously, Paine signed this letter "P.M.H."—his initials in reverse.[40]) Scarcely had the ink dried on Paine's reply when George Mathiot clarified his position on the Hydro Electric Light. Doubtless he was embarrassed that Paine and his supporters had exploited his brief and narrowly focused report. Mathiot now went on record unambiguously: "I never doubted the erroneousness of Mr. Paine's assertions as to the main features of his affair."[41]

Taking seriously the burden of educating inventors in first principles, *Scientific American* published a detailed, multi-issue article on illuminating gas, authored by J. B. Blake of Boston, a gas engineer. In one of the last installments, Blake dissected Paine's claims and sought to explain his popularity: "Mr. Paine's statements, although quickly repulsed by scientific persons, being considered so grossly absurd as to be unworthy of further thought, were upheld by many unacquainted with the subject, and therefore gained believers very rapidly, as do all new and novel inventions; and it would seem the more preposterous the statements, the more converts are gained; so craving is the public mind for new things, and so assiduously does it seek excitement."[42]

In mid 1851, in the midst of these testy exchanges, the beleaguered Paine received a gift: a British patent on the Hydro Electric Light. (He never received a U.S. patent.) This action had required disclosure of the invention's parts and mode of operation.

Scientific American reproduced the patent specifications and drawings, uncharacteristically without comment.[43] Curiously, Alfred Newton, the patent agent who had taken out the patent on behalf of Paine, was the London representative of Munn & Co. A week later, *Scientific American*, which was in a somewhat awkward position, nonetheless reaffirmed its negative opinion of the invention and furnished reasons to doubt the patent's claims.[44]

Having read the abstracts of every British electrical patent from this period, I conclude that the examiners were poorly prepared for this technical work or were negligent in carrying it out. For example, they sometimes patented the same invention repeatedly, and even accepted abstruse and nonsensical specifications. That Paine's invention was patented indicated only that he had an able agent in London and had paid the $500 application fee. (In mid-nineteenth-century America, the fee varied from $15 to $30.)

A British patent was a prestigious validation and a potentially salable property. However, the British patent marked the beginning of the end for Paine's Hydro-Electric Light. The confusing drawings and obscure specifications could not conceal the fact that the invention was unable to produce light for next to nothing. Not long after the patent was made public, *Scientific American* penned the invention's obituary: ". . . it was of no economical value whatever; and we do hereby assert that, for all *practical beneficial* purposes, it is extinguished now and forever."[45] Investors, perhaps now a bit wiser, did not flock to Paine's door. Nonetheless, an unrepentant Paine retained the posture of a misunderstood inventor persecuted by close-minded scientific authorities.

_⸻⌇⌇⌇⸻_

If nothing else, Paine was resilient. Just as the Hydro Electric Light was slipping into the shadows, he brought forth the "benzole" or "atmospheric" light. This invention rested on the claim that air would be carbureted when bubbled through a mixture of alcohol, water, and benzole (an impure form of benzene used, like turpentine, as a solvent). According to Paine, burning this carbureted air created a bright white light. The U.S. Patent Office's file on this invention is uncommonly thick and includes testimonials from men of science for or against the light's technical feasibility. This record not only furnishes further insights into conflicts that can arise between scientific authorities but also reveals the secret ingredient in Paine's Hydro Electric Light.

Scientific American took notice of the new invention in May 1851, merely observing that "Mr. Paine has been amusing [Worcester newsmen] with some of the volatile hydro-carbons."[46] Clearly there was nothing new here. Paine got the same reaction from the Patent Office when he submitted an application for the benzole light; it was almost immediately rejected for lack of originality. Indeed, the Patent Office had received a letter from Henry Adams of Boston insisting that Paine had purloined the invention from him.[47]

During the following year, Paine submitted revised applications and engaged in lengthy exchanges with the examiner, seeking to convince him that the benzole light was an invention. Paine at last asserted that the invention's novelty resided in the use of *moist* air in addition to benzole and alcohol.[48] In an unusual move, the Patent Office insisted that Paine conduct experiments in Washington to demonstrate the efficacy of moist air because theory suggested that the moisture, upon evaporating, would reduce the temperature of the gas and diminish its ability to burn. But Paine protested for months that he could not afford to comply.

In another odd move, Commissioner Ewbank appealed to Joseph Henry to experiment with Paine's benzole light so that the Patent Office could "arrive at a correct decision."[49] Henry was reluctant to waste his time on a project whose results he believed were certain to be negative. Nonetheless, he acceded to Ewbank's request and enlisted J. Bryant Smith, an assistant to Benjamin Silliman Jr., to perform the experiments in New York. In May 1852, after learning the results, Henry reported to Ewbank that "the flame from the mixture was less brilliant, that it sooner declined, and that the mixture itself cooled more rapidly, with the same amount of air passed through it, than the alcoholic solution."[50]

Ewbank, apparently dissatisfied with the report, asked Henry if he had been present during the trials. He also requested that the experiment be repeated in the presence of the patent examiner Leonard Gale or himself.[51] In a brief and evasive reply, Henry—taking umbrage at having his word questioned—reasserted, on the basis of "theory and direct experiment," that the addition of water was ineffective. Henry (not a chemist) perceived a need to ratchet up the argument from authority and enlisted Robert Hare. The latter placed at the end of Henry's letter a signed paragraph stating that the addition of water would adversely affect the production of heat or light.[52] Two days later, Henry wrote a more expansive letter, admitting that he had not been present during the entire two days of the trials, but insisting that "the results are in accordance with what I anticipated, and have no doubt as to their correctness."[53] He also supplied Ewbank with a certificate attesting to the experiment's validity. The next day, Gale drafted a letter rejecting Paine's application, but on Ewbank's orders the letter was never sent, perhaps because more experiments were in the offing.[54]

Shortly after threatening Ewbank that he would petition Congress for relief from what he believed was the Patent Office's oppressive, unjust, and illegal treatment, Paine at last agreed to an experiment in Washington.[55] The inventor's partner and agent, a Mr. Pedrick, performed the demonstration at the Patent Office in the presence of witnesses, including Leonard Gale and Jonathan Lane, and a report was prepared. The conclusions were favorable to Paine's claims and in accord with sworn statements furnished to the Patent Office by Paine's supporters: 'The light having burned full 7 hours . . . was regarded as sufficient to show the advantage of the water in the alcohol & benzole compound."[56] Gale submitted the report to Ewbank, and on July 13, 1852, Paine received U.S. Patent 9,119 for an "Improvement in Benzole-Lights."

No doubt Henry fumed over the stunning rejection of his authority on matters scientific. This defeat underscored an irksome reality: a scientific authority had no statutory power to affect an invention's fate. In the United States, only the Commissioner of Patents could grant patents. In view of the conflicting opinions and evidence on the benzole light, as well as political pressure on the Patent Office to be more liberal in its evaluations, Ewbank gave Paine the benefit of the doubt.[57]

And doubts did remain. After all, in the trials of Paine's benzole light, a number of factors had not been held constant and might account, singly or in combination, for the divergent results on the effects of moist air. Among these were differences in the purity and relative quantities of ingredients, in the particular apparatus employed, and in environmental conditions. And we cannot dismiss the possibility that Henry, having previously labeled the Hydro Electric Light a fraud and its maker a humbug, was unwilling to give a thorough and sympathetic trial to another Paine invention.

In any event, the benzole mixture was almost certainly the secret ingredient in Paine's Hydro Electric Light.[58] After all, instead of bubbling air or hydrogen through turpentine to carburet it, Paine could have passed the gases through the benzole-alcohol-water liquid to produce the white light—or even benzole alone.[59] No one, after all, had chemically analyzed the purported turpentine employed in his public demonstrations. It could just as well have been the ingredients of the benzole light, concealed by the complexity of the apparatus' plumbing. The timing of Paine's patent application suggests that the benzole light was a spinoff from the Hydro Electric Light—the only part that might have worked.

‿⌒⌒⌒‿

Whether or not Paine gained riches from his Hydro Electric and benzole lights, he was soon pursuing different inventions, including improved carriage wheels and ventilators for railroad cars. He eventually accrued more than two dozen American patents, and in some instances sold the rights to companies. He had, it would appear, rehabilitated his reputation as an honest inventor. And from time to time Paine delivered at the Worcester city hall lectures on the "Vicissitudes of an Inventor's Life, as Illustrated by my own Experience."[60]

Paine was no doubt a hero to many mechanics and inventors. Regardless of public exposés and refutations of the Hydro Electric Light that were based on established principles, some people would maintain that Paine had accomplished what the experts said was impossible, but that his ambitions to furnish a source of cheap light had been thwarted by the gas companies and their scientific apologists. However, even true believers might have entertained doubts after Paine's later escapades with an electromagnetic motor.

Although Paine was granted five patents for improvements in electromagnets and electromagnetic engines in the year 1870 alone, the specifications were unfathomable in relation to accepted principles, and none was consequential.[61] Nonetheless, this outpouring of inventiveness propped up Paine's assertions that he had discovered how to get around the old constraints on using galvanism to operate machinery. From a tiny battery of four ordinary Bunsen cells his motors could, he claimed, generate great power for only pennies a day—comparable to the price of steam.[62] Paine's secrets—not all of which he had disclosed in the patents—would liberate magnetic energy on an astonishing scale.

Paine did gain some early endorsements, a fact that *Scientific American* pondered with exasperation. The editor concluded that Paine's "absurd proposition has been received, yea, swallowed whole, by persons who have heretofore enjoyed reputation for common sense, if not sagacity, in things scientific. But this easy credulity in the present case, shows that they have been over-rated. They belong to that large class of individuals, intelligent and sound in ordinary matters, but in whose minds there runs a vein of lunacy upon the perpetual-motion question; the result of careless and deficient training in scientific principles. From this class, Mr. Paine will draw followers and money."[63] And he did. The indefatigable inventor pitched his motor to potential investors in New Jersey, and they succumbed, buying shares in the Paine Electromagnetic Engine Company. With the company's alleged capital of $3 million, Paine began work on an even larger motor. This was completed in mid 1871 and put on exhibit in New York. Just as Paine promised, with a diminutive battery the motor drove a circular saw almost all day long—or so it appeared.

Paine's motors, like his Hydro-Electric Light, aroused controversies in the popular and technical press, including *Scientific American*, with no resolution in sight. Suspicions were aroused by Paine's claim that 3 grains of zinc consumed in a battery would yield the equivalent of 67 million foot-pounds of force.[54] After all, experiments by Joule, Faraday, and others had established that the actual conversion factor was less than Paine's figure by many orders of magnitude. P. H. Vander Weyde, medical doctor, polymath, and lecturer, discounted Paine's claim, insisting that it was tantamount to having "found perpetual motion." Vander Weyde also went step by step through conventional calculations to expose Paine's conceptual and computational errors.[65] Paine, in his reply, gave no ground, insisting that "the forces developed by the action of a single Bunsen cell, *if* utilized and converted into power, *would* drive the largest ship afloat."[66]

The first person to overtly raise the specter of fraud was the young civil engineer Henry Rowland, who in later decades would become one of America's most distinguished physicists, occupying the chair of physics at Johns Hopkins University. In a letter to *Scientific American*, Rowland recounted a visit to Paine's office. Before the motor was turned on, Rowland and Paine briefly debated fundamentals of physics. The inventor, who apparently had little regard for Joule or Faraday, asserted that his experiments had disproved the conservation-of-energy law. Rowland, no doubt reeling in disbelief, then turned his attention to the motor, which was mostly enclosed in a large iron shroud resting on the floor. When Paine engaged the battery, the saw, connected to the motor by belts and shafting, sprang to life, shaking the building and cutting boards with ease. After Rowland peppered Paine with penetrating questions, Paine threatened to toss him out the window.

Upon leaving Paine's office, Rowland discovered that the room directly below, which Paine had also rented, contained a steam engine. Barred from entering the locked room, Rowland was unable to prove exactly how Paine had pulled off the deception. But Rowland did say that "unless the greatest development of modern science [the conservation law] be overthrown, this machine cannot but derive its power from an extraneous source."[67] An indignant supporter of Paine wrote that the extraneous source might be "the newly discovered force of psychic power."[68]

Apparently annoyed by the seemingly interminable controversy, J. E. Smith, who had had his own suspicions after viewing Paine's motor, sought to end the matter by offering Paine $500 to demonstrate his motor to Smith, several other witnesses, and the editor of *Scientific American*. Paine would have to meet certain conditions, including running the motor while it was detached from its base plate and allowing inspection of all circuitry. "Now, Mr. Paine," taunted Smith, "you cannot avoid placing either me or yourself in a truly ridiculous position. I pays my money, and you takes your choice."[69]

Vander Weyde soon added another $500 to the pot and invited those interested "in the advancement of theoretical science, or in the progress of industry, or in the preservation of human life secured by a safe substitute for the steam boiler, to come forward and follow our example."[70] In a brief letter that brought an end to coverage of the motor in *Scientific American*—"Exit Mr. Paine"—the besieged inventor declined to perform for the $1,000 prize, calling it a "sportsman's bluff."[71] In Henry's characterization of humbugging two decades earlier, the humbug's "deception frequently begins in folly and ends in fraud." But sometimes it goes farther and ends in farce.

_⁓⁓⁓_

Henry Paine, clearly a skilled mechanic, a prolific inventor, and a shrewd manipulator, apparently did not distinguish between artifact and artifice. As *Scientific American* had observed, humbugs could tap into a reservoir of credulity for inventions that challenged authorities. Moreover, history had shown that men of science were fallible;

perhaps the latest inventor's claim would also survive their naysaying. But, one wonders, how could people familiar with the principles of electricity, like Dr Channing and Professor Doremus, lend their names to fraudulent schemes? This strange phenomenon yielded years later to a social explanation that I cannot better. According to Henry Morton, a man of impeccable scientific credentials, "an ordinary sense of politeness hardly allows one, when visiting a man's workshop on his own invitation, to treat him as if he were a cheat and his statements as falsehoods, even if we so believe them to be." Instead, the observer is likely to utter something that could be represented, outside that social context, as an endorsement. Once these words are publicized and taken up by the press, Morton added, it is difficult to issue clarifications. Having learned this lesson the hard way, Morton offered an effective remedy: refuse the invitation unless the inventor is willing to reveal every "interior detail of the machine."[72]

Paine's electrical inventions remind us that the power of scientific authority has limits. Thus, although the Hydro Electric Light and the miraculous motor were debunked early and often, Paine was able to raise money, attract tenacious supporters, and obtain patents. Apparently, the ideal of a scientific authority passing unequivocal, consequential judgments on inventions—an ideal advocated by *Scientific American*, Henry, and others—was sometimes difficult to realize in practice owing to a variety of social factors.

Despite the earnest efforts of Henry and others, the boundaries of institutional, authoritative science remained fuzzy and fluid, and no hard-and-fast educational or occupational criteria determined who was or was not an authority on a specific matter. As a result, the public and the press were often unable to give appropriate weight to statements from varied sources claiming to represent scientific viewpoints. And, of course, heated controversies with colorful characters sell newspapers.

In addition, scientific men held tenaciously to principles that later research might overturn. When brought forcefully into the open, this vulnerability could weaken any scientist's claim that he spoke transcendent truths. To some, it might have appeared that invoking certain principles to undermine an inventor's claim was no more than an arrogant profession of faith that those principles would hold for all time. Paine also pointed out a related logical flaw in arguments from scientific authority. In his view, men of science were in effect saying that unless they could attribute a given technological effect to a known cause then it must be humbug. In the midst of the Hydro light controversy, Paine countered with the argument that seemingly inexplicable effects could also result from legitimate causes not yet known to science.[73] By emphasizing the vulnerabilities of scientific authority, Paine had an apparently reasonable rationale for continuing to insist that he had discovered new science that could explain how his inventions worked.

Moreover, the eagerness of the press to publicize the achievements of local inventors, which catered to parochial pride and the public's insatiable thirst for novelty, could prolong the life of a fraudulent invention. Also, such controversies can be framed as the lone inventor battling the scientific establishment in a David-against-Goliath morality play. Such dramas, often laced with a leitmotif of struggle between social classes, can provoke sympathy for the underdog and occasionally translate into political support or even state funding. (Recall, for example, the cold fusion controversy of recent times.)

In some cases, especially when many players are involved, scientific authorities lack the power to derail inventions they deem impossible. That difficulty was aggravated in the case of Paine, who, Henry admitted, was the most plausible humbug he had encountered.[74] Although inventions that flouted core scientific principles might enjoy some time in the sun, in the end they could only fail as technologies. Paine's electrical inventions have long been forgotten. Nonetheless, between an invention's birth announcement and its death notice—a period that may last many years—fascinating power struggles sometimes take place.

As we have seen, the electromagnet that Joseph Henry devised around 1830 led almost immediately to the invention of motors, magnetos, and telegraphs. It also spawned hundreds of lesser-known inventions drawing current from batteries. Some of these were brought to market, and a few—including clocks, "fire alarm telegraphs" and hotel annunciators—enjoyed appreciable adoptions. However, many of the novelties languished until, decades later, they were commercialized, sometimes after reinvention, in altered societal and technological contexts.

The relatively obscure electromagnetic inventions of 1840–1880 represent a remarkable burst of creativity, for the electromagnet was imagined as an essential component in countless devices and technological systems. A wave of inventions following in the wake of a new component's appearance is a common pattern, one also seen in more recent times after the commercialization of the transistor, the laser, and the computer microchip. This general process is sometimes called "component-stimulated invention."[1]

The engine of creativity lies in inventors' envisioning connections between a component's performance characteristics and potential devices that might exploit these for carrying out functions in ongoing or anticipated activities. Creating this kind of linkage is, according to Arthur Koestler, a bisociative act that often occurs unconsciously.[2]

The rapid adoption of the telegraph, the Morse-Vail register in particular, was the major vehicle for disseminating information about the electromagnet's performance characteristics. In any telegraph office, onlookers could watch as the register's small electromagnet—whose motions were controlled by someone at a distant place—impressed dots and dashes in a moving paper tape. Moreover, telegraph operators, who also had the job of maintaining a station's equipment, became knowledgeable in the basics of electrical technology. Not surprisingly, telegraphers sometimes conjured new applications for the electromagnet, such as Edison's justly famous stock ticker.

But a person did not have to be a telegraph customer or a telegrapher to learn about the electromagnet and its capabilities. Beginning in the 1840s, descriptions of telegraph systems were widely reported and illustrated in newspapers, magazines, and books. And scientific instrument makers, including Daniel Davis in Boston, had commercialized a variety of components (such as Morse-Vail registers) and display devices (such as electric bells).[3] Commercial telegraph apparatus was also conspicuous at international expositions, beginning with the 1851 Crystal Palace Exhibition in London. Thus, by the early 1850s, numerous people in Western countries were aware of the telegraph, its basic mode of operation, and the pivotal role played by the electromagnet.

Smitten with the possibilities of using electromagnets to build new devices, many people became inventors. Their inventions were usually aimed at augmenting or replacing existing technologies, including those actuated by human hands, but many were far-reaching novelties inspired by the electromagnet's unique combination of performance characteristics. To wit, an electromagnet can create magnetism of potentially great strength, can be turned on and off rapidly, can be actuated at distances, long or short, and can produce reliably and repeatedly precise motions of small amplitude. I submit that people familiar with the electromagnet could not help but internalize these performance characteristics in some form. Matching them to the requirements of a device that could solve a perceived problem in some realm of activity, inventors envisioned new electromagnetic devices and technological systems. Clearly, opportunities abounded in industrializing countries for people in many occupations to indulge in these bisociative acts. And during this heady time for independent inventors, patent-management systems supplied financial incentives for apparent technical success in problem solving.[4]

The electromagnet fomented a conceptual reorientation by vastly increasing the ways in which inventors could potentially extend humanity's powers over the material world. With no other technology was it possible for people to produce, instantaneously, action at a distance. The sampling of inventions in the next section, touching on activities as varied as textile manufacturing and church services, testifies to the electromagnet's power to move minds as well as machines.[5]

⁓

Nautical technologies presented many opportunities for electromagnetic invention. Thomas Taylor's "electric annunciator," for example, would enable a pilot to convey orders to the engineer or helmsman by means of a five-position (stop, ahead easy, ahead full speed, back easy, back full speed) electromagnetic indicator.[6]

J. P. Joule, of energy-conservation fame, proposed a regulator that automatically steered a ship on a steady course, employing a compass that operated as a switch. If the ship deviated from the desired course in one direction, it closed a circuit. An electromagnet in turn actuated a steam valve that caused the rudder to correct the course. A second electromagnet came into play when the ship deviated in the other direction.[7]

A more sinister seagoing invention was J. A. Ballard's "electric torpedo boat," which could be steered toward a target ship. From the shore, the "pilot" directed the movement of this explosive-laden steam-powered vessel by means of an electrical cable. Electromagnets opened and closed valves that controlled the rudder. A separate wire set off the explosion when the target was reached.[8]

Alexander Bain, Scotland's most prolific inventor of electromagnetic devices, conceived a depth sounder that slowly submerged a switch at the end of an insulated cable. Upon touching bottom, the switch caused an electromagnet on deck to release a hammer, which struck the bell.[9]

During the Atlantic telegraph project, Charles Bright deployed an "electrical log" for measuring a ship's speed, information that was needed for safely paying out the cable. The device consisted of a waterwheel at the end of an insulated cable that was lowered into the water. At each revolution of the wheel, a circuit was broken, which caused an electromagnet to advance a register on deck that indicated the ship's speed.[10]

Boilers of steam engines in ships and factories could explode when water levels got too low, and they worked poorly when overfilled. These problems occasioned electrical solutions for regulating the water level. A. Achard's sophisticated version monitored the water level with a float that operated switches connected to electromagnets controlling the flow of water. It also rang a bell when the level was too high or low or when the battery failed.[11]

As the example of the steam boiler hints, industries—especially those seeking to mechanize—supplied many problems that electromagnetic devices might help to solve. French inventors, sometimes working in concert with factory owners, took the lead. By the early 1850s, for example, a Jacquard loom had been outfitted with electromagnets that moved particular threads into place as dictated by a textile's design.[12]

The expense and labor of engraving a printing plate brought forth W. Hansen's electrical remedy, which employed two electromagnets, one to raise and the other to lower the stylus. The stylus was connected to a mechanical linkage whose lateral movements were controlled by an operator who moved a "feeler" over the surface of the drawing to be copied, which had been rendered in nonconductive ink on a conductive surface. When conductive areas of the drawing were touched, one electromagnet suspended the stylus above the plate; when the feeler touched the lines of the drawing, the other electromagnet pressed the stylus downward, engraving the plate.[13]

Electromagnetic devices were also expected to enhance safety and security in buildings. Mr. T. Allan of London, who had been suffering thefts from his factory, outfitted the window with a switch and ran a wire to an electromagnetic bell in his home nearby. When the window was next opened, "the signal bell set up a violent ringing," allowing Allan to hasten to his factory and catch the burglar in the act.[14] Across the Atlantic, where burglary was an equal vexation, the U.S. Patent Office issued 18 patents for electromagnetic burglar alarms before 1873. John Rutter came up with a fire alarm by putting a mercury thermometer in a circuit with an alarm bell. When the room temperature became dangerously high, the rising mercury in the tube made contact with two platinum wires, which closed the circuit and thus sounded the electromagnetic alarm.[15]

Devices were also proposed for enhancing comfort at work and at home. Samuel and Cromwell Varley devised a "thermo-regulator" that could, among other feats, close a damper when a room's temperature rose beyond a preset value. A mercury thermometer was fitted with a float that, upon rising, closed the circuit to the electromagnet-controlled damper. Théodose Du Moncel's regulator maintained a comfortable room temperature by using electromagnets to manage air flowing from hot and cold vents. George Sternberg's system (figure 15.1) operated a damper in his parlor and could also regulate the temperate of a liquid by adjusting the gas feeding a Bunsen burner.[16]

Chronometry was another fertile field, and many designs were fashioned for electric clocks (figure 15.2). In one design, an electromagnet periodically advanced a ratchet

Figure 15.1
Sternberg's "electro-magnetic regulator." Source: *Scientific American* 23 (1870): 126.

Figure 15.2
An electric clock. Adapted from Guillemin and Thompson 1891: 754.

wheel, which drove a gear train and also wound a spring that, in turn, activated the electromagnet and caused the hands to move.[17] In addition to purchases by wealthy individuals, a few electric clocks—usually large ones—were installed in public places, including the Philadelphia Merchants' Exchange and the gallery of the General Post Office in London.[18]

Perceiving the need to communicate accurate time to many cities, especially for regulating the movement of trains, Alexander Bain came up with a master clock that could periodically advance any number of slave clocks in other places. The working of this system was made possible by a switch on the master clock that periodically sent an electrical impulse, via parallel circuits, to electromagnets that advanced the slave clocks' hands. Bain developed, patented, and sold such systems beginning in the 1840s.[19] They were widely adopted by schools, railroads, factories, and government offices; moreover, they enabled standard time to be reproduced over telegraph lines—eventually throughout the world.[20] In New York City, to help people set their watches and clocks, a "time ball" was dropped periodically by telegraphic signal from Western Union's main building.[21]

The French clock maker and magician Jean-Eugène Robert-Houdin (from whom the magician known as Harry Houdini took his stage name), contrived electric clocks and also invented electromagnetic devices for his shows. In one illusion, Robert-Houdin claimed that he could instantly turn a strong man into a weakling. Resting on the stage floor was an ordinary-looking wooden box, which Robert-Houdin lifted effortlessly. Another man, dared to repeat this unremarkable feat, was embarrassed by his failure to budge the box. The illusion's secret was simple: behind the scenes, Robert-Houdin's confederate had turned on a strong electromagnet, implanted in the stage floor, which gripped the bottom of the box, onto which a sheet of iron had been fastened.[22]

Robert-Houdin was not alone in using electromagnets to create illusions. Enterprising spiritualists used them to produce otherworldly clicks and pounding sounds. A medium named Colchester had even deceived First Lady Mary Todd Lincoln. Joseph Henry, an arch-enemy of spiritualists, related Colchester's secret to President Lincoln: the clicking noises were caused by an electromagnet controlled by a switch on the medium's arm that responded to a flexed biceps. Henry had learned the secret from a maker of telegraph instruments—the very same man who had outfitted Colchester.[23]

Electromagnets were also coupled to musical instruments. Building on his system for synchronizing clocks, Alexander Bain proposed that a pianist or an organist could simultaneously play several instruments in several cities: the keys could be depressed by electromagnets connected electrically to a master instrument.[24] In pipe organs of churches and theaters, the keyboard had to be installed very close to the pipes because of the need for a responsive mechanical linkage. However, the electric organ freed the keyboard to be placed elsewhere. The keys were switches that controlled the flow of air by means of electromagnetic valves on each pipe. The "electric organs" in New York's Chickering Hall and Grace Church were built by Hilborne Roosevelt, a cousin of Theodore.[25]

⌐⌐⌐

Local telegraphs of many kinds also arose during this period, most of which had special-purpose accessories in place of Morse-Vail keys and registers. Hotels' annunciator systems enabled guests to summon a valet or to signal another need to a central operator. The system made use of special sending units for guest rooms and had a display panel for the operator that indicated the room number and nature of the request; some systems even incorporated fire alarms. Electrical annunciators were commercialized and enjoyed some adoptions in competition with mechanical annunciators.

The idea arose early on that individuals and firms might construct private telegraphs for their exclusive use in intra-city communication. R. Hoe & Co., makers of printing presses in New York, installed a 2-mile telegraph line to connect their office and their factory. This telegraph made it possible to monitor the progress of orders and respond to questions requiring timely answers.[26] Private telegraphs were popular in Great Britain too; one merchant kept track of doings at the docks from his country home, and Lord Kinnaird laid down a line from his castle to a town 8 miles away so that he could send orders to tradesmen.[27]

One company in New York solicited business for a "domestic telegraph." A small instrument was placed in a house and connected to a district telegraph. By moving the appropriate lever or button, one could summon a policeman, a fireman, or a messenger. The editor of *Scientific American* subscribed to the service, at $2.50 per month plus 15 cents an hour for the messenger's time. When he tested the system by calling for a messenger, less than 3 minutes elapsed before the man rang his doorbell. There followed a hearty endorsement in *Scientific American*: "... this district or domestic telegraph is a most useful and valuable institution, promotive of comfort, convenience, and safety of families. That it will soon come into general use, cannot be doubted."[28]

Private telegraphs crisscrossing Manhattan eventually created urban clutter, with poles hosting dozens of wires sprouting on nearly every street corner. The exposed wires were an irresistible attraction for mischievous boys, who playfully tossed kite tails or strips of cloth weighted with rocks over the wires, creating a scene that at times resembled "the limbs of an African prayer tree, with its burden of rags, tags, and strings hung on by pious wayfarers."[29]

In 1845, the Boston physician William Channing (1820–1901), who had written a book on electromedicine, publicized his vision of a remarkable electrical technology that could make cities safer by speeding up the response to fires.[30] Comparing a city to a living organism, he argued that a "Central Station" could collect information from peripheral sensors located throughout the city. Like the brain, the central station could process the incoming information and send signals to peripheral locations, producing mechanical effects. This system would allow the location of a fire to be swiftly relayed, through the central station, to nearby fire brigades. Presumably, more rapid responses would spare property and lives.

Channing's inspiring vision, known as the "fire alarm telegraph," languished until 1847, when its potential value was dramatized by a fire that destroyed more than 100 buildings in Boston's North End.[31] Not surprisingly, Boston's mayor called for the installation of a fire-alarm telegraph. Wasting no time, Channing proposed a detailed plan to the City Council, which accepted the proposal and furnished $10,000 for its implementation. To help bring this project to fruition, Channing teamed up with

Moses Farmer (1820–1893), a local telegraph engineer.[32] Farmer had already embarked on what would become a long and lucrative career in inventing. Eventually he acquired about 100 patents, which included telegraph accessories, electrochemical processes, electric lights, motors, and telephone apparatus. Farmer and Channing developed and installed the hardware for Boston's fire alarm telegraph.[33] In 1855, Channing reported in a lecture at the Smithsonian Institution that it had been working well for 3 years.[34]

The original installation of the Channing-Farmer system required several entirely new components. First there were the "Signal Boxes" through which someone could inform the central station of a fire. Firmly attached to a building, each consisted of a sturdy cast-iron box with a hinged door. Inside was a crank that, when turned, sent a coded electrical signal that identified the signal box's location (e.g., District 1, Box 5). To ensure that the operator at the central station could read the location correctly, the signal was sent redundantly by turning the crank six times. By 1855, Boston had 46 signal boxes. To promote rapid responses, they were placed no more than about 800 feet apart, usually opposite a street lamp. The signal box was also equipped with an electromagnet, which could tap out return messages from the central station, and a key for sending more detailed information. To prevent false alarms, the signal box remained locked until needed; it could be opened by a nearby resident whose address was posted on the box.

When a signal was received at the central station, located next to City Hall, a bell rang—perhaps to awaken the operator—and a Morse-Vail register recorded the signal box's location. To sound the alarm and summon fire brigades, the operator used a special keyboard. It incorporated a clockwork mechanism that rotated conductive metal strips on a cylinder below each key. By depressing the proper keys, the operator could ring alarm bells, announcing the district number at regular intervals and sending coded messages such as "the fire is out." Here in one man's hands was the power to simultaneously ring bells all around Boston.

The "striking machine" (figure 15.3) weighed 800 pounds and cost $165. As of 1855, the system encompassed 24 bells in churches, firehouses, and schools. When the signal arrived, an electromagnet released a detent that set in motion a weight-driven gear train, which then swung the heavy hammer against the bell. By querying the central station from a signal box, firemen learned the location that had sounded the original alarm and then hastened to the fire.

To guard against sabotage, snow and ice, and other mishaps, Boston's system used heavy iron wire that was held aloft by sturdy brackets on tall buildings and was well insulated and protected from lightning. The circuits had redundant connections between the central station, the signal boxes, and the alarm bells so that the system still worked in the rare event of a broken wire. At the central station, another new apparatus periodically tested the continuity of all circuits.

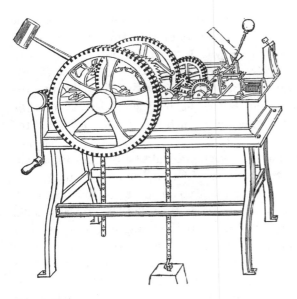

Figure 15.3
A bell ringer for a municipal fire-alarm telegraph. Adapted from *Scientific American* 7 (1852, n.s.): 227.

Powered by the central station's Grove battery and by local batteries in each belfry, the fire alarm telegraph came in around the budgeted figure but was doubtless subsidized by the inventors' labor. Farmer estimated that future systems would cost about $150 per mile. (Boston's system had 49 miles of wire.)

Enthusiasts maintained that the new system made it impossible for Boston to suffer another devastating fire. But this was wishful thinking. The "Great Boston Fire" of 1872 ravaged 65 acres, destroyed around $80 million in property, and took at least 20 lives.[35] The problem—in this disaster and others—was not a failure of the fire alarm telegraph but insufficient men and machines to fight huge conflagrations. Although fire alarm telegraphs reduced false alarms, whether the *early* ones had much effect on the fate of fires is a matter requiring more research. However, officials in many large cities were convinced of the technology's effectiveness, and so similar systems proliferated in Western countries.[36]

The electromagnet made possible other technologies whose beneficial effects were indisputable, including devices to control the movement of trains. Early railroads often had but one track for trains traveling in both directions, along with several "turnouts" for passing. Careful scheduling of departures for fast and slow trains worked only up to a point, because on both one-track and two-track railroads trains made many unplanned stops. Indeed, a catastrophic crash could be imminent if a train suffered a breakdown, ran into a rockslide or a snowbank, or had a close encounter with a big beast. Newspapers and magazines often ran sensational stories of horrific wrecks. For example, in 1854 a passenger train and a train loaded with gravel collided on the Great Western Railroad in Canada. About 60 people perished. *Scientific American* lamented that the line lacked a railroad telegraph which might have prevented the collision.[37]

Many people rapidly figured out that the telegraph could become an important adjunct to a railroad, and, conveniently, many telegraph lines ran along tracks.[38] In principle, telegraph operators at stations on a rail line could do a better job of scheduling trains and also send alerts if one did not arrive on time. But these measures did not eliminate the danger of two trains colliding when one train's engineer, unaware that another train sat motionless ahead, proceeded at full speed. (Trains do not stop on a dime.) An obvious solution was to furnish trains with portable sending units that could be tapped into the telegraph line in an emergency. Another was to equip trains with special telegraph systems (perhaps using the rails or a separate metal ribbon between them as conductors) that would allow continuous communication with stations. Neither invention was entirely satisfactory, and I doubt that many were installed. In the meantime, concern for safety was growing with the proliferation of rail lines. (By 1870 the United States had nearly 50,000 miles of intercity track—more than all the European countries combined.[39])

The telegraph system that best reduced the risk of wrecks, and in modified form remained in use throughout much of the twentieth century, was "block signaling." It was first proposed in 1842 by Charles Wheatstone's telegraph collaborator, William Cooke. The basic principle is that "trains are kept apart by a *certain* and *invariable* interval of *space*, instead of by an *uncertain* and *variable* interval of *time*."[40] A track was divided into blocks whose boundaries corresponded to station locations, and no train was permitted to enter a block occupied by another train. Mechanical indicators, operated by electromagnets and placed close to the tracks so engineers could see them, displayed either a "danger" signal or an "all clear." Because operators knew when a train had passed their station, and were conversing with each other by telegraph, they could send the appropriate signal to the indicators and thus to the engineers. The state of every signal was also displayed electrically at all stations on the line.

Innumerable railroad telegraphs, with specially designed components, were commercialized for operating the block system. In England, six different systems had been installed on various lines by the late 1870s. Yet older timetable-based systems survived on many railroads because block signaling required new equipment and well-trained operators. The old system was cheaper, but its toll of human lives was incalculably greater.

By 1880, automatic railroad telegraphs were being installed on some lines. In this system, trains' movements caused the signals to change through electromagnetic devices. Besides lessening the chores of station operators, the automatic system permitted blocks to have arbitrary lengths unrelated to the locations of stations.

As I noted earlier, another function of railroad telegraphs was to send a time signal for synchronizing all clocks on a line. But at any given place, train time usually differed from local time because of the crazy quilt of time zones. A step in the direction of standard time was taken by several railroads in New England, beginning in 1851. Station clocks were coordinated with Boston time through signals received from Harvard University's observatory in Cambridge.[41] Eventually, the need for standard time on railroads was the impetus for establishing worldwide time zones. A young German working in the Swiss patent office, a man familiar with telegraphs, railroads, and time zones, pondered how to determine when two clocks widely separated in space displayed the same time, and so arose Albert Einstein's special theory of relativity.[42]

Another railroad problem that electrical technology helped solve was the lassitude or carelessness of switchmen. With only a hand-operated lever, a switchman could divert a train from the main line to another track. Occasionally, after a train passed, a switchman forgot to return the lever to its original position. An American, Thomas Hall, came up with a technology consisting of three major electrical components (in addition to the insulated cables that connected them, and a battery): a switch on the junction of the tracks that opened and closed an electrical circuit in response to the lever's position, a bell, and an electromagnetic indicator of the lever's position that could be seen by the arriving train's engineer (figure 15.4). To alert the switchman that the lever was not in the main-line position, the electric bell sounded a continuous alarm. Hall's "improved switch signal and alarm" enjoyed some adoptions, beginning in 1867.[43]

Grand Central Depot in New York, built in 1871 by the railroad tycoon Cornelius Vanderbilt, was the predecessor of the current Grand Central Terminal. More than a showpiece of Victorian architecture, it had the daunting function of coordinating the movement of thousands of passengers and about 130 trains coming and going every day. Imposing order on so many people and machines was the responsibility of the "dispatcher," who, from his perch high above the tracks, received reports on the whereabouts of approaching trains, sent departing trains on their way, and signaled baggage handlers when to stop loading. The dispatcher could accomplish these tasks because the depot was wired with telegraph-inspired apparatus for receiving and sending information. Pressure-sensitive switches under rails indicated the passage of a train from one block to another, bells allowed simple communications within the depot, and an array of buttons enabled the dispatcher to issue instructions to engineers. Signals on the tracks had a disk, painted red on one side, that could be turned,

Figure 15.4
Hall's railroad "switch and alarm." Source: *Scientific American* 16 (1867): 277.

by means of an electromagnet, to face toward or away from an oncoming train, indicating "stop" or "go."[44] Even with its sophisticated electrical control system, however, Grand Central Depot was pronounced obsolete almost immediately after opening, so rapidly did numbers of trains and passengers increase.

Despite their limitations, railroad telegraphs, in all their varied incarnations, made possible the operation of busy stations such as Grand Central Depot and appreciably improved railroad safety on local and long-distance lines.

The preceding examples underscore a theme mentioned in earlier chapters: spanning a developmental distance, long or short, almost always required the knowledge and skills of an instrument maker, a mechanic, or a mechanical engineer. Indeed, virtually all electromagnet-containing technologies can properly be called *electro-mechanical* devices. Thus, beyond the creative act of envisioning that the electromagnet could perform an essential function in a new technology, the inventor had to possess a mechanical turn of mind and metalworking skills, or had to call upon the services of others who did. Often the result was a complex mechanical system embodying only a trivial amount of electrical expertise—the sort that could be easily gleaned from a textbook. Making a similar point, W. Bernard Carlson observed that "the electrical industry in the late nineteenth century was based as much on craft knowledge as on scientific theory."[45]

No device better illustrates the dependence of electrical technology on mechanical expertise than the "copying telegraph." The idea of such a device, which occurred to many people, exploited Alexander Bain's earlier invention of an electrochemical telegraph printer. In Bain's printer, which was commercialized in the 1840s, a paper tape was treated with dye. When current was applied by a metal stylus to the moving paper, the dye changed color.

Bain himself was aware that his printing telegraph might be developed into a machine for making facsimiles of letters and drawings.[46] However, an Italian abbot, Giovanni Caselli, proposed the most intricate and technically successful system; he called it the pantelegraph.[47] The basic idea was beguilingly simple: the sender writes a message on a piece of metal foil with nonconductive ink. A stylus, in a circuit with the foil, is then gently drawn over the foil in a series of parallel, closely spaced scanning lines. During each pass, electrical continuity is made when the stylus touches the foil and broken when it contacts the ink. Thus, the spatial pattern on the foil is translated into electrical current that goes on or off for varying durations. This signal is then sent along a telegraph line. At the receiving end, an electrical stylus scans the dyed paper at the same time and in the same manner, and thus appears—line by line—a "fac simile" of the original letter or drawing.

Caselli made a prototype pantelegraph in 1856, but was unable to work out the details. In France, with a subsidy from Napoléon III, Caselli hired the instrument maker Paul-Gustav Froment, who helped to turn the preliminary design into functioning hardware. Besides devising a reliable scanning device, the men had to perfect a mechanism for precisely synchronizing the scanning between sending and receiving units. They solved this problem by using a large pendulum at each end, both of which oscillated in tandem and regulated the scanning motions of the sending and receiving styli (figure 15.5). The pendulums were controlled by electromagnets receiving electrical signals from a single clock. Scanning at a resolution superior to that of most twentieth-century televisions, the pantelegraph's transmission of handwriting—at 15 words per minute—was, however, much slower than an ordinary telegraph.

To fully describe the pantelegraph's mechanical parts and the intricacy of its operation is far beyond my abilities. Suffice it to say that a complete description required 12 pages, including seven drawings, in an 1888 book, and that the American patent specifications were unusually lengthy, with five pages of text plus two pages of drawings.[48]

In February 1865, the pantelegraph was put into operation between Paris and Lyon on France's government-owned telegraph system. This application was underutilized, lasting only until the end of the decade, and the company Caselli founded to

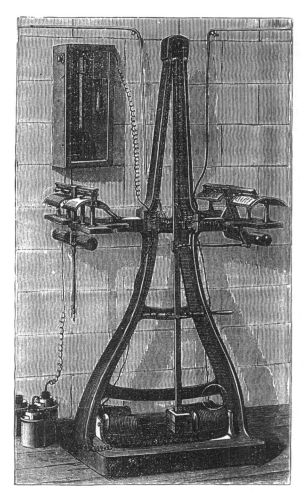

Figure 15.5
Caselli's pantelegraph. Source: Prescott 1877: 755.

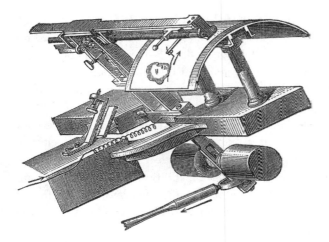

Figure 15.6
Detail of Caselli's printer. Adapted from Prescott 1877: 755.

Figure 15.7
A message printed by Caselli's printer. Adapted from Prescott 1877: 752.

manufacture and market the invention eventually folded. In 1961 a pantelegraph system from a French museum was restored to working condition and exhibited for several months. It operated "flawlessly, six hours a day."[49] The pantelegraph is an extreme example of the mechanical virtuosity required to realize many nineteenth-century electrical technologies; even today it stands as a noteworthy achievement of the mechanical arts.

The electromagnet stimulated people in various occupations to invent devices for exercising power over mechanical systems of human design. Although many electro-magnetic inventions would eventually be called "labor-saving," inventors tended to regard these technologies as means to *extend* human capabilities, sensory and mechanical, so as to engage the world—including other people—in new ways. Indeed, inventors underwent a dramatic change in how they viewed the possibilities for using electricity in everyday life.

Before radio beacons, radar, and the global positioning system, traveling by sea was treacherous. Although the mariner had onboard technologies for determining locations, such as sextants, compasses, chronometers, sounders, and charts, their use was far from foolproof. During the nineteenth century, mariners became increasingly reliant on coastal aids such as lighthouses, buoys, and lightships. Although a coastline densely dotted with lights did not eliminate shipwrecks, it helped mariners establish their locations, avoid obstacles, and make it safely to port.[1]

As maritime commerce and naval activities accelerated, given added impetus by steamships and international telegraphs, groups concerned with safety at sea, including merchants, ship owners, captains, marine insurance companies, and naval officers, became strong advocates of improved coastal lighting. In response to these pressures, national lighthouse organizations, if not already in existence, were established to expand the number and effectiveness of navigation aids.

In the United States, the Light-House Board was given jurisdiction over aids to navigation in 1852. The Light-House Board was appointed by the president, reported to the Secretary of the Treasury, and secured appropriations from a miserly Congress one project at a time. At its central depot on Staten Island, it manufactured some lighting equipment, conducted experiments, and tried out apparatus offered by vendors.[2] Among the board's founding members was Joseph Henry, who chaired the committee on experiments until his death in 1878.

Britain had a private guild, called Trinity House, that had, over several centuries, acquired increasing power over navigation aids. In 1836, Parliament granted it complete control over English and Welsh lighthouses. Trinity House was under the jurisdiction of the Board of Trade, from which it had to request funds for new projects. Although Trinity House's board lacked permanent members with technical or scientific expertise, it appointed scientists (including Michael Faraday and John Tyndall) as advisers.[3]

France had the Commission des Phares, a part of the civil administration that was staffed mainly by engineers and at least one scientist. By buttressing proposals with scientific and technical arguments, it was able to tap state coffers with relative ease for the funds needed to try out, and sometimes to adopt, new technologies. Because the commission's role included the creation of new science and new technology, it could take advantage of the government's perennial interest in fostering these activities. Moreover, it was a matter of national pride for France to maintain international leadership in areas of traditional excellence, such as applications of electricity and lighthouse technology.

France's ideal for a coastal lighting system, closely met in practice, was that no mariner plying French coastal waters would ever be beyond sight of a lighthouse in fair weather. Following a scheme of lighthouse grading devised early in the century by Augustin Fresnel, principal lighthouses with very bright (first-order) lights were established on prominent headlands; lesser lights (second to sixth order) were placed between them to mark rivers, changes in coastal contours, and hazards. Early lighthouses projected their lights with reflectors, which resulted in much wasted light that vanished skyward; in 1822, Fresnel invented a system of complex and costly lenses that directed more of an oil lamp's light seaward. France's coastal lighting system set the standards for lighting technology and for coverage toward which other countries strove.[4]

As countries dotted their coastlines with navigation aids, the number of lighthouses—many having Fresnel lenses—rose dramatically. From 1860 to 1885, their numbers increased by 68 percent in England, by 84 percent in the United States, and by 35 percent in France. Despite the proliferation of lighthouses, increases in shipping almost ensured that ships would continue to run aground, collide, and sink, especially when visibility was poor. From about 1860 to 1880, there were about 1,200 wrecks at sea per year.[5] With so many new lighthouses being built, there were ample opportunities to try out, even adopt, new technologies that might be more effective in foggy and hazy conditions.

Lighthouse organizations experimented continuously with illuminants, reflectors, lenses, buoys, foghorns, whistles, and construction techniques. Lighthouse scientists also intensified their research on illuminants and optical systems, and added new topics such as the effects of atmospheric conditions on the transmission of light and sound. As a result, lighthouse science—empirical and theoretical—expanded dramatically in France, England, Scotland, and the United States. Among the important contributors was Joseph Henry, whose work led to improved acoustic technologies.[6]

Aids to navigation, lighthouses in particular, were an important arena of international competition—especially between the major colonial powers, Britain and France. Not surprisingly, lighthouse organizations kept abreast of developments elsewhere, gathering information and sponsoring official visits, which sometimes led to trials of new technologies already adopted in other countries. Not surprisingly, lighthouse organizations were constantly receiving proposals for new kinds of lights. And so it

was that the lighthouse became a prominent venue for displaying and evaluating virtually every new bright light. However, few of the new lights were adopted, because
they failed to meet the demanding performance requirements of lighthouse illumination. Oil lamps remained the dominant lighthouse illuminant.

_⌒⌒⌒_

The lighthouse was not the only anticipated market for new lighting systems. Other
potential locations for safer and brighter lights included train stations, government
buildings, theaters, concert halls, and factories. An especially important market was
city streets. In 1858, *Scientific American* expressed a growing consensus among urban-
ites: ". . . the best lighted cities are . . . the most free from crime, because robbers can
skulk in dark streets, and pursue their nefarious practices with comparative impu-
nity."[7] By the beginning of the Civil War, gas street lamps had been installed in parts
of most American cities, but in brightness and coverage they were far from ideal.

Announcements of new lighting technologies appeared often in the technical and
popular presses. Many an inventor claimed that his system was a cheaper illuminant
than gas. However, such comparisons were problematic because the cost of gas varied
so much from city to city and country to country—usually correlated with the price
of coal, which was burned in large retorts to make "coal gas." Yet for a while these
claims would draw attention to an inventor and his lighting system, which might
help to attract investors. A few of the new lights came to market and found specialized
applications, such as limelight in theaters. But in the middle of the nineteenth century
none posed a serious challenge to the burning of gas for illuminating rooms, streets,
and other spaces. The first electrical technology to compete for lighthouse and other
markets was the battery-powered carbon arc.

Before the late 1830s, when constant batteries appeared, carbon arcs remained
mainly in the hands of experimenters and science lecturers. In public demonstrations,
the carbon arc's brilliance left onlookers dazzled and amazed, but the heavy drain on
the battery limited the displays to a few minutes at most. This was not the sort of
performance that invited people to think up new applications. Although "constant
batteries" could sustain a heavy drain for longer periods, they slowly lost power, and
the light faded. Nonetheless, the new batteries did stimulate inventors.

Beyond selecting a long-lived battery, inventors faced two immediate technical
challenges. The first was the challenge of forming durable carbon rods. Soft charcoal,
such as Humphry Davy had used, was not suitable for more than a brief exhibition.
A common solution was to use carbon-rich residues that had accumulated in coal-gas
retorts, for they could be easily shaped into rods. Soon experimenters were confecting
rods by combining materials such as sugar, powdered coke, pitch, and coal dust.[8]
Many rods gave reasonably effective service, but a lack of homogeneity could cause

flickering. The second challenge was finding a way to automatically maintain a constant distance between the rods as their tips slowly wore down (the positive one twice as quickly) and changed shape. To electrical experimenters, the solution (evident by the 1840s) was to use electromagnets. Wired in the light circuit, an electromagnet could move one or both rods, usually through a clockwork mechanism, in response to changes in the amount of current drawn by the arc. Invented and patented in profusion, these devices were known as "regulators," and, as with batteries, each variety carried its inventor's name (figure 16.1).[9] With the availability of arc light regulators (the first electrical feedback mechanism), some inventors believed, the electric light would be judged practical by consumers.[10] (In later years, some regulators, including the Duboscq regulator shown in figure 16.1, moved both rods.)

Despite the ongoing expense of refurbishing the battery, entrepreneurs set up demonstration systems of arc lights in several countries.[11] Arc lights were decisively superior to candles, gas, and oil lamps in one performance characteristic: they were brighter by orders of magnitude. In many applications, however, the light's full brightness greatly exceeded the illumination requirements. Thus, advocates identified locations where lights were used intermittently or where one or just a few blinding lights might be appropriate, such as city squares, theaters, and lighthouses.[12] William Grove, inventor of a constant battery as well as the fuel cell, suggested in 1849 that the electric light could be "well and easily applied" in lighthouses.[13] And he wasn't the only Englishman thinking along those lines.

One of William Grove's countrymen, W. Edwards Staite, became an early promoter of the "electric light" (a term that meant arc lights until the advent of Edison's incandescent lamp). Staite designed several lamps and regulators (figure 16.1, left), and concocted recipes for making the rods. His first patented recipe used compressed coke powder treated by a saturated sugar solution and baked in an airtight vessel.[14] With occasional collaborators, he received five patents for arc lighting inventions during the period 1846–1849.[15]

Another unique feature of Staite's technology was the use of a sealed glass lamp to enclose the carbon rods. According to Staite, his lamp had decisive performance advantages over gas lights: "The absence of all smoke and flame, and noxious gases—the non-consumption of oxygen—the impossibility of its igniting surrounding substances—and the simplicity of the apparatus."[16]

Staite founded a company and took his light show on the road. Held before large crowds and gatherings of the elite, his public exhibitions never failed to captivate observers, for this brilliant and penetrating light was white like sunlight, in contrast with the yellowish light afforded by candle, oil, and gas flames. Objects could be seen

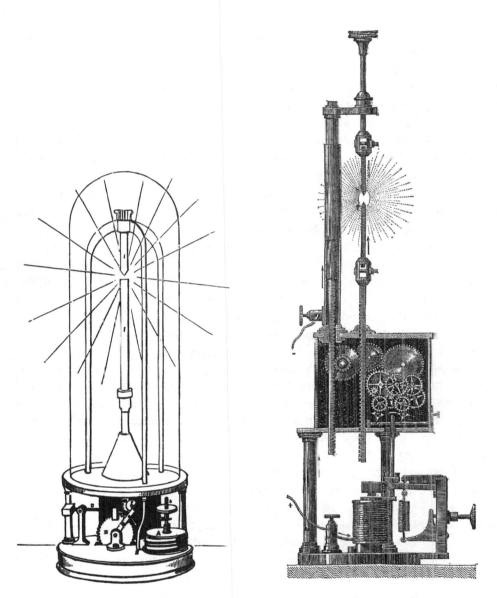

Figure 16.1
Left: Staite's arc light, 1848. Source: King 1962b: 337; courtesy of Smithsonian Institution. Right (not to same scale): Duboscq's arc light, 1870s. Source: *Scientific American* 33 (1875): 35.

in their natural colors. After a demonstration at Worsley Hall, Queen Victoria and Prince Albert were said to be "much gratified and interested."[17]

Beyond impressing onlookers and receiving positive newspaper coverage, Staite's light appeared to be a technical success. According to a note in *Chemical Times*, it met the two technical desiderata of the electric light: the rods were of sufficient purity and the flow of current was regulated to maintain a steady light. Thus, Staite was credited with having "applied a well-known phenomenon of electricity to a practical purpose."[18] The editors of *Mechanics' Magazine*, given a private showing, were surprised that a small battery was able to sustain a steady light for so long. They concluded: "Lighting by electricity has been a favourite dream of many; but Mr. Staite is unquestionably the first scientific experimenter who has reduced it to (what we may almost venture to call) a practical certainty."[19] *Scientific American* reported "that this one light completely eclipses ten gas lights," adding: "The gas companies had better look out. The dissatisfaction of the public with their mismanagement may have begotten a rival destined to eclipse many more than merely ten of their gas lights."[20] This was the first of many occasions when advocates of electric lighting sought to take advantage of widespread hostility toward the purveyors of gas.

Potential real-world applications for his light came easily to Staite. He suggested, for example, that it could be used for a kind of telegraphy, perhaps placed in glasses of different colors and actuated by color-coded telegraph keys. Thus, for signaling trains at night, "Mr. S. proposes to have fixed, at required distances from the stations, a signal post, on which two or more lamps may be fixed—say, one enclosed in a red glass, one in green, and one in white . . . The red light may indicate 'danger,' the green light 'caution,' and so on." Such a system would "be of great service in preventing accidents at night."[21] Staite puffed that his invention was "suitable for all purposes of lighting—private dwellings, shops, churches, or streets; and which shall not only be the best and purest light ever seen, but *the cheapest*."[22]

First-order lighthouses were an obvious application. After one of Staite's exhibitions in London, this use was "discussed and its advantages set forth."[23] The first trial by Trinity House took place on the South Pier at Sunderland in June 1850. Erected on a platform above the existing lighthouse, Staite's light was turned on at 10 p.m. It was so bright that onlookers had to avert their gaze. A woman at Ryhope, 3 miles away, was able to read a previously unopened letter, and lighthouse officials 7 miles out to sea noted that the electric light put the existing light on the North Pier to shame.[24] W. C. Wilkins, a Trinity House engineer, offered Staite "any assistance that may be in my power to further its introduction for lighthouse purposes."[25]

Michael Faraday, a consultant to Trinity House, arranged for trials of new lights.[26] To put it mildly, he was a tough sell, in one case issuing this warning to a would-be demonstrator: "Above all I most earnestly beg of you to produce no imperfectly prepared apparatus or any thing that requires excuse."[27] Trinity House and other prospective adopters of the electric light were especially concerned about the expense of keeping up the battery. Staite answered that "there is no light so cheap as that evolved by voltaic currents of electricity."[28] He supported this astonishing claim by calculating that the cost of materials consumed per hour, including one-seventh pound of zinc, would be far less than the cost of candles (7s 6d) or gas (6s 8d). (Staite also noted that zinc sulfate, a by-product of the battery's operation, itself had some value.) It is doubtful that any prospective purchaser took these figures at face value; Faraday certainly didn't. In any event, Trinity House declined to buy Staite's light, and so did everyone else.

Scientific American penned an epitaph for Staite's enterprise in 1851: "... a few weeks ago this Electric Light became insolvent, it was executed by a number of indignant creditors, and its body consigned to that place where it had threatened to send all its old but sturdy opponents."[29] Staite died in 1854, apparently without having profited from his inventions.

The French lighthouse organization experimented with battery-powered arc lights in its central workshops between 1848 and 1857. However, the electric illuminant was not adopted, owing to inconsistent results and the excessive costs of battery maintenance.[30] In the United States, Joseph Henry was categorically opposed to carbon arcs in lighthouses, and none were installed during his lifetime. Despite entreaties from inventors, in no country did a lighthouse ever boast a permanent battery-powered arc light.

The expense and the difficulties of battery maintenance were surely important factors in the electric light's rejection. Moreover, lighthouse organizations still entertained doubts about the carbon arc's ability to satisfy other performance requirements, such as operating continuously—sometimes for 16 hours at a time in winter—without flicker or failure. Arc lights were eventually emplaced in some lighthouses, but not until the batteries were replaced by magnetos.

_⌒⌒⌒_

In the mid 1850s, the French scientist Antoine Becquerel examined the expense of the battery-powered electric light and determined that 60 Bunsen cells consumed a kilogram of zinc in the first hour, after which the power and the light dropped off drastically. Also taking into account the need to refresh the acid, Becquerel calculated that for an equal amount of light (350 candles/hour) the electric light in Paris would be four times more costly than gas but about as costly as oil and candles (ignoring labor for refurbishing the battery).[31] William Grove's similar studies yielded an estimate of $.75 per hour for a 30-cell battery equivalent to 1,444 candles.[32] These relative costs, of course, varied over time and place, and also depended on the kind of cell in use. But no matter who computed them or how, the result was always the same:

". . . the cost of light . . . from electricity, with our present means of producing it, must be greatly in excess [of ordinary illuminants]."[33]

But cost is only one performance characteristic that potential consumers consider. Many expensive technologies, even in pre- or non-industrial societies, had their beginnings as prestige goods for elite consumers, often performing important symbolic functions in social and political contexts.[34] Only later did these technologies find utilitarian applications.

The electric light's technical competence invited entrepreneurs and potential consumers to weigh this technology's unique performance characteristics (the brightest and whitest artificial light) against the performance characteristics, utilitarian and symbolic, of illuminants already in use for specific applications. The components needed to build arc lighting systems were readily available. Batteries had been commercialized in large numbers and in many varieties for electrometallurgy and telegraphy. In the 1850s, instrument makers, especially in Germany and in France, began to offer regulators and carbons. Thus, someone could assemble a system without enormous expense and exhibit the light to prospective customers, such as a business or government; at the very least, a demonstration would receive newspaper and magazine publicity underscoring the arc light's unique performance characteristics. These diverse trials did establish niche markets for battery-powered arc lights, some of which endured long after the advent of steam-powered generators. Battery power was pronounced practical for particular applications because the consumer might place a high value, not only on brightness and whiteness, but also on compactness, portability, simplicity or safety of operation, and the ability to awe observers.

_⁓⁓⁓_

One of the earliest applications was in photography. In the era of daguerreotypes, exposures were painfully long, which could try the patience of a person sitting for a portrait. In addition, any movement could result in a blurred image. To reduce exposure time, photographers needed bright illumination, such as sunshine supplied by skylights and directed through mirrors. However, cloudy days shortened the workweek, and in northern climes the workday in winter was brief. In view of these difficulties, photographers tried out every technology of artificial illumination and adopted some of them—at least for brief periods—even though all had performance deficiencies in the studio setting. Trials with arc lights were under way by 1850.[35]

Photographers learned that the electric light shortened exposure times for portraiture, yet it also produced ghastly images with washed-out faces and sunken eyes. To ameliorate this, they employed reflectors and blue-glass diffusers, which yielded a more flattering image. Because photography did not require a steady light for long periods, the arc could be managed without a regulator; one needed only an adjustable, insulated mount for the carbons. Also, the infrequent episodes of battery maintenance were hardly onerous for people already skilled in manipulating chemicals.

One of the better-known photographers who used the electric light was Mathew Brady. Trained by Samuel Morse in the daguerreotype process, Brady is best known for his Civil War photographs. However, neither on battlefields nor in his New York and Washington studios did Brady take many images himself; he was more of an entrepreneur and organizer of other photographers. In 1857, to advertise his studio's skill in handling the new illuminant, Brady held an exhibition, in his gallery on Broadway, of "photographs by electric light."[36] I wonder if Brady used this technology to make any of his famous portraits of Abraham Lincoln.

In 1857, during a 3½-hour lecture on light by R. Ogden Doremus at the Academy of Music in New York, Brady used a large camera and electric lighting to produce photographs before a "spellbound" audience of 3,000.[37]

By 1850, it had been learned that the electric light could stop motion. An 1849 American textbook on photography mentioned its use for "producing pictures instantly" by (e.g.) capturing the image of a person in the act of placing a foot on a doorstep.[38] William H. Fox Talbot showed that a brilliant flash of the electric light could stop the motion of a rotating disk.[39] The electric light was also used to illuminate scenes that may have been impossible to photograph otherwise. For example, Felix Nadar took a Bunsen battery, a carbon arc, and a camera into Paris's catacombs and sewers. Overcoming stench and steam for 3 months, he secured 100 eerie images.[40]

───⌇⌇⌇───

Lecturers had long used the "magic lantern" to surprise and delight audiences, sometimes exhibiting the microscopic world. Glass plates, gelatin, and mica that held images of plant cells, ant eyes, oral flora, and other exotica were projected through a system of lenses onto a wall or screen. By 1850, the magic lantern, commercialized in many varieties, was a staple of science lectures and a necessity in many classrooms. Famed lecturers, including Michael Faraday in England and R. Ogden Doremus and Henry Morton in America, used it to great effect.[41]

As a teaching technology, the magic lantern's weakness was the light source. The limelight, although very bright, involved heating a piece of lime to incandescence in an oxygen-hydrogen flame. Among other shortcomings, this technology required great care in preparing, storing, and using dangerous gases.[42] The new magnesium light, also bright, was very expensive, because it burned a rare metal. An ordinary oil lamp's light—safe, simple, and cheap enough—was very dim.

Beginning in the 1850s, makers and users of magic lanterns began experimenting with carbon arcs. Many lecturers were already exhibiting the electric light. Most of them also used batteries for other demonstrations, so the upkeep of batteries required no new skills. Appreciating that the arc light was a suitable illuminant for their products, some makers of magic lanterns offered the electric light as an accessory. In 1870, *Scientific American* could write: "The electric light affords, probably, the strongest and best illumination for the magic lantern."[43] It was also used in projection microscopes.[44]

‿⌒⌒⌒‿

A long-standing cultural imperative in the world of warfare was the need to direct a beam of light toward a ship, a fort, or advancing troops. By bringing potential targets out of darkness, the electric light could help soldiers aim their cannons and artillery. It might also deter the advance of armies into well-lit areas. Not surprisingly, experiments with the electric light were undertaken in many countries. The French were perhaps the first to use it in battle: during the Crimean War, French sailors used a parabolic mirror to project a beam of light from the deck of a ship to the point of attack.[45]

In England, in 1863, the electrician J. H. Hearder assembled an electric light apparatus with an 80-cell zinc-iron battery and a parabolic reflector 20 inches across. From the Citadel at Plymouth he was able to direct a fairly tight beam toward structures and other features of the landscape; the light's efficacy was observed at these places and by third parties nearby. The keeper of the Eddystone lighthouse, perched on a rock 16 miles at sea, reported that the light coming from the shore was as brilliant as the brightest star. As Hearder aimed his light at a ship, then at a nearby cliff, and then at various buildings, witnesses furnished uniformly favorably reports that, Hearder concluded, "went to prove the intense illuminating power of the concentrated ray, its perfect manageability, and the extraordinary distance at which its effects were appreciable."[46] Hearder also mentioned that the battery retained only one-sixth its original power at the end of the tests.

The first large-scale use of the electric light in warfare took place during the Franco-Prussian War of 1870–71. Threatened by German troops advancing on Paris, the French command cobbled together electric lights for the redoubts and forts outside the city wall. Employing Serrin and Foucault regulators, Bunsen batteries, and telegraph wire, these lights enabled the defenders to scan for German troop movements. Both the French and the Germans also used a few magneto-powered lights in this conflict.[47] Although the Germans took Paris, the military importance of the electric light—however powered—had become indisputable.

‿⌒⌒‿

Lighting large spaces was the arc light's forte. And, conveniently enough, large spaces were owned by people and institutions that could afford to use expensive lights in spectacular displays. Thus were lit, for example, the Chamber of Deputies in Brussels, Westminster Bridge in London, and the Rhein Bridge in Germany.[48] The electric light

was not just an illumination technology, for its unique visual performance also signi-
fied and advertised the power of those who employed it. They could command the
electric genie to emerge from acid-filled jars and turn night into day. Although some
of these installations were ephemeral, their effects on social consciousness were not.

An 1865 drawing of a town square illuminated by electric light (figure 16.2) invites
us to infer that this novel technology, while furnishing light for mundane activities,
was itself an irresistible attraction. An awesome display of temporal power, the electric
light left in darkness the church spires barely visible in the distance. Perhaps God was
giving way to forces of nature that man might control. For a few people marveling at
the artificial sun, the electric light might have raised troubling questions about who—
or what—would affect their future.

Electric lights were exhibited in many European towns, often by entrepreneurs
seeking to sell their system to a local government for outdoor lighting. In 1855, an
electric light tried in Deal, England, was said to have "a most transcendent and vivid
appearance, and is a vast improvement upon the gas lights."[49] The Arc de Triomphe
in Paris was the site of a demonstration (by Lacassagne and Thiers, inventors of a new
regulator) in which four arc lights shone down upon the Champs Elysées in several
directions.[50]

The electric light also found its way into theaters. As early as 1846, a parabolic mirror
projecting the light onto a silk screen had been used in a Parisian play to mimic the
rising sun. A decade later, Jules Duboscq, an instrument maker who with Léon Foucault had devised a sturdy regulator, was hired by the Paris Opera to create special
lighting effects. For nearly 20 years Duboscq used battery-powered carbon arcs with
stunning and memorable results. To simulate lightning, he attached a small, unregu-
lated arc to a concave "magic" mirror that, when shaken by hand, imitated the zigzag
trace of a bolt from the sky. Actors used the same mirror, with wires emerging from
a sleeve and a finger-operated switch, to create phantasmagorical effects. In a production of *Moses*, Duboscq's light caused a rainbow to appear after the Red Sea closed
behind the fleeing Hebrews.[51] These kinds of short-duration applications could be
quite economical because wear-and-tear on the battery was minimal.

Napoléon III, a patron of science and technology, was fond of sponsoring extravagant festivities, some of which served to exhibit, for the admiration of domestic and
foreign observers, new French technologies, including the electric light. In June 1867,
the emperor held a grand ball at the Tuileries in honor of the tsar of Russia. The *New
York Times* reported that the ball was attended by "the most distinguished set of guests
that has ever been brought together in one assembly," including France's emperor and
empress, the tsar, the king and the prince of Prussia, and Count Bismarck. From a
ballroom balcony guests could gaze at the "fairy spectacle" below: "The grand avenue,
the gardens, the fountains, even the trees themselves, were one mass of brilliant illumination." In addition to the 70,000 or so gas burners at work, "the fountains were

Figure 16.2
A battery-powered arc light illuminates a town square. Source: Zöllner 1865. Courtesy of Dibner
Library, Smithsonian Institution.

tinted with the intense rays of the electric light—varying every moment and producing an inconceivably beautiful illusion."[52]

In 1868, in Paris, Napoléon held his national *fête*, which celebrated the greatness of France and that of the emperor. The events of that day included military processions, a regatta on the Seine, a high mass at Notre Dame, free performances at theaters and the Opera, and pantomimes. In the evening there was a fireworks spectacular above Paris, a city aglow with artificial lights on grand avenues and fountains illuminated by the electric arc.[53] These were just two of many occasions in which the electric light was an adjunct to French statecraft.

In America the electric light played scarcely any role in entertainment public celebrations, or political life, but there were some exceptions. In Boston, an electric light accompanied the inauguration of an immense new organ boasting 5,474 pipes. One evening in the autumn of 1863, Boston's elite gathered in the Music Hall to hear renowned organists take turns playing the new instrument. Announcing the beginning of the program, gaslights were turned down. Suddenly the electric light, streaming from a gallery, fell upon the great organ, causing its burnished pipes to glisten and sparkle. A few minutes later, the light was turned toward eight flags while the organist played "The Star-Spangled Banner."[54] Another conspicuous exception occurred in February 1871, when Washingtonians held their own Mardi Gras. Attended by at least 50,000 people, the activities included a procession of horse-drawn vehicles, a mile-long foot race, and another race in which goats pulled miniature trotting gigs driven by young boys. When darkness fell, the broad avenue from the Treasury Building to the Capitol was packed with celebrants who basked in limelights and electric arcs. The Washington carnival did not take hold as an annual holiday, but it did make the front page of the *New York Times*, which described the event as an "interesting exhibition of unsurpassed and unobjectionable frivolity."[55]

In France the electric light was rapidly judged practical—indeed, economical—for several commercial and industrial activities. In 1854, the Napoléon Docks were lit up so that stevedores could work after sunset. After 4 months it was calculated that the nightly cost of the light was no more than 38 francs—less than 5 centimes per man. The paltry expenditure permitted 800 men to handle cargo at night. The electric light also allowed construction to continue day and night on a Paris mansion.[56] Those in charge were apparently unconcerned that this work schedule was disrupting the lives of laborers.

Battery lights were used extensively during the construction of the Spanish Northern Railway. In more than 9,400 hours of use, ten lamps furnished illumination for a dock and various excavations. In tunnels, they did not foul the air as much as torchlight did. Most surprising, the electric light was said to be "60 per cent. cheaper than torchlight."[57]

In Russia, an electric light powered by a 48-cell battery was installed on a locomotive on the Moscow and Kursk Railway. The tsar had expressed his opinion that such a light, projecting 500 yards ahead, would reduce collisions at night. However, trials of electric lights on American trains had shown that the batteries suffered from the constant shaking, which increased the costs of maintenance. Beyond experiments such as the tsar's, I have found no evidence that any railroad used battery-powered electric headlights on its locomotives.[58]

Perhaps the most novel and specialized demonstration of the electric light took place under water. Jules Duboscq came up with a small, lightweight version of his regulator that would fit into a tightly sealed glass tube of great strength. The light had a lens, which allowed the diver to direct the beam wherever he desired. Current was supplied through insulated copper wires connected to a battery on ship or shore. Lit by the turn of a screw, the light would operate for 2 hours. Duboscq's portable submarine light was tested in 15 feet of water and reportedly worked well.[59]

The commercial viability of technically competent arc light systems, in stark contrast to the lack of robust battery-powered motors on the market, allowed experimentation with the new illuminant. The large number of trials that took place in the middle of the nineteenth century showed that battery-powered electric lights could perform adequately in various activities. In Napoléon III's national festival and in similar events having mainly social and political functions, the extravagance of the electric light was more than tolerable: it was highly appropriate because the activities had symbolic performance requirements. In utilitarian applications where a very bright light was desirable, such as special effects in theaters, photography, and the magic lantern, the electric light enjoyed performance advantages over its competitors, for the latter were also expensive, sometimes dangerous, and perhaps difficult to employ. And in the lighting of construction sites and docks, the electric light was regarded by some as a practical illuminant—even in economic terms.

Although the battery-powered electric light was not widely adopted, the trials and experiments had far-reaching effects. The lighting of a town square, a bridge, a tunnel, a factory floor, a fort under siege, a legislature, or a lighthouse—however short-lived—exhibited much more than the carbon arc itself: it conspicuously advertised, and perhaps fomented, widely felt needs and cultural imperatives for brighter light in particular kinds of public and private spaces. It markedly recalibrated the performance requirements of artificial illuminants. Inventors, I suggest, inferred from these exhibitions that a large latent market—including governments—was waiting to be tapped if only they could create new and less expensive electric lights. Given these incentives, inventors responded with new technologies, including huge, steam-powered magnetos.

17 If at First You Don't Succeed . . .

As his short-lived cable in New York's East River showed, Samuel Morse had already realized that large telegraph networks would have to include underwater segments. Even before the Baltimore-Washington line was done, Morse was contemplating intercontinental communication through deep-sea cables. Writing to Secretary of the Treasury John Spencer in 1843, Morse drew "the practical inference . . that a telegraphic communication on my plan may with certainty be established across the Atlantic! Startling as this may seem now, the time will come when this project is realized."[1]

Morse was just one of many people who prophesied that the Old World and the New World would be linked by wire, perhaps soon. In 1852, giving lively voice to this vision, *Scientific American* claimed that "the earth will be belted by the electric wire, and New York will yet be able to send the throb of her electric pulse through our whole continent, Asia, Africa, and Europe, in a second of time."[2] This expectation, already a cultural imperative, was founded on Victorian faith in the inevitable, unrelenting progress of science and technology—a faith intensified by telegraphy itself. After all, the rapid adoption of land lines during the 1840s on both sides of the Atlantic had generated immense profits for investors, stimulated commerce, and quickened communication for governments and for citizens. The telegraph even held out the promise of better relations between countries. Many potential investors and consumers believed that oceans were merely temporary obstacles that could, with sufficient resources and ingenuity, be overcome.

However, neither Morse nor other visionaries foresaw the number and size of the obstacles. The Atlantic telegraph entailed a developmental distance unprecedented for an electrical technology. The problems were seemingly endless, their resolution taking a decade. The enterprise also consumed vast sums from investors, required generous subsidies from the American and British governments, and fostered the creation of various electrical and mechanical technologies as well as new electrical-engineering science—even a mega-corporation. Owing to the difficulty of this undertaking and its far-reaching effects on human behavior, the Atlantic telegraph is often hailed—

properly, I believe—as one of Victorian society's most remarkable technological achievements.

The Atlantic telegraph came about in a manner most unusual for an early electrical technology. The project was driven, not by an inventor, a scientist, or a government agency, but by an American businessman: Cyrus Field. Like other entrepreneur-promoters lacking technical expertise, Field turned to men of science for advice. Although virtually all authorities pronounced the project technically feasible, there was a lack of unanimity on the design features of the cable itself and on its likely signaling performance. A man of action committed to pursuing his project expeditiously, Field had to choose among options presented by experts. E. O. Wildman Whitehouse, the telegraph man whose opinions prevailed early on, wielded considerable power over the project's plan and execution.

We know little about how nineteenth-century entrepreneur-promoters made decisions when presented with conflicting advice, but the experts' relative social power certainly had an effect. Decisions made under this degree of uncertainty sometimes have undesirable and expensive outcomes. Cyrus Field's project is a case in point. The first "functional" cable failed in a few weeks, with a loss of about $1 million. Whitehouse was discredited, and power shifted to other scientific authorities, including Professor William Thomson of Glasgow. Despite the debacle, Field mustered investor and government support to try again (and again), but the cable itself—the priciest part of the enterprise—had to be redesigned. The new cable embodied a different weighting of mechanical, electrical, and financial performance characteristics, and at last allowed effective and profitable telegraphy across the Atlantic.[3]

⁓⌒⌒⌒⁓

In 1845, Ezra Cornell had laid a 12-mile cable under the Hudson River that connected Fort Lee, New Jersey, with New York City. This cable appears to have incorporated some design features of Morse's original—and faulty—Baltimore-Washington underground line: a lead sheath and two copper wires covered with cotton. But Cornell's cable had inside it a layer of India rubber for additional insulation. In contrast to Morse's cable, Cornell's was manufactured without defects and sent signals for several months until an ice floe severed its lead shield.[4] This was perhaps the first working underwater cable of nontrivial length.

Soon several cables were laid in Europe, including an 1851 line across the English Channel connecting Dover and Calais.[5] Supported by demand from expanding telegraph networks, British companies took a leadership role in creating and commercializing cable technology, which they parlayed into a monopoly. One important innovation was the use of a newly discovered material: gutta percha, a rubber-like substance obtained from trees in the British East Indies and sold almost exclusively by the Gutta-Percha Company of London. Gutta percha is a superb insulator and its mechanical properties—waterproof, easily molded when warm, firm but not brittle when cold—also commended it for cables.[6] A common design had a core of several strands of copper wire covered by one or more layers of gutta percha. This might then be coated with jute or tarred hemp, and spirally wound strands of braided iron wire might be added for protection.

Because the Atlantic cable would be long and laid in deep water, its design involved a tricky balancing act among varied electrical, mechanical, and financial performance characteristics. For example, iron armor augmented strength but also increased cost, volume, and weight (the Dover-Calais cable weighed 12,000 pounds per mile). A thicker copper conductor reduced electrical resistance while adding weight and cost. An effective design for the Atlantic cable—one that achieved acceptable tradeoffs between performance characteristics—eventually grew out of power struggles among scientific authorities and bitter experience gained from failed cables.[7]

The prime mover of the Atlantic telegraph was Cyrus Field (1819–1892) a New York merchant who made his first fortune before the age of 30 selling, among other products, pricey paper.[8] This ambitious and energetic man soon tired of the paper business and sought new challenges. His brother Matthew, a civil engineer, approached Cyrus on behalf of Frederick Gisborne, an engineer and electrical inventor who had become mired in a difficult telegraph project across Newfoundland. Gisborne's plan included an eventual link by ship from Newfoundland to Ireland, which would cut days from transatlantic communication. One evening in early 1854, Gisborne discussed his troubled project at Field's home, hoping to entice him into investing. After Gisborne left, Cyrus, alone in his library, traced on a globe the path of the projected Newfoundland line. It appeared to Field, as it had to many before him, that Newfoundland—the easternmost land mass in North America—was indeed the gateway to Europe. Why not, he thought, continue the Newfoundland line beneath the sea all the way to Ireland, enabling telegraphy to England, the Continent, and beyond?

Excited by the possibility of an Atlantic telegraph, Cyrus Field knew that he dared not commit to such a fantastic project without first securing the approbation of scientific men.[9] The day after meeting with Gisborne, Field sent off two letters. Of Samuel Morse, generally believed to be *the* American authority on telegraphic practice, he asked whether it was feasible to send a signal through 2,000 miles of submarine cable. Initiating a long-term relationship, Morse met with Field and answered his question affirmatively, certain that the laws of electricity allowed such transmissions. However, both Morse and Joseph Henry (who was not consulted) believed that propelling a current through a lengthy cable would require a high-tension battery.[10]

Field's second letter went to Matthew Maury, Superintendent of the Naval Observatory and one of the world's leading oceanographers.[11] As it happened, during the previous year an American, Lieutenant Commander O. H. Berryman, had made provisional soundings and sampled the sea floor from Newfoundland to Ireland; along the way he also acquired data on wind and currents. Maury related to Field Berryman's encouraging findings for a submarine cable: the sea floor was a plateau deep enough to escape anchors and icebergs but sufficiently shallow to allow laying of a cable. Moreover, except near the shores the bottom was free of rocks. (A few years later, prodded by Field, Maury sent Berryman on a more thorough survey, the results of which were equivocal; however, a British government survey in 1857 furnished firm evidence of the "telegraphic plateau."[12])

Armed with the endorsements of two authorities whose assessments carried weight, Field approached other moneyed men in New York with a proposal to form a company for completing the Newfoundland line. His first investor was Peter Cooper, who was persuaded that this project would shower humanity with benefits galore. With Cooper's aid, Field recruited additional investors. These men organized the New York, Newfoundland, and London Telegraph Company, which in March 1854 secured subsidies from the Newfoundland government in the form of "a guarantee of the interest on £50,000 in bonds as well as grants of land and a right-of-way across the island for the overland telegraph line."[13]

The Newfoundland line included a submarine link to Nova Scotia, about 75 miles away. In the fall of 1854, Field sailed to London to order the cable from the Submarine Cable Company, owned by the brothers Jacob and John Brett. This was the company that had laid the Dover-Calais cable. Field broached the possibility of an Atlantic cable to John Brett, who was sufficiently interested to purchase shares in the new company.

With Field, Cooper, and Morse aboard, the bark *Sarah L. Bryant*, towed by a steamship, began paying out the cable during the summer of 1855. Although the voyage began in calm seas, partway across the strait the vessels encountered a severe gale that threatened to capsize the bark. To save it, the captain ordered that the cable be cut. Field learned from this experience that a sailing ship was not maneuverable enough to lay a cable.

Cyrus Field, who personified the entrepreneurial spirit, was not discouraged by failure. He went to England again and ordered a new cable. A steamship successfully laid it the following summer; and in 1856 Matthew Field's crew, which sometimes reached 600 men, completed the 400-mile land line across Newfoundland. Costing more than $1 million, the project had completed the crucial first link.[14] Now the time had come to cross the Atlantic, but first there were a few matters of electrical engineering to consider.

Latimer Clark, an English telegraph engineer, had encountered a curious phenomenon while signaling through underground lines. The transmissions, he found, were attended by an appreciable "retardation" and stretching of the signal.[15] Clark was painfully aware that this effect might threaten the economic viability of long submarine cables by limiting the transmission rate: signals sent too rapidly would be impossible to decode. Not surprisingly, Clark turned to Michael Faraday, hoping that he could diagnose the problem and suggest a solution. After witnessing Clark's impressive demonstrations employing long cables both above ground and immersed in water, Faraday identified the problem. A buried or submarine cable, with its two conductors (copper wire and metal sheath) separated by gutta percha, formed a condenser (capacitor). Thus, retardation came about because it took time for the signal—an electrical pulse for a dot or a dash—to charge the cable. The extent of retardation depended on the cable's electrostatic capacity, which in turn was determined by its length and mode of construction.[16]

About this time, William Thomson (1824–1907), a brilliant young professor of mathematics at Glasgow University who had broad interests in natural philosophy and in applications of science, learned about the retardation problem.[17] Faraday's analysis had been qualitative, but Thomson built a mathematical model based on heat flowing through a metal rod that took into account both electrostatic capacity and electrical resistance. From the model he deduced several laws governing signal transmission.[18] Of greatest concern was the "law of squares," which dictated that retardation would increase in direct proportion to the square of the cable's length. Thus, retardation in a 200-mile cable would be four times that in a 100-mile cable.

Thomson's analyses led him to propose that a large-diameter conductor of the purest copper would reduce electrical resistance while a proportionally thicker layer of gutta percha would decrease the cable's electrostatic capacity. To lessen the stretching of the signal, he recommended the use of a mechanism to send current only in brief, symmetrical pulses (perhaps lasting a twentieth of a second); he also recommended immediately clearing the line of residual electricity by reversing the current's polarity. Thomson also claimed that the Atlantic line could probably be worked with a low-tension battery, provided that the receiver was a galvanometer of utmost sensitivity—an instrument he would eventually design. Thomson's conclusions, his law of squares in particular, were challenged by E. O. Wildman Whitehouse, a British surgeon and electrical experimenter who was collaborating with the Bretts' firm and its young telegraph engineer, Charles Tilson Bright.[19] On the basis of his own experiments, Whitehouse insisted that the degree of retardation was a function of distance alone, not its square.[20] Thus, a slender and lightweight cable, not the thicker, heavier one Thomson advocated, would serve for the Atlantic telegraph. After several exchanges between these authorities in the *Athenaeum*, Thomson reworked Whitehouse's data and argued that they were actually in conformity with his law of squares.[21]

On the one hand, Thomson's design sought to enhance a cable's most important use-related *electrical* performance characteristic: transmission rate, in words per minute. A thicker cable, so long as it had a correspondingly heavier iron wrapping, would also be stronger. On the other hand, Whitehouse's design favored lower capital costs as well as ease of handling and paying out—*but not strength*. Clearly, at the beginning of the Atlantic telegraph project, arguments could be marshaled in support of either design.

Significantly, neither Faraday nor Whitehouse nor Thomson claimed that retardation would render an Atlantic cable impracticable. Others were less certain. The British telegraph engineer F. R. Window calculated that an Atlantic cable could transmit only one word per minute.[22] Thomson estimated a transmission rate nearly as dismal: about three to four words per minute.[23] Yet, playing cost accountant, he forecast that, at 30 shillings for a 20-word message, the Atlantic telegraph could yield a respectable return on investment. When he offered this forecast in 1856, Thomson knew neither the size of the ultimate investment—no one did—nor the costs of operating and maintaining the system.

In October 1856, Morse witnessed an experiment on retardation in which Whitehouse and Bright connected in series ten 200-mile segments of underground cable, making a total line of 2,000 miles. Using Whitehouse-designed senders and receivers, they transmitted between 210 and 270 signals per minute—roughly equal to 10 or 12 words.[24] In a letter to Field, Morse claimed that this experiment "most satisfactorily resolved all doubts of the practicability as well as practicality of operating the telegraph from Newfoundland to Ireland."[25] That sanguine claim, however, assumed that submarine and underground cables would perform identically. In any event, it must be emphasized that the electrical authorities whom Field consulted, though they had differing views on the design of the cable itself, all gave the project the go-ahead.

‑‑‑‑‑‑‑‑‑‑

The Newfoundland cable had been financed entirely by American investors, but Field realized that the Atlantic link would require a large infusion of British capital. After several meetings with government officials, Field struck an auspicious deal. The British government would provide ships for taking new soundings across the projected route and for laying the cable, and would then pay £14,000 per year for use of the cable.[26]

Field now organized the Atlantic Cable Company, capitalized at £350,000. With a government-guaranteed return on investment of 4 percent plus dividends from additional profits that the cable might yield, Field was able in just a few weeks to obtain subscribers for the entire stock offering—at £1,000 per share—in London, Liverpool, and Manchester.[27] Field's pitch to potential investors was irresistible. In Liverpool, for example, he promised that the cable would transmit at 10 words per minute; he also brought along John Brett and Charles Bright to endorse the project's technical feasibility.[28] The American entrepreneur assumed a huge risk himself by subscribing 25 percent of the stock.

In appointing scientific authorities to the new company, Field seemingly covered all bases. The company had Charles Bright as chief engineer, E. O. Wildman White-house as electrician, and William Thomson on the board of directors.[29] Morse, who had given Field's companies free use of his patents and had gone to London at his own expense to conduct experiments, was offered an honorary place on the board as well as stock at par value. This was a decidedly stingy offer, and Morse was displeased. Although refusing the offer, he nonetheless maintained cordial relations with Field, for the American inventor earnestly wanted this grand enterprise to succeed.[30]

Field lobbied Congress to match the British government's generous subsidy. Secretary of State William Seward championed the legislation, arguing "that after the telegraphic wire is once laid, there will be no more war between the United States and Great Britain."[31] Congress reluctantly went along, apparently not chastened by the Page locomotive boondoggle just a few years earlier or convinced by arguments that a cable both of whose ends were controlled by London would be counter to the best interests of the United States. President Franklin Pierce signed the act with Field attending, on March 3, 1857. The American government would provide a $70,000 annual subsidy for 25 years by purchasing telegraphic services, and would furnish Navy ships to help lay the cable.[32]

By this time, the cable itself was being manufactured by three English companies: the Gutta-Percha Company applied three coats of insulation to seven strands of copper wire, and two firms that originally made cables for mines (called "iron ropes"), Glass, Elliot & Company and R. S. Newall & Company, put on the iron sheathing. Between the iron and the gutta percha was a layer of tarred hemp treated with preservative.[33] The design of this cable, which was only five-eighths of an inch in diameter, privileged flexibility, light weight, and low cost.[34] Because Field insisted that it be ready for laying in the summer of 1857, the cable's *electrical* performance characteristics would have to be learned after it went into operation, for there was no opportunity for testing it under realistic conditions.[35] At this stage of the project, Field's belief that the cable could transmit at a profitable rate rested on the experiments of Whitehouse and Bright and on Morse's endorsement of their findings. That Thomson's arguments had been marginalized was manifest in the company's official history, published in the *New York Times* in August 1857.[36]

The 2,500-mile cable cost £178,835, weighed about 2,000 pounds per mile, and was completed on time. A much more heavily armored cable was attached to the shore ends to resist damage from anchors and rocks. Field had also contracted for the design

and manufacture of paying-out machinery having demanding performance requirements, such as the ability to release the cable at a controlled rate using a brake for
preventing a dangerously rapid descent.[37] The many new technologies and inexperi-
enced men—not to mention the fledgling engineering science of cable design—made
this project a high-risk gamble, but the company's public position was one of unqualified confidence.[38]

_⌒⌒⌒⌒_

Too large to be carried by one vessel, the cable was divided equally between the *U.S.S.
Niagara*, the largest steam-powered warship afloat, and the *Agamemnon*, a huge British
man-of-war.[39] Both ships had to be greatly modified, at government expense, to
accommodate the cable and the paying-out machinery. The plan was for both ships
to head west from Valentia, Ireland, accompanied by escort vessels furnished by both
governments. The *Niagara* would disgorge her cable first, then the *Agamemnon* would
splice on her segment and resume the journey to Trinity Bay, Newfoundland. Crews
of 60 men worked 3 weeks to load the cable segments, which in all weighed more
than 5 million pounds.

In the months preceding the start of the cable-laying, festivities and publicity in
both countries raised expectations to unreasonable heights. Morse, Field, and numerous dignitaries and prophets waxed poetically about the benefits that would follow
the ability to send messages, almost instantaneously, between the Old and New
Worlds. Caught up in cable fever, many people earnestly believed that humankind
was on the threshold of a new age, perhaps even perpetual peace. Would electrical
wonders never cease? Field promised the honor of sending the first message to Queen
Victoria, titular head of the world's largest commercial and colonial empire.

Before a cheering crowd, the cable's shore end was secured on August 5, and the
expedition began its westward journey with Field, Morse, and Thomson aboard;
Whitehouse claimed to be ill and remained on land.

Only 5 miles out, "the cable caught in the deck machinery and parted."[40] Still in
shallow water, it was retrieved and spliced, and the journey resumed at lower speed.
The *Niagara* maintained constant contact with the shore station at Valentia to ensure
that the cable retained electrical continuity. A few days later, the cable went dead but
then revived. This brief failure was troubling because no one could pin down the
cause.

The erratic performance was all but forgotten a few days later when the cable broke.
Morse described the somber scene: "All hands rushed to the deck . . . the men gathered
in mournful groups, and their tones were as sad, and voices as low, as if a death had
occurred on board."[41] The immediate cause of the break was all too apparent: the brake
on the paying-out mechanism had been improperly managed.[42] Lacking equipment

to retrieve the cable from nearly 2 miles down, Field and crew returned to Ireland; the remaining cable was offloaded and stored.[43] Field managed to frame this failure as a limited success, for 335 miles of cable had been submerged. Writing to his family, he denied being discouraged; rather, he was certain "that with God's blessing, we shall connect Europe and America with the electric cord."[44] After all, the cable had parted owing to an accident that could be avoided with an improved paying-out mechanism. Field assigned the redesign of this equipment to William Everett, Chief Engineer of the U.S. Navy. Everett's new machine included a water-cooled brake that engaged gradually when triggered by the tension of a runaway cable. The trigger point was set at 3,000 pounds, half the cable's purported tensile strength.[45]

Another new technology was a super-sensitive galvanometer designed and patented by William Thomson. Eventually known as a "mirror galvanometer," it worked as follows. Several small magnets were attached to the back of a tiny mirror, which could reflect a beam of lamplight reaching it through a slit in a screen. The mirror was suspended by a silk thread into a coil of wire. When a current passed through the stationary coil, its magnetic field caused the magnet-bearing mirror to rotate in proportion to the current's strength. The reflected spot of light, which moved correspondingly, was aimed at a graduated scale from which one could read the current's strength. The distinctive performance characteristic of this galvanometer was that a trifling current—undetectable by any other device and incapable of actuating telegraph registers—could move the light spot.[46] Whitehouse, however, was determined to retain his register, whose operation required high tension.

In preparing for the next expedition, Field ordered 700 more miles of cable, which was laboriously coiled in the holds of the *Agamemnon* and the *Niagara*. The *Agamemnon*'s hold was too small, so some of the cable had to be stored on deck. Before setting off, members of the expedition practiced splicing, and repeatedly tested the new machinery and telegraph equipment. All was now ready.

The plan this time was for the *Agamemnon* and the *Niagara* to sail together to the midpoint of the Atlantic; there the cable ends would be spliced and the ships would head to opposite shores. Cable laying began on June 26. After only a few miles, the *Niagara*'s cable broke. The ships returned to the starting point, the cable was spliced, and again the *Agamemnon* and the *Niagara* headed their separate ways. The next day, however, after 80 miles of cable had been payed out, Thomson reported that the circuit was dead. The ritual of rendezvous, splicing, and resuming the voyage from the starting point had to be repeated. For two days everything went smoothly until after 300 miles the cable broke again. Because communication between the two ships could not be easily maintained at this distance in the absence of a functioning cable, the defeated squadron returned to Valentia.[47] News of the latest failure sent shares of the Atlantic Cable Company plummeting to as little as £200.

Field now had to convince the company's directors that another attempt should be made immediately, insofar as the ships still held enough cable. For some people, a second try seemed futile in view of the unexplained failures. However, after heated discussions punctuated by the resignation of one vice chairman, the directors permitted Field to proceed. The ships were provisioned, and on July 17 they embarked for the mid-Atlantic.[48]

To reduce the chances of a severed cable, the trigger point for the brake was set at 1,700 pounds. This larger margin of safety proved satisfactory. Although the cable had a few obvious flaws, these were corrected before it was sent overboard, and no breakage occurred. However, this expedition was again plagued by unexplained losses and resumptions of the signal. Despite additional problems, including a patch of bad weather and the *Niagara*'s briefly running off course, the ships reached their respective destinations after laying a cable 1,950 miles long.[49]

On August 5, Cyrus Field telegraphed the Associated Press from Trinity Bay, Newfoundland, that the laying of the cable had been completed, and that "the electric signals sent and received through the whole cable are perfect."[50] Americans greeted the news with unrestrained enthusiasm: "New York went wild with rejoicing. A hundred guns were fired on Boston Common and the bells of the city continued their joyous peal for an hour. For weeks the national jubilation continued and the Atlantic Telegraph remained the most important subject of conversation."[51] Field began to receive accolades from heads of state and other notables. Basking in the glory, he credited his co-workers, including men of science, mariners, and long-suffering company directors.[52] On the cover of *Harper's Weekly* of August 21, however, there was room for only Field's portrait. The latter image also graced a book on the history of the Atlantic Telegraph, including its recent triumph, which was advertised for $1 in the *New York Times* on August 14. In another ad, souvenir pieces of spare cable mounted in brass, silver, and gold were sold as "Charms or Watch Keys" at prices ranging from 15 cents to $5 each.[53]

The ritual exchange of banalities between Queen Victoria and President James Buchanan was delayed for more than a week while Whitehouse's electrical system, augmented by Thomson's mirror galvanometer, was tested and tweaked.[54] Eventually the queen's 98-word note reached Washington and the president's 149-word reply arrived in London. Happy news of this exchange led to another round of celebrations in both countries. So exuberant were the festivities in New York that fireworks destroyed the landmark cupola on City Hall.[55] Joseph Henry, writing to Field, spared no praise:

"This is a celebration such as the world has never before witnessed. It is not alone to commemorate the achievements of individuals, or even countries, but to mark an epoch in the advancement of our common humanity."[56]

Not widely publicized, however, was the cable's dismal transmission rate: scarcely 0.1 word per minute for the queen's note and about 0.2 word per minute for the president's.[57] Some messages did surpass the one-word-per-minute mark, but not by much. Evidently, transatlantic communication was going to be very expensive. Cable men perhaps wondered whether high rates would render this line impractical for all but governments and the most well-heeled customers.[58] That issue turned out to be moot, since the cable never entered commercial service.

While celebrations continued, communication gradually worsened until effective signaling ceased in September; by the end of October the cable was dead.[59] Distraught cable men looked far and wide for possible causes, but one seemed particularly salient: by using an induction coil to send messages through the balky cable, Whitehouse had fatally damaged the thin layer of gutta percha insulation. Thomson and other electrical authorities blamed Whitehouse because they knew that high tension can break down insulation. According to Thomson's calculations, the place of failure was about 270 miles from Ireland. Whitehouse insisted that the failure was attributable to a mechanical flaw very near the Irish coast, but tests disproved his conjecture. Although the causes of the cable's demise were eminently disputable, there was no denying that Whitehouse had influenced its design and dictated its mode of operation.[60] For several reasons, the board of directors summarily fired him.

~~~~~~~

To examine problems of cable design and operation, Great Britain's Board of Trade appointed a committee of prominent scientists and engineers, including the telegraph experts Charles Wheatstone and Latimer Clark. In effect, this committee became the project's de facto scientific authority. After two years of investigating and deliberating, the committee declared its firm support for future attempts, for the problems seemed soluble. These men were no doubt encouraged by the dozens of submarine cables that were operating successfully, including several more than 500 miles long; yet some cables had failed, including one under the Red Sea in 1860. The recommendations included close monitoring of cable manufacture and use of high-purity copper for the conductor. After comparing successes and failures of previous cables, the commission noted a telling pattern: ". . . in almost all cases small cables had been found liable to mishaps, while the heavier the cable . . . the greater had been its durability."[61] Clearly, Thomson's design ideas were now being taken very seriously.

Another development, this one organizational, contributed to later efforts: the Gutta Percha Company and Glass, Elliot & Company joined forces in the Telegraph

Construction & Maintenance Company. Capitalized at £1 million, the combined corporation was able to both manufacture and lay cables.

Although planning for the next cable expedition slowed during the Civil War, Field was still pressing both governments for renewed support. The U.S. government indicated a willingness to induce investors by putting on the table an offer of guaranteed interest—2 percent for 25 years—contingent upon a match from Great Britain.[62] Britain at first gave no commitment to Field's entreaty, in part because she was exploring other options for a transatlantic telegraph. Another American telegraph promoter, Taliaferro Shaffner, had organized the North Atlantic Telegraph Company, which projected a line from Labrador to England through Greenland, Iceland, and the Faroe Islands. The alleged advantage of this scheme was the shortness of its submarine segments, none longer than 800 miles. In support of this effort, the British government undertook soundings along the proposed route, and the project appeared feasible notwithstanding the constant threat of icebergs.[63] But Shaffner's line did not come to fruition.

Eventually the British government more than matched the American subsidy by agreeing to buy telegraphic services for £20,000 per year and promising investors an 8 percent return on new capital of £600,000. But, having been burned by the failure of the heavily subsidized Red Sea cable in 1860, the British government insisted that the cable work.[64] Certain that it would, Field and the company's directors sold more stock, but they had to sweeten the deal by promising an 18 percent dividend, which assumed that the cable would be in use 16 hours a day, 300 days a year. At 30 pence per word, the total annual income would amount to £413,000—assuming a transmission rate of 11.5 words per minute.[65] Nearly half the stock offering was subscribed by the Telegraph Construction & Maintenance Company.

The new cable embodied design principles that William Thomson had enunciated nearly 10 years earlier.[66] The conductor contained more copper per mile to reduce resistance, and the insulating layer of gutta percha and tar-impregnated hemp was somewhat thicker, which lowered electrostatic capacity. Yet the steel shroud was perhaps this cable's greatest innovation. It was designed not only for strength—while remaining flexible—but also for increased buoyancy. It was composed of 10 strands of soft steel wire (no. 13), each wrapped in yarn impregnated with a waterproof preservative. Weighing 3,575 pounds per mile, the cable was 1.1 inch in diameter. Significantly, its greater volume contributed to a relatively low specific gravity, and so it sank more slowly than its denser predecessors despite being nearly twice as heavy. The cable's tensile strength was 15,500 pounds—more than twice that of the 1858 cable. The colossal shore segments were 2.5 inches in diameter and heavily armored.

To reduce the risks of mid-ocean splicing, it was decided to make a single cable 2,700 miles long. The obvious problem was that such a monstrous cable would require an

Figure 17.1
The *Great Eastern*. Source: Dibner 1959: 68.

equally monstrous ship, but by 1865 such a ship became available. The *Great Eastern* had been designed by the English civil engineer Isambard Kingdom Brunel and built by the Scott Russell Company at Millwall on the Thames (figure 17.1). At 693 feet long and 120 feet wide, the *Great Eastern* was by far the largest ship afloat. Driving the enormous screw propeller and two 56-foot paddle wheels were ten steam engines, together producing 10,000 horsepower. The capacious vessel was intended to carry 14 million tons of cargo *and* all the coal needed to travel from England to Australia.[67] She was launched in 1858 and put into service the following year.

Although a technical marvel, the *Great Eastern* was a commercial catastrophe. A group headed by Daniel Gooch finally purchased the ship at auction for only £25,000—less than her salvage value. When a representative of the Atlantic Telegraph Company came calling on Gooch, he was delighted to offer use of the ship in exchange for a block of cable stock and membership on the company's board of directors.[68] The great white elephant of the sea now had a noble and potentially lucrative calling, and the Atlantic Telegraph Company grew ever larger and more complex.

The hold of the *Great Eastern* was outfitted with three large tanks to store wet 8 million pounds of cable, distribute its weight, and prevent it from sliding around in choppy seas. It also held 8,500 tons of coal, hundreds of barrels of provisions, and 18,000 eggs. Installed on deck was paying-out machinery of an entirely new design that promised an accident-free descent. Also on deck was a menagerie, including oxen, pigs, sheep, and flocks of fowl, to help feed the 500-man crew.[69]

As in all previous expeditions, Thomson was aboard, for he was one of the men assigned to test the cable throughout the voyage and, upon arrival in Newfoundland, send messages eastward. The chief electrician was C. V. de Sauty, a company man who possessed vast experience with submarine cables.[70] Samuel Canning, the Chief Engineer, oversaw the cable-laying operation.

Accompanied by two escort vessels, the *Great Eastern* began her voyage to the New World in mid July, 1865. Twice over the next few days the alarm gun sounded as electrical continuity suddenly waned. Miles of cable had to be hauled up by a small engine on the ship's bow. The faulty sections, whose leakage of current had been caused by stray pieces of iron piercing the insulation, were cut out and the loose ends spliced. After each repair, the paying-out process resumed at a good clip with a strong signal.[71]

When the alarm sounded a third time, the cable was in deep water, 1,200 miles from Ireland. Again it would have to be taken up partway, cut, and spliced. With so much weight to lift, however, the feeble hauling engine strained for hours yet made little progress. The cable chafed and parted, its sea end settling more than 2 miles to the bottom. Aboard the ship were grapnels and a 5-mile rope, and so Canning decided to retrieve the lost cable. Three times the cable was snagged, only to return to the bottom when the rope broke. On the fourth attempt, the rope gave way again, and now it was too short for another try. The *Great Eastern's* disheartened crew returned home.[72]

Although the paying-out machinery operated flawlessly, poor performance of other mechanical equipment had once more doomed the project. Nonetheless, these problems seemed remediable through use of better grappling and retrieving equipment. On the plus side, the new cable had performed well, transmitting a strong and intelligible signal just as Thomson had anticipated. Moreover, it was brawny enough to endure routine paying out and hauling in.

⎯⎯⌒⌒⌒⌒⎯

During the winter of 1865–66, Field had to raise funds for another expedition, slated for the summer of 1866. Because it was necessary to found a new company to comply with British law, the financial dealings became even more convoluted. Nonetheless, with optimism still high that the next attempt would succeed, the parties worked out the details by March. The new corporation, the Anglo-American Telegraph Company, issued £600,000 in stock, which was promptly subscribed, and contracted for another cable.[73] Minor changes were made to the cable's design (e.g., the steel strands were galvanized), which rendered it slightly lighter, stronger, and more corrosion-resistant. In addition, the engineers secured a much more robust hauling engine and better grappling equipment. And the *Great Eastern* underwent a minor overhaul, including removal of a thick coat of barnacles from her steel hull.[74]

After the splice was made to the shore-end cable, the heavily laden *Great Eastern* retraced her westward path. While the cable was being payed out at 6 knots, the electricians constantly exchanged messages with the Valentia station. During this tedious process, the operators at both ends became adept at sending and decoding signals, which carried news of the Austro-Prussian War and stock-market quotations. Much to the relief of all on board, the voyage was marred by no major problems, not even bad weather; despite having begun on Friday the 13th.

On July 27, 1866, after connection was made to the shore-end cable in Newfoundland, Field telegraphed a brief message: "We arrived here at nine o'clock this morning. All well. Thank God, the cable is laid, and is in perfect working order."[75] The cheery news spreading around the world elicited another round of congratulations for the indefatigable promoter Field, who again sung the praises of all who had contributed to the project's success, including the investors. And contribute they did: the total cost of all five attempts was approximately £2.5 million—more than $300 million in today's dollars.[76]

Two days after the connection was made, messages were already being sent at 5.5 words per minute.[77] The provisional charge for a 20-word message was £20 in gold or $150, with 20 shillings for each additional word. Remarking on this steep tariff, which might yield as much as $4 million per year, the *New York Times* hoped that the cable would not become "a vehicle of extortion."[78] On the first day it was open for business, the New York office sent 20 messages and collected in cash $3,046.[79] Before the Atlantic telegraph's first week of operation had ended, Field was able to report to the press "We are now receiving messages . . . at the rate of over twelve and a half words per minute . . . the electricians are delighted with the perfectly distinct character of signals."[80] At last the brash promise Field had made to potential investors a decade earlier—a transmission rate of 10 words per minute—had come to pass. As the engineering science of cable telegraphy advanced in the decades ahead, the signaling speed on this cable would continue to increase.[81]

In late September of 1866, after 30 arduous attempts to grapple the lost cable of 1865 and with provisions running dangerously low, the elusive prize was brought aboard the *Great Eastern* and spliced; the remainder of the cable was then laid without incident.[82] Now there were two functioning Atlantic cables. With ample unused capacity, on November 15 the company reduced the tariff by 50 percent. By 1875, the rate had descended to $1 per word.[83] These cables earned substantial profits. As demand increased over the decades, new cables were laid.

After its stellar service in the Atlantic, the *Great Eastern* became a workhouse for the cable industry, laying more long lines. She was eventually replaced by special-purpose ships, beginning in 1874 with the *Faraday*, a huge vessel designed by William Siemens and built in England.[84] Just a few years later, submarine cables made possible a communication network that included all major cities of the northern hemisphere.

By the end of the century, Africa, South America, and Australia had been drawn into this worldwide web. And so was realized *Scientific American's* bold prophecy of 1852. In 1892, when ten cables linked North America to Europe, a telegram from New York to London cost 25 cents per word, much cheaper than before but still not within easy reach of ordinary Americans.[85]

_⌒⌒⌒_

Not long after the Atlantic telegraph began operating, Latimer Clark sought to show that a submarine cable could be worked on a low-tension battery, consistent with William Thomson's views. First, Clark directed the men in Newfoundland to join the two cables. Next he assembled an unusual single-cell battery: a silver thimble containing sulfuric acid and a piece of zinc. Then, with this diminutive power source, Clark actuated the mirror galvanometer through the combined cable—around 3,700 miles.[86] Such demonstrations underscored Thomson's theoretical perspicacity as well as Whitehouse's engineering shortcomings.

Why did Field at first follow Whitehouse's advice on cable design? The facile answer is that a Whitehouse-type cable would be cheaper. Cost no doubt influenced the decision, but surely Field also believed that it would work. To gain a better understanding of this decision, let us reprise Field's role in this project.

Field was an entrepreneur and a promoter with training in neither natural philosophy nor telegraphy. Captured by the vision of an Atlantic telegraph, he nonetheless knew that others would have to design, construct, and install the system. Thus, he turned to scientific authorities, who, he learned to his dismay, had divergent views on cable design and operation. How could a businessman judge which men of science had the better understanding of signaling through long submarine cables? Choosing among conflicting expert opinions is a difficulty potentially faced by any entrepreneur or promoter whose project is unprecedented and encroaches on phenomena little explored by scientists.

Thomson was a young professor of mathematics—a "savant"—whose analyses were expressed in dense mathematics, essentially opaque to laymen. However, the implications of his work were all too clear: a Thomson-inspired cable would be expensive and perhaps so heavy that it could not be easily transported and payed out.[87] Moreover, because Thomson was a relative newcomer to electrical science and the practice of telegraphy, his power to persuade was limited. In contrast, Whitehouse was actually experimenting with cables and designing new telegraph instruments, and his ideas justified not only a less costly design but also one that seemingly created fewer ancillary problems. Field accepted the advice of the man having more experience with actual cables, a man whose ideas were supported by Faraday and Morse.[88] In this power struggle, Thomson, when set against Whitehouse, Faraday, and Morse, was the much weaker authority.[89] Field's reliance on Whitehouse's advice in 1857 and 1858 was, it would appear, eminently reasonable.

Entrepreneurs and promoters are impatient people who value decisiveness. They are also eternal optimists, apt to make decisions under conditions of great uncertainty and expecting to work around obstacles. Field was no different: he wanted to get on with uniting the continents. However, this haste doomed the early efforts because no one had worked out every devilish detail, electrical or mechanical. Although potential modes of mechanical failure were considered, some possibilities (e.g., breakage of the cable in deep water) were apparently judged so improbable that no remedies were at first provided. As a consequence, each cable-laying expedition became a costly experiment whose failings indicated the need for new technologies, such as more sophisticated paying-out machinery and powerful hauling engines. Not until one set of problems were solved did others—wholly unanticipated—appear.

Field was not unique in pushing a project that had to invent the requisite engineering science and technology along the way. After all, this kind of protracted learning process—"trial and error" would not be a misnomer—afflicts the development of many a complex technological system. However, in the creation of electrical technologies, the Atlantic telegraph presented a range and a magnitude of difficulties never before encountered. That this enormous developmental distance was eventually traversed testifies to the ability of Field—"the most obstinately determined man in either hemisphere"—to raise capital, despite failure after failure.[90] Field was no doubt persuasive in arguing that success was just around the corner, that the last major technical hurdles had been surmounted. But one must not forget that investors, already hooked on telegraphy, were prepared to believe that trans-Atlantic communication was destined to come and that its use would be a necessity in the affairs of countries and businesses. Moreover, given the inducement of high dividends, they expected a functioning cable to be a profitable investment. And so, even in the face of failures these players made available more financing.[91] Field was also fortunate that, during the 10 years his project took, others were also experimenting with long submarine cables, generating additional principles of engineering science that aided in redesign of the cable and associated machinery. Such visible progress also helped to inspire confidence in Field's enterprise. And it didn't hurt that the *Great Eastern* became available in time for the 1865–66 expeditions—a matter of serendipity that enabled a single vessel to lay the massive cable.

In attributing the eventual success of the Atlantic cable to factors such as the tireless efforts of an entrepreneur-promoter, investors willing to plunge in anticipation of huge profits, the coeval growth of necessary engineering science and technology, and serendipity, one is apt to underplay the many kinds of generous government support that made this Herculean enterprise possible. Most obvious was the commitment of both the United States and Great Britain to buying blocks of telegraphic services, which guaranteed moneyed men a return on their investments—if the cable worked. A less obvious subsidy was the underwater surveys that agencies of both governments undertook. Then there were the vessels—the *Agamemnon*, the *Niagara*, and their escorts—furnished for the 1857–58 expeditions and sundry vessels supplied in 1865–66; these vessels also required modifications, provisioning, fuel, and numerous navy personnel. And let us not forget *laissez-faire* economics: by allowing joint-stock companies to combine with other joint-stock companies, the British government enabled establishment of large, vertically integrated corporations that could not have existed otherwise. British officials believed that generous support for Field's project was warranted because a cable link would serve national interests. And as a tool of empire—colonial and commercial—it did.[92]

Not all promoter-driven or government-supported projects end as happily. Some developmental distances turn out to be so great that confidence in the project's technical feasibility wanes and support is terminated. Today we have neither aircraft carriers made from an ice-sawdust composite nor nuclear-powered airplanes, for both projects stalled in the face of obstacles regarded as insuperable. And despite countless billions of public funds poured into nuclear fusion, that technology may never produce electricity commercially.

During much of its first century, the U.S. government, distancing itself from the decadence of European monarchies and husbanding its limited financial resources, built no grandiose monuments to flaunt the country's might or glorify its leaders. In contrast to Buckingham Palace and Versailles, the White House was a rather modest residence. Two of America's most iconic structures—the Statue of Liberty and the Washington Monument—were not completed until the 1880s, and the federal government initiated neither project. Before the dedication of the Statue of Liberty, the most dramatic architectural symbol of the United States and its representative form of government was the Capitol, with its distinctive dome.

In comparison with European legislative seats, the old Capitol, rebuilt after the War of 1812 largely by enslaved laborers, had been an undistinguished structure. And, with the rapid growth of the country during the first half of the century, its offices and chambers became too cramped for the many senators and representatives from the new states. In a 14-year project (1852–1866) that tripled its size, the Capitol was refurbished and extended north and south, becoming "a sparkling jewel glittering with the finest materials, art, and architecture money could buy."[2]

Plans for the enlarged Capitol indicated that the old wooden, copper-clad dome—a leaky fire hazard—would, if retained, look ridiculously small and squat. Thus, in 1855, while work on the additions was still under way, Thomas Walter, Architect of the Capitol, designed an iron dome more in keeping with the proportions of the elongated edifice. The drawing of the dome hanging in Walter's Capitol office attracted the praise of legislators who passed by. Within a few months, Congress appropriated $100,000 toward construction of the new dome—a sum that ended up as little more than a down payment. Capping the cavernous rotunda, itself a ceremonial space where Abraham Lincoln would lie in state amidst John Trumbull's War of Independence paintings, the projected dome's grandeur would certainly evoke universal admiration. Intended to outclass the most magnificent domes of Europe, which Walter had studied firsthand, this indulgence in political architecture became the consummate "symbol of American self-government and democracy."[3] Work on the dome continued intermittently during the Civil War, demonstrating Lincoln's resolve that the nation endure united.

Topped by a bronze sculpture titled *Freedom*, the intricate, magisterial dome reached nearly 300 feet above street level, towering above all other Washington buildings and affording stunning views for anyone able to trudge its 365 steps. No less impressive was the dome's interior: at its "eye," 180 feet above the rotunda's floor, was a canopy adorned with a fresco by Constantino Brumidi, a political refugee from Italy who had previously created artworks for the Vatican. His decoration of the dome's canopy, which covered 4,664 square feet, was entitled *The Apotheosis of Washington*.[4] Costing $40,000, the fresco showed the first president ascending gloriously to heaven. Surrounding George Washington and his angelic entourage were groups of figures representing America's central concerns: war, commerce, agriculture, marine, mechanics, and science. The latter group included Benjamin Franklin, James Fulton, and Samuel Morse, depicted in the company of an eighteenth-century electrostatic generator and a battery of Leyden jars. Lighted during the day through windows, how could this fresco awe and inspire nighttime visitors? Because chandeliers would block views of the elevated artwork, the dome's upper reaches would have to be illuminated by a mass of individual gas lamps arrayed around its perimeter.

Lighting hundreds of gas lamps presented a tall challenge, for each burner would have to be lit individually. The original design for dome lighting called for a less labor-intensive system: carrier tubes that could ignite many burners at once. Requiring two parallel, independent systems of gas pipes, this solution was cumbersome, inelegant, and potentially dangerous.

_⌒⌒⌒_

Lighting multiple gas lamps spread over large spaces or perched in hard-to-reach places was framed by some inventors as a problem that electrical technologies could solve. That electricity could ignite inflammable gases with a spark or a red-hot wire had been known since the latter decades of the eighteenth century, and several instrument makers had even marketed hydrogen-burning lamps with spark ignition.[5] But these products had generated little interest. In the middle of the nineteenth century, as gas illumination became nearly universal in cities large and small, inventors perceived an opportunity to replace the ubiquitous lamplighter with electrical technology.

A general solution had been proposed in France as early as 1850, some technologies had been patented in England, and by 1860 there were several installations in both countries. To light a hard-to-reach part of a hall at Edinburgh University, electricians had rigged up an electromagnet on each lamp that swung a platinum wire over the burner. When current was passed through the wire, it got hot enough to ignite the gas.[6] But the flow of gas was still controlled manually. Somewhat earlier, however, a more complete system had already been envisioned. In 1852 *Scientific American* exhorted inventors to solve the problem with telegraph technology: "Very simple machinery only would be required to turn the cock of the gas-pipe, and, simultaneously with the escape of the gas, to apply the electric spark."[7] Apparently, someone with a rudimentary knowledge of electricity and a facility with mechanisms could create an electric gas lighter. Although far from the frontiers of science, it would be on the cutting edge of lighting technology—and no doubt pricey.

A system along these lines came to fruition a few years later. The driving force was Samuel Gardiner Jr. (1816–1880), an inventor-*cum*-electrician living in New York.[8] Gardiner's first inventions (1853–1855) bespeak skill in designing metal mechanisms. Inspired by the California gold rush that began in 1849, all were ore-processing machines for winning gold from its rock or sand matrix. One machine crushed and pulverized ore, a second removed magnetite (a highly magnetic iron oxide), and the third amalgamated gold with mercury. Gardiner was clearly comfortable designing devices with gear trains and other moving parts.[9]

Gardiner's machine for removing magnetite particles employed a series of permanent magnets that rotated through a moving slurry of gold-bearing sand. With each revolution, brushes swept off the magnetite particles adhering to the magnets. However, anyone who has played with magnets and iron filings knows that complete cleaning of the magnets would have been impossible. Seeking to improve the ore separator, Gardiner might have explored the potential of electromagnets, whose on-off actions were visible in any telegraph office. Perhaps the magnetic ore separator was a bridge to his electrical inventions. Another possibility is that he was following *Scientific American's* advice, for his gas lighter reflected some acquaintance with telegraphs and electromagnets, and would both turn on and light the gas.

To build the patent model, Gardiner turned to L. A. Hudson, a philosophical instrument maker in Syracuse, New York, who also made magnetos for physicians.[10] Gardiner's original idea was to light a series of gas lamps with sparks. Hudson informed Gardiner that a battery could not produce the required spark, proposing instead the use of a hot wire, which Gardiner adopted. If Hudson's account is accurate, it indicates that Gardiner, like Morse, was captivated by an electromagnet's performance characteristics but began his project ignorant of important electrical principles. Nonetheless, by 1860, Gardiner was calling himself an "electrician."[11]

The electric gas lighter (figure 18.1) combined features of the telegraph key and the telegraph register. It had two keys that controlled battery current in separate circuits. The first sent current to the electromagnet, which in small steps turned a stopcock that admitted gas to the burners. The second circuit contained a number of fine platinum wires, connected in series, each mounted with insulators slightly above and adjacent to a burner's orifice. Strokes on the first key gradually allowed the gas to flow. After the second key was pressed, the platinum wires became red hot and ignited the

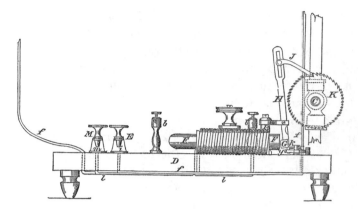

Figure 18.1
Gardiner's electric gas lighter. Adapted from *Report of the Commissioner of Patents for 1857. Arts and Manufactures*, volume 1 (Government Printing Office, 1858), p. 233.

gas. Since the stopcock could rotate 360°, pressing the gas key several more times eventually extinguished the burner. The mechanism for rotating the stopcock may have been inspired by the paper-tape driver in a Morse-Vail register. (Compare figure 12.1.)

A section of the original patent application that did not survive revision attests to Gardiner's belief that this invention would have many applications: "The object of this invention is to light the whole or any number of the gas-lights employed in a public or other building or district simultaneously or without the necessity of lighting each separately by hand, and to enable the gas chandeliers and other suspended gas-lights in theatres and other places, and other lights in positions not easily reached by hand or mechanical agency, to be lighted without difficulty."[12] This was a heady and ambitious vision for 1857.

In establishing the invention's technical feasibility, Gardiner did not require the approbation of men of science, for he had incorporated common electrical effects into a novel arrangement of metal parts similar to those of telegraph apparatus. On the basis of well-known electrical and mechanical principles, the electric gas lighter could work. Because the invention was in many respects prosaic, U.S. patent examiners had difficulty seeing in it a glimmer of originality, and so they rejected the application.[13] After two appeals and many revisions, however, Gardiner gained the patent.

The invention was simple in principle, but Gardiner and Hudson had to develop a detailed design and hardware that worked in practice. Fashioning the first versions of the gas igniter apparently did not require a major capital outlay. Like Thomas Davenport and other independent inventors of that era, Gardiner cobbled together funds for making the patent model and experimenting with prototypes, presumably by using his own resources and drawing upon his social network.

Gardiner's approach to profiting from the basic invention was somewhat novel, for he neither started a company nor sold the patent. Thus, he avoided the need to convince capitalists and entrepreneurs that his invention might make money for them. Instead, Gardiner worked as a contractor, selling and installing a system tailored to the unique requirements of every customer and built with his patented components and hard-won technical expertise. Becoming a contractor was a pioneering strategy for commercializing an electrical invention. Fraught with risks, it nonetheless preserved the inventor's independence and reduced the number of players who had to assess the technology. And, as Gardiner made improvements along the way in switches, burners, and other components, he secured more patents to solidify his monopoly position.

Samuel Gardiner had no difficulty buying parts. Manhattan was a manufacturing mecca. John Norton, a maker of "magnetic telegraph apparatus" with a business at 447 Broome Street, was probably a major supplier, for he had extravagantly endorsed Gardiner's invention in a sworn deposition to the Patent Office. Though he acknowledged only that his favorable opinion was based "upon actual personal experiments with the invention," Norton no doubt was already making parts for Gardiner.[14]

To alert potential customers that his electric gas lighter might be practical for them, Gardiner had to display the system working in a real-world application, perhaps enlisting the press to publicize a success. Identifying a target market was easy. Obviously, the most promising applications were large public spaces where gaslights abounded, such as theatres, places of worship, streets, hotel lobbies, factories, and legislative chambers. Precisely these sorts of consumers, some of which had experimented with battery-powered arc lights, might be able to afford an expensive accessory for their gaslights. Perhaps more importantly, they might assign a high value to having a state-of-the-art lighting technology.

In the spring of 1857, probably before the patent application was first submitted, Gardiner and Hudson installed a system in the Broadway Theatre. The work was arduous, involving the modification of each burner in the chandeliers, stringing up and connecting wires, affixing electromagnetic mechanisms to the stopcocks, and setting up a large battery. In its first public trial, the electrical system operated the theatre's lights on two consecutive nights in May. This "experiment" was reported with some enthusiasm on page 4 of the *New York Times*: "When the invention has been perfected it will be of great practical utility. Whole cities can be lighted instantaneously, and all the danger and trouble attending the illuminating process in public places of amusement entirely avoided." But the *Times* added a caveat: "At present . . . it is defective. It fails to ignite all the burners, with absolute certainty, and is far from being instantaneous in its operation. . . . The escape of gas which occurred before the majority of the burners were lighted last night, created anything but a pleasant sensation."[15] This was not the kind of publicity that Gardiner hoped to garner. In contrast, *Scientific American*, in a half-page spread, praised the installation effusively, perhaps because the patent agency that owned the magazine handled Gardiner's applications.[16] Instead of describing it as an experiment, *Scientific American* stated that Gardiner's system had been "successfully introduced in the Broadway theatre . . . and arrangements are being made for introducing it in one of the principal theatres in Philadelphia." Because it would have been difficult to extricate the apparatus from the theater, especially from the modified lamps and stopcocks, I suspect that this demonstration was less an experiment than an actual installation.[17]

After sketching how the Gardiner system worked, *Scientific American* favorably compared its performance characteristics to those of the lamplighter. In addition to saving gas, it was fast: about 30 seconds to turn on the entire house, versus two men laboring for an hour. This rapid working allowed the lamps to be turned off and relit between acts. Moreover, the hazardous lighting of stage lamps with torches, which had caused many a catastrophic fire, was eliminated. The apparent clincher was the unsupported claim that "one of the insurance companies has offered to insure theatres for 25 per cent less premium where this apparatus is employed."

Characteristically, *Scientific American* was silent on other performance characteristics that might have concerned would-be buyers. Although noting that the battery contained 30 Smee cells, each in a one-gallon container, the article made no mention of the purchase price, the costs and difficulties of battery maintenance, or the liabilities of a series circuit. Nor did it acknowledge that every major installation would have to be overseen by an electrician.

Aware that their system had shortcomings, Gardiner and Hudson labored to modify the design. In effect, each new installation was a development project underwritten by a purchaser of means. Between 1858 and 1875, Gardiner received a dozen patents for lighting gas by electricity (including one reissue), many of which involved alterations to major components.[18] The patents included a lava tip for the gas burners (lava is heat resistant and also an electrical insulator), a tiny platinum coil as the heater, a mix of parallel and series wiring for large installations, and visual indicators of the amount of gas released to each set of burners. Gardiner even came up with a meter, so that individual customers could be charged for the electricity they consumed.[19] Eventually these inventions would constitute a sophisticated system, but even in its earliest installations the electric gas lighter appears to have worked well enough to attract buyers.

Consumers weight a product's anticipated performance characteristics—utilitarian, financial, and symbolic—in relation to alternatives and to their own applications and priorities.[20] In the late 1850s, there was precious little information on the new technology's maintenance requirements and costs. Even so, it was not hard to predict that—in view of the need to refurbish the battery and perhaps employ an electrician, and in view of the amortized purchase price—the electric lamplighter would probably not be cheaper than its human counterpart. However, for some consumers these financial shortcomings would be outweighed by symbolic performance characteristics, not least of which was the cachet of progress that electrical technology in any form was coming to embody by virtue of the telegraph, which was affecting more lives every day. Also, because the lighter operated many gas lamps with near simultaneity, it dramatically demonstrated mastery of the magical electrical force. Because it advertised its owner's discernment and wealth, Gardiner's technology was an appropriate product for elite consumers.[21]

On the eve of the Civil War, *Scientific American* reported that Gardiner's system was "being rapidly introduced." Among the installations were those at Tiffany & Co, at "Mr. Belmont's picture gallery," and at the Academy of Music, all in New York. The gas lighter also entered several homes, including Gardiner's. At the front door were two ivory knobs that controlled the chandelier in the front parlor; in the next room a set of keys operated a second chandelier, and so on. The cost of installing the system in a three-story house was said to be $100, but this hefty expense included a special feature supposed to scare off burglars: if any door or window was opened at night, it would light every lamp in the house.[22] Perhaps Gardiner's most significant pre-Civil War installation—but not necessarily sale—was to Congress. Installed in the old Senate chamber (soon to become the Supreme Court's home) early in 1858, the system operated 1,500 burners in a huge chandelier.[23] According to *Scientific American*, these varied installations were "working successfully and giving perfect satisfaction."[24]

Not long after the gaslights in the old Senate chamber were electrified, an inventor named Archibald Wilson planted the seeds of eventual competition to Gardiner's system. Wilson's invention employed a tiny battery, an induction coil, and parallel wiring to spark gaps on each burner, but it lacked a mechanism for turning on the gas.[25] The *American Gas-Light Journal* tilted in favor of Wilson's system because, in principle, it would be much easier and cheaper to maintain on account of its minuscule battery.[26] And because Gardiner's heating coils were wired in series, tracking down a break in the circuit would be difficult.[27] Years later, an entrepreneur employing the Wilson system would challenge Gardiner for the profit and prestige of lighting the Capitol. In the meantime, Lincoln was elected president and the rebels bombarded Fort Sumter.

The outbreak of the Civil War had many effects on American farms and businesses, and likely dried up demand for Gardiner's gas lighter, despite its technical, commercial, and symbolic successes. The war also stimulated inventors, who, in meeting or even anticipating military needs, offered new designs for necessities such as portable telegraphs, camp stoves, and artificial limbs. Samuel Gardiner invented and patented a novel bullet that promised to multiply the users of prosthetic devices: one second after penetration, the bullet exploded in the body. Over the objections of the Chief of Ordnance, the Secretary of War bought 110,000 exploding bullets from Gardiner for $35 per thousand, but apparently only 35,000 were dispensed to the troops. Many people, including General Ulysses S. Grant, considered these bullets barbarous.[28]

ˏᴖᴖᴖ

From the start, the project to extend the Capitol had been mired in bickering, political and personal. The Architect of the Capitol, Thomas Walter, and the chief engineer, Montgomery Meigs, wanted to create a monument not only to the country but also to themselves, and this sometimes devolved into disputes about construction details and controversies over credit. But what Walter and Meigs had in common was the conviction that every detail of the building should be first-rate, from heating and ventilation systems to ornamentation. In responding to funding requests for this ambitious project, members of Congress were divided and conflicted: they did not want to betray the public trust by spending money on mere embellishments, yet they also wanted a Capitol that would befit an America already aspiring to world leadership. And so a reluctant Congress funded, piecemeal over more than a decade, a showcase of sophisticated art and technology.

Despite many distractions, including the garrisoning of troops in the Capitol at the start of the war, the extension project lumbered ahead.[29] In 1863, the year of the Emancipation Proclamation, the dome's exterior shell was finished. This required, among other tasks, hoisting and wrestling into position pieces of cast and wrought iron weighing a total of nearly 9 million pounds. The following year, the dome was painted white. An early historian of the Capitol wrote that the dome was "more imposing to the eye" than those of St. Peter's in Rome and St. Paul's in London; indeed, he asserted, "There is no dome in Europe more graceful in its lines and proportion."[30] In 2001, William Allen, architectural historian of the Capitol, noted that the dome's "commanding presence over the city changed the perception of the nation's capital forever. Impressed at the sight of the great white dome, Europeans no longer sneered and Americans gained a welcome sense of national pride."[31] This magnificent aesthetic performance did not come cheap. The final cost of the dome alone was a breathtaking $1,047,292. Apparently this was a small price to pay for an America architecturally proclaiming its importance in the community of nations.

Soon the dome would be as impressive inside as outside. In December 1862, anticipating that the project would be finished soon, Walter invited Gardiner to submit a bid for dome lighting, and two days later had the bid in hand. Because this project was unprecedented in scope and challenges, no one could predict exactly how many burners would be required (one early estimate was a mere 300), the size of the battery, or various other details. In view of these pervasive uncertainties, Gardiner had framed the bid in terms of unit costs for unspecified numbers of components and for associated materials and hardware. To these totals he had added percentages for his profit; presumably he would also profit on every part sold to the government. The prices seemed "just and reasonable" to Walter, but the decisive factor was his belief that there was "no other method by which the inner dome of the Capitol can be lighted than that of Mr. Gardiner."[32] This judgment obviated the standard practice of soliciting bids.

Benjamin French, Commissioner of Public Buildings and disbursing agent, agreed with Walter's judgment and awarded Gardiner a contract to light the dome.[33] The contract was based on the terms set forth in Gardiner's proposal, which afforded ample opportunities for costs to escalate. Who, for example, could question the size of the battery needed (at $29+ per cell) or the number of thumbscrew fasteners (at $1 each)? Gardiner was to be paid periodically on the basis of detailed invoices. Although the contract represented a meeting of the minds in 1862, it contained ambiguities that would later erupt into disputes between Gardiner and the government. In any event, the U.S. government had once again become a patron of pricey new electrical technology.[34]

An early invoice is evidence of the incompleteness of Gardiner's proposal. Items appearing for the first time in that bill included a solder furnace, varnish, amalgamating brushes, gas regulators, and a hydrometer (for testing the battery acid).[35] Clearly, Gardiner had vastly underestimated what it would take to build the system. His experience had been mainly in lighting a plethora of burners on chandeliers. Installing and testing the dome lights—hundreds of dispersed burners at several heights—was far more complicated. Gardiner was learning that the developmental distance for a cutting-edge technological system is difficult to forecast.

Gardiner believed that under the contract he could buy what was needed, hire assistants, and bill the government for expenses plus his profit. But it wasn't that simple, especially in wartime. In early 1864, vexed by delays in getting checks, Gardiner wrote to the Secretary of the Interior demanding immediate payment for overdue bills.[36] In skirmishes over the next few years, the government refused to pay Gardiner's charges for his labor and sometimes that of his assistants, arguing that the percentage increments for profit had included labor. In several more letters to the Secretary of the Interior, most of which went unanswered, Gardiner persisted in arguing his interpretation of the contract. Benjamin French, who had authorized the contract, strongly supported Gardiner's position.[37]

In the meantime, a new player increased the possibilities of conflict. Now serving as Secretary of the Interior was James Harlan, a heavy-handed administrator who, upon taking the job in mid 1865, began firing employees, including everyone above the age of 60.[38] Another casualty was Thomas Walter, the long-suffering Architect of the Capitol, who was reassigned by Harlan just months before the enlarged Capitol was declared done. Walter resigned and was replaced by a young man named Edward Clark. Harlan, like his predecessor, put off Gardiner's requests for payment.

Harlan asked Clark to explain the apparent cost overrun of the dome lighting project, which he believed should have been no more than $5,000. Clark, in a dilatory response, held Gardiner blameless in the escalation, noting the large number of burners required, which Clark himself had authorized.[39] No doubt these exasperating experiences taught Gardiner that sympathetic players in a government bureaucracy could come and go—a lesson learned earlier by Samuel Colt.

With the interior of the dome nearing completion, Constantino Brumidi toiled for 10 months on the brilliant fresco covering the large canopy. Below him, working on narrow ledges without ropes or scaffolds were Samuel Gardiner, L. A. Hudson, J. Richardson, and David Small, whose combined labor would total about two-and-half man-years.[40] Because the dome was a new structure, there were no existing gaslights to be retrofitted. This gave Gardiner the opportunity to emplace an entire lighting system, including gas pipes, stopcocks, and his own lava-tipped burners with their dainty platinum coils. The gas pipes fed three tiers of lamps at heights of 45, 80, and 165 feet above the rotunda's floor, which served 1,050 burners inside the dome and 60 more outside and above the dome.[41]

Even though Gardiner's earlier installations had been operating for several years, some people questioned whether an electrical system could work properly in the dome's massive iron carapace. Some also fretted that on such a large scale it might not function at all. These issues were not by then questions of science, but of electrical engineering. Gardiner and Hudson hoped to lay them to rest by thoroughly insulating the wires and electrical components and installing a sufficiently hefty battery.

The zinc-carbon battery was gargantuan. Two hundred cells, each housed in a glass jar 14 inches tall and 13 inches in diameter, filled a room measuring 36 by 45 feet. Evidently, Gardiner was taking no chances that a dearth of power would prevent all the lamps in a tier from lighting at once. Current to the electromagnets and the coils was distributed through 5 miles of copper wire (at $3 per pound) insulated with two layers of linen covered in critical locations by rubber tubing. The electrical controls included a new dial-plate switch, developed especially for the dome project, with eleven keys and corresponding vernier indicators that put the battery into the circuit in groups of twenty cells.[42] Surrounding the dial-plate switch were ten additional keys, with indicators, that sent current separately to the electromagnets and heating coils. The keyboard was placed in a passageway just outside the rotunda's floor and occupied a mere two square feet.

The lighting project required an outlay estimated at between $20,000 and 30,000.[43] This expenditure would have seemed exorbitant in comparison with the annual salaries of federal employees. (A gardener earned $1,440, a watchman $1,000, and a water-closet cleaner $538.) Nonetheless, the expenses were on a par with those of frescoes, sculptures, and other decorative elements: Emanuel Leutze's painting *Westward the Course of Empire Takes its Way* cost $20,000, as did the casting and installation of the dome-capping statue, *Freedom*. And, no matter the cost, the electric gas lighter would compliment the Capitol's state-of-the-art heating and ventilation systems, features that advertised America's growing prowess in industrial technologies. In view of the Capitol's obvious political functions, Gardiner's system would have to perform well symbolically. Specifically, it would have to be perceived as the most sophisticated system available, and would have to impress all who saw it in operation. In this context, parsimony was no virtue.[44]

ᴖᴖᴖ

In mid 1865, as the lighting project and Brumidi's fresco neared completion, Congress was going through one of its periodic paroxysms over alleged contracting irregularities and waste in federal projects. The target this time was the interminable extension of the Capitol. Although most congressional accusations of "malconstruction and fraud" concerned building materials, Harlan and French independently appointed commissions to evaluate the dome lighting project.[45]

Harlan's commission consisted of Joseph Henry, Charles Page, and James Simpson, the latter a lieutenant-colonel in the Army Corps of Engineers. Henry was a logical choice to chair the commission, not only for his expertise in matters electrical, but also because he had been consulted throughout the extension project on acoustics and ventilation, and had even helped test the strength of building materials. Yet serving on the same commission with Page must have been uncomfortable for Henry, since the two had been estranged for many years. It is doubtful that they interacted much, if at all, in carrying out the commission's charge.

After inspecting the lighting work in progress, but 6 months before submission of the final report, Page wrote to Harlan, offering two major conclusions.[46] The first was that "the apparatus and materials furnished by Mr. Gardiner for this purpose are of a creditable and substantial character and give promise of a successful experiment." The second conclusion, from the man who had spent $20,000 in federal funds on his ill-fated electric locomotive, was somewhat surprising: "... the prices charged by Mr Gardiner for materials, fixtures, and other items are much too high, and ... the continuance of purchases at his rates would be inconsistent with public interests and economy." In support of this indictment, he noted that Charles Chester of New York could supply batteries for $10 less than Gardiner's price of $33 per cell.[47] Page also advised that pursuit of the project was justified, but "under such supervision as will save the department from the commission or imputation of extravagance."

The final report, not submitted until April 1866 and compiled of parts written in different hands, included Henry's familiar mantra on the impracticality of galvanic power: "... electricity is the most expensive of all the motive powers ... and therefore should never be employed when a desired effect can be as readily and efficiently produced by other less expensive means."[48] For Henry, economy and efficiency trumped all other performance characteristics, regardless of a technology's political and social functions. Thus, the report insisted, galvanism was an extravagant means to rotate stopcocks, for the gas could have been controlled by hand and lighted mechanically. Follow-on costs for maintaining the battery would, the commission estimated, amount to $5,000 a year—including $2,500 for Gardiner's salary.[49] In witnessing the operation of the lighting system, the report claimed that it took 15 minutes to ignite the burners. The commission cautioned against "further application of electricity ... to light other parts of the Capitol until the efficiency and expense of the present experiment have been fully and satisfactory shown."

French's commission rendered a more positive assessment. The members were "four eminent electricians" who were to visit the Capitol and "examine the work thoroughly, and carefully, and test it practically, and make a report to me."[50] Indeed they were a distinguished group, chaired by Samuel Morse. The other members were Edward Knight (a civil and mechanical engineer who had authored a comprehensive mechanical dictionary), a Professor Nicolas Pike, and the telegraph promoter Taliaferro Shaffner.[51] Their report succinctly described the technology and its operation. It even speculated that, although battery maintenance would annually consume 50 gallons of sulfuric acid, 600 pounds of zinc, and 80 pounds of mercury (for amalgamating the zinc plates), no other apparatus could light the lamps as effectively and economically. As for the system's functioning, the report affirmed that "its efficiency and practical performance leave nothing to be desired; its permanency may well be admitted from the solid and honest character of the work; and the economy with which it performs its duty is beyond dispute." The report mentioned how the demonstration wowed observers: "The workmanship of the dial and appurtenances, and the electro-magnetic engines, is substantial and elegant, and deservedly attracts great attention from experts and intelligent casual visitors, from whom the effect of the manipulation of the keys elicits murmurs of applause."[52] In short, the system worked well and, significantly, was a properly impressive technology for the Capitol. After reading the report, French must have breathed more easily. His confidence in Gardiner vindicated, French recommended that he extend the system to the ground level of the rotunda, using power from the dome battery, and this was done in 1867 at a cost of $3,000.[53]

Before receiving his commission's report, Secretary Harlan had invited a small group to attend a demonstration of the dome lights. Carried out on the evening of January 24, 1866, in the long shadow of Lincoln's assassination and relief at the South's surrender, the display was a dignified counterpoint to the tension and discord in Congress over Reconstruction. Members of the party included Secretary Harlan, Commissioner of Patents T. C. Theaker, Thomas Walter and Edward Clark (past and present Architects of the Capitol), and Joseph Henry. A beaming firsthand account in *Scientific American*, perhaps supplied by Gardiner himself, recorded the occasion. Entering the dark rotunda, the visitors waiting "for the artificial light to dawn for the first time upon the splendid interior . . . were amply rewarded for their journey on a stormy evening, the space above them showing like an immense vault through whose open mouth the heavens were visible, peopled with the fraternizing demi-gods of ancient and modern times."[54] The emergence of Brumidi's inspiring fresco out of darkness, effected by no more than a gentle touch on several keys, was a remarkable spectacle, even to observers familiar with telegraph technology—with the likely exception of Henry.

The conflicting assessments of Gardiner's system doubtless resulted from observations made at different times during the installation's shakedown period. But they probably also reflected disparate agendas at work. Henry was an idealist on the federal payroll who had eschewed profiting from his early electrical inventions and who selectively regarded with disdain men who profited from theirs. In contrast, Morse and Shaffner had prospered by commercializing technology that incorporated some of Henry's electromagnetic principles. Page and Henry, at odds with each other nonetheless shared an antipathy toward Morse.[55]

It seems clear that once the system was pronounced done, federal officials were satisfied with it. In 1867, Benjamin French affirmed that the dome lighting was "a complete success," praising Gardiner as "an honest, upright gentleman."[56] And long after the controversies over the Capitol extension project had subsided, Edward Clark remained a staunch supporter. In 1873, confirming the forecasts of French's commission, Clark penned the following words in a testimonial for Gardiner: ". . . your Electric Gas Lightning Apparatus has been in use at the U.S. Capitol, for about eight years, and has always proved effective. I know of no instance of a failure in lighting. I can therefore recommend it as being safe and reliable."[57] What is more, Gardiner's gas lighter in the dome and the rotunda remained in use for decades, although in 1879 the huge battery was replaced with a dynamo.

_⌁⌁⌁_

Shortly after the dome project was finished, Gardiner and Hudson had a falling out. It was about money, to be sure, but Hudson was also chafing over the disproportionate credit he believed Gardiner was getting for their technology. Although Gardiner held all the patents, had negotiated the contract, and had organized and managed the project, Hudson had made technical contributions all along and had worked on the dome to the point of exhaustion and financial hardship. Aggrieved, he sent Harlan a lengthy and choleric history of the collaboration, requesting that his own name be engraved as the "designer" on a plate to be affixed to the dome-light keyboard; this doubtless did not happen.[58]

Gardiner continued petitioning the government to pay overdue bills. In 1866 he had accepted "with protest" a check for $6,831.91 as final payment for the dome lights. Pressed by Hudson and by laborers threatening to sue for back wages, Gardiner filed a claim with the government for an additional $3,890. Clark did not support this claim, and it was denied.[59] Rebuffed by the bureaucracy, Gardiner petitioned Congress for relief and again got nothing.[60] As late as 1872, he sought federal compensation for the use of his lighting patents.[61]

Whether Gardiner's work as a federal contractor enabled him to earn a respectable middle-class living I cannot say. Although he profited from the sale of every part, received the checks, and decided how to spend the money, he apparently came up short at the end. Yet he did ensure that the lighting system had high-quality materials and excellent workmanship, as befitted a project of national significance. Perhaps his fine work would lead to a $2,500-per-year-sinecure as the "Electrician U.S. Capitol," a position he sometimes affected when signing letters.[62] That was not to be. His endless petitions to the government, seeking payments that he believed were years in arrears, and fending off Hudson's accusations of malfeasance, had palpably weakened his influence in Washington.

Gardiner suffered another setback in 1870, as competition arose for Capitol lighting projects. In a despairing letter to Clark, Gardiner reported that a forged telegram had lured him to New York with the prospect of a new project.[63] Upon his arrival, he had been wrongly arrested. Writing from jail, and requesting a loan of $150 to make bail, Gardiner was certain that this stratagem had been concocted to enable passage of a bill to purchase Wilson's gas lighter for the new Senate chamber. Conveniently, Gardiner was out of town when the vote took place, as were his two most ardent Senate supporters.

Although Gardiner insisted that the Wilson system was unreliable (installed in Steinway Hall two years earlier, it had already fallen into disuse), the bill was passed and another New York contractor, A. L. Bogart, was hired to do the work for $4,500. Bogart completed the project by the end of 1871, and Clark reported that the Wilson system worked well. Moreover, owing to the small battery (whose plates were immersed only during use), the installation would be inexpensive to maintain. Of course the gas had to be turned on manually, but it was still an electric gas lighter. Soon Bogart was proposing to replace the Gardiner system.[64]

In the early 1870s, demand was growing for electric gas lighters, and more competitors were entering the business, able to buy off-the-shelf components from several manufacturers, eventually including Western Electric.[55] These products enjoyed sales well into the early twentieth century, suggesting that many people were favorably disposed to lighting gas lamps by battery-supplied electricity. But, like the Wilson system, most electric gas lighters used spark ignition and operated the valves by hand.

As tourists made their way to Washington, their guidebooks extolled the Capitol's architectural and artistic triumphs and called attention to the cost of this or that feature. The million-dollar dome was touted as the largest and most beautiful in the world. Of course no tour of the dome was complete without stops at Gardiner's battery room and keyboard. There they could view, according to Townsend's guide of 1873, the "strongest battery in the world," which powered "an almost miraculous apparatus." Press the keys and "at a wink the great hollow sphere is aflame. You can see the spark-spirit run on tip-toe . . . , planting its fire-fly foot on every spear of bronze." But, the blurb continued, "thirty thousand dollars is dear even for a miracle.'[56]

_⌒⌒⌒_

The descriptions of Gardiner's gas-lighting technology in the Capitol dome foreshadow, in some respects, Thomas Edison's later electric light and power system. If the coils had been energized in the absence of gas, which was doubtless done many times in the course of putting up and testing the system, they would have glowed in the dark—1,050 incandescent lights. Even more suggestive was the parallel structure of pipes and wires that delivered gas and current to each burner. Although Edison was only 19 years old when the lights went on in the dome, perhaps in the late 1870s the Wizard of Menlo Park, a *Scientific American* subscriber, perused back issues and drew inspiration from Gardiner's technology. After all, Edison did acknowledge that his electrical distribution system was modeled on that of gas lighting, a juxtaposition that Gardiner and Hudson had previously created.[67]

And there is another interesting parallel to be drawn between the work of Gardiner and Edison. In 1858, Gardiner and Levi Blossom invented an incandescent lamp with a platinum element. A glance at the patent drawings suggests that they had come up with a prototype light bulb. However, the text indicates that the platinum coil was placed inside a signal lamp, an enclosure having internal reflectors and sides of colored glass. Clearly, the ventilation holes in the glass would have precluded a vacuum—and none is mentioned. The only novelty claimed for the "improved electric signal-light" was the placement of a platinum coil in a lantern, which could be employed for signaling on ships, military installations, buoys, and even lighthouses.[68]

Platinum elements had long been used to demonstrate electricity's heat- and light-producing effects. The problem was that, if white heat was sustained, the metal would melt. However, if the Gardiner-Blossom lantern had been used as anticipated, only in short bursts for transmitting a code, then melting might have been prevented. This curious lantern was perhaps the first technically competent design for an incandescent lamp, albeit one intended for a very specialized use.

19 Machine-Age Electricity

The course of technological change often takes weird twists and turns. Sometimes a cutting-edge technology is an utter flop in the marketplace, as the first picturephones were; sometimes a technically undistinguished product, such as the Walkman, can be embraced by millions of consumers and achieve iconic status. And sometimes even a component invented for a dead-end or even a discredited technology can become a part of an important system.

An invention reminiscent of Paine's Hydro-Electric Light led to the development, in the late 1850s, of the first steam-powered magnetos used in electric lighting. Commercialized in England and France, they were installed in a handful of lighthouses in both countries and served well. However, other countries adopted electric lights in only a few lighthouses or not at all. This case of "differential adoption" offers an opportunity to examine how institutional consumers evaluate a new technology's practicality.

The giant magnetos developed mainly for lighthouses did enjoy some successes in other applications, gradually reducing the exclusive dependence of the electric light on batteries. However, the greatest effect of this new technology was to alert inventors and manufacturers that a vast potential market might be tapped if only they could develop generators that were more efficient, more compact, and less costly. The result, before the end of the 1860s, was the invention—by many people in many countries—of the "dynamo-electric machine" or "dynamo," which met these new performance requirements.

The dynamo appeared to render the electric light capable of challenging other illuminants in potentially lucrative applications: factories, public buildings, street lighting, and, of course, lighthouses. Dynamos also seemed able to substitute for batteries in electrometallurgy and telegraphy. With large markets anticipated, dynamos were rapidly commercialized by numerous firms.

The story of magneto-powered lights begins in the middle of the nineteenth century with Floris Nollet, a descendant of Jean-Antoine Nollet, Benjamin Franklin's nemesis in electrical theory. A physics professor in Brussels, Nollet (1794–1853) was one among many people experimenting in mid century with new illuminating technologies that might compete with gas and oil lamps. He was particularly taken with the possibility of using the brilliant limelight in lighthouses. For such a scheme to be at all feasible, Nollet would have to produce, on site, copious quantities of oxygen and hydrogen, whose combustion would render limestone luminous.[1] To evolve these gases in abundance, Nollet planned to use a powerful magneto for decomposing water. In collaboration with the engineer Joseph Van Malderen, he designed a number of large magnetos, yet his first design was hardly novel.[2] In fact, the mechanical arrangement of its parts—two large permanent magnets with coils rotating between them—was, like Charles Page's telegraph magneto, simply a Saxton magneto times two.

In 1853 Nollet patented a huge magneto that promised to generate greater power by placing together, on the same shaft, five dc compound magnetos similar to those John Woolrich had built in England for electrometallurgy.[3] This would be termed a "five-disk" machine, which denoted the five radially arrayed, magnet-coil complexes. The use of many coils in each disk contributed to a smoother output of current—an advantage for electrolysis.

Although Nollet died before acting on his plan, a company was formed to exploit his patents and refine the magneto's design.[4] The result was the development of the world's largest and most powerful electrical generator, which was driven by a steam engine. Despite this technological achievement, the company expired, laid low by the accusation that it had bilked investors, including Napoléon III, through the fraudulent lure of cheap limelight. But all was not lost, for a new firm, the Alliance Company, was founded in 1855 to refine the Nollet-Van Malderen magneto and explore possible markets.[5] The following year, one machine was installed in the gas-lighting plant of a Paris hospital for generating hydrogen and oxygen, but this application apparently fizzled.

To attract businessmen, the Nollet-Van Malderen lighting system presumably would have to perform better in utilitarian applications than gas and oil lamps. To its detriment, the electrolysis system involved costly and complex apparatus, and it was doubtless much more expensive to operate and maintain than conventional lights; its only performance advantage was the limelight's brilliance. And critics pointed out that if coal were used to produce gas rather than to fuel the steam engine, more usable light probably could be obtained. Nollet's lighting scheme was, at best, a curiosity that the commercial world could and did ignore. There was, however, another way to use a magneto for furnishing light.

Seeking advice on potential applications for the huge magnetos, the Alliance Company consulted Frederick Holmes, a British chemistry professor and engineer.

Holmes suggested a much more plausible way to use them in a lighting system: they could provide current for arc lights. Adding to this plan's attractiveness was the availability, from several instrument makers, of arc lamps with built-in regulators that had already proved their worth in battery lights. For several years, however, the Alliance Company did not act on Holmes's advice.

In England, Holmes had also been designing large, multi-disk magnetos, and he was firmly committed to using them in arc lighting. Already in 1853, not long after W. Edwards Staite failed to commercialize his battery lights, Holmes had shown the fitness of magnetos for this task. His plan was to perfect, manufacture, and sell the large machines to Trinity House, and in 1857 Holmes made his first overture

Michael Faraday, still scientific adviser to Trinity House, knew that a magneto had performance advantages over a battery in powering arc lights, but doubted that the electric light could best the traditional oil lamp in all-around performance. No doubt its first costs, including purchase of the electrical equipment and steam engine, would be steep; and it would require more space and more workers. Despite these misgivings, Faraday was willing to give Holmes's system a fair trial.

_⌒⌒⌒_

The Holmes magneto incorporated 120 permanent horseshoe magnets, each weighing 50 pounds, and 160 coils.[6] An image from a London exhibition in 1862 (figure 19.1) suggests that Holmes's machine stood at least 3 meters tall, but later models were slightly more compact. The magneto furnished direct current for one arc lamp, light from which was projected through a small Fresnel lens (the beehive-like object visible in the figure).

Perhaps to keep the machine from shaking itself to pieces and to prolong the life of wear-prone parts, Trinity House insisted that the armature spin at a leisurely 100 rpm.[7] In generating sufficient power for an arc lamp, under the constraint of low speed, Holmes's magneto had to rely on size and brute force. However, at the dawn of generator-powered lighting it was part of a state-of-the-art electric light.

In May 1857, Holmes exhibited his system at Trinity House's Blackwall wharf and lighthouse. With Faraday's blessing, Trinity House approved an installation in the upper lighthouse (the "high light") at South Foreland, on the cliffs near Dover where, years earlier, electrical blasting had helped to build a roadbed for the Southeastern Railway. The light's initial trial began in early December 1858; it was discontinued for several months to allow modifications, but was resumed the following March. Involved in these trials, Faraday stated that they had "established the fitness and sufficiency of the Magneto-Electric Light for Lighthouse purposes, so far as its nature and management were concerned."[8] Despite this technical success, and despite endorsement by its scientific expert, Trinity House itself objected to additional electric lights on the

Figure 19.1
Holmes's magneto-powered lighthouse light. Adapted from frontispiece of Holmes 1862. Courtesy of Dibner Library, Smithsonian Institution.

grounds of their great capital expense, but was overruled by the Board of Trade. Thus, Holmes's systems were installed in both lights at South Foreland, a second pair was added at Dungeness, and a single light was established at Souter Point.[9]

By the time of the Souter Point installation, Trinity House permitted Holmes to run his magnetos at 400 rpm. Smaller than the original South Foreland machines, each magneto still had a footprint of 2.4 square meters, stood 1.7 meter tall, and weighed 3 tons.[10] Because this machine produced alternating current, both carbon rods wore down at about the same rate.

Across the English Channel, a similar adoption pattern transpired in France. After 3 years of tests using Alliance magnetos, a lighthouse at La Hève was electrified in December 1863.[11] Two years later, the second lighthouse at La Hève received an electric light, and in a few more years another was installed at Gris-Nez. The Nollet-Van Malderen magneto had at last found permanent employment. And, unlike its English counterpart, the French Lighthouse Commission embraced the electrical illuminant with enthusiasm.

Trials of arc light systems in England and France were watched with interest in America. *Scientific American* published accounts of the European goings-on and highlighted the favorable technical findings.[12] In addition, Joseph Henry, chairman of the U.S. Light-House Board from 1871 to 1878 and, since the board's inception in 1852, head of its active Committee on Experiments, sought the latest information on electric lighting. Although informed about European activities and conducting his own experiments with generators, Henry did not advocate the electric light or any new illuminant that failed to meet stringent performance requirements. Henry laid these out explicitly in responding to an inventor's query: "First the light must be of sufficient brilliancy. . . . Second. It must be persistent so as to continue in full force for at least sixteen (16) hours. Third. It should be economical, not exceeding in cost the same amount of light as derived from lard oil. Fourth. It should not require more skill or cost in the way of attendance than the light now in use."[13]

In Henry's view, the arc light fell short. His summary judgment was unequivocal: "I can say from a theoretical as well as practical knowledge of the whole subject that in the present condition of the invention it [the electric light] cannot be substituted for oil-light."[14] The American scientific authority had spoken once more, but his views clashed with those of French lighthouse scientists who were no less versed in the theory and practice of lighthouse illumination. Clearly, these opposed judgments stemmed not from alternative physics in the two countries but from assigning different weights to the electric light's performance characteristics.

Seeking firsthand information on European developments, the Light-House Board dispatched George Elliot of the Army Corps of Engineers on a tour of European lighthouses. Elliot's detailed technical account included cost data on various illumination systems in England and France. He claimed that these countries "are in advance of us in using both the gas and electric lights in positions of special importance."[15] Although he recommended that electric lights be tried in a few American lighthouses, Elliot concluded that the greatest recent advance was the use of kerosene in place of colza (rapeseed) or animal oil.

Elliot's report, which was favorably covered at length in the *New York Times*, more than implied that the American system of navigation aids was inferior to those of France and England.[16] The Senate took notice and asked the Appropriations Committee to investigate. The committee's report asserted that the "Board does not investigate and keep pace with the improvements made in light-house illumination."[17] One example cited was the board's reluctance to construct any experimental electric lights. Henry had heard this accusation before, but was unmoved. Apart from the great expense of electric lights, he contended that in a dense fog no artificial light could penetrate very far. Therefore, he argued, "Recourse must . . . be had to some other means than that of light to enable the mariner to recognize his position on approaching the coast when the land is obscured."[18] And that is why Henry devoted his attention to refining acoustic aids such as foghorns and whistles, which were deployed in large numbers by the American board—and by Europeans. With Henry as Chairman until his death in 1878, the Light-House Board ignored Elliot's suggestion to outfit a lighthouse with an arc lamp.[19]

rnnn

Whereas Holmes seems not to have successfully promoted his magneto-powered lights to consumers beyond Trinity House, the Alliance Company aggressively sought new markets, and sometimes found them.[20] Elite consumers were alerted to the possibility of substituting magnetos for batteries in temporary installations such as the French emperor's winter festival. An early purchaser was the Societé Générale d'Electricité, a pioneer in street lighting. With the magneto one could also envision relatively permanent installations—a sales pitch that was especially effective with the French monarchy. In 1867, the yacht of Prince Napoléon was outfitted with an Alliance system whose beacon brightened the vessel's path, making it possible to enter dangerous ports such as Casablanca at night.[21]

Not to be outdone, the Austrian emperor also bought a system for his yacht. On one voyage, he steamed through the Suez Canal, "lighting marvelously the shores."[22] This vessel may have been guided to the canal's Mediterranean entrance by its lighthouse, in which an Alliance electric light had been working since the canal's opening in 1869.

An engineer of the steamship line Compagnie Transatlantique became aware of the Alliance light, whose purchase had been promoted as a way to lower insurance rates. Purportedly, ships outfitted with bright beacons would be less likely to collide with obstacles, the shore, and other ships. On the basis of this premise, nearly a half-dozen of the company's ships were supposed to carry Alliance systems. However, most captains resisted the new technology, complaining that it was "a cumbrous luxury, expensive, awkward in use, and of doubtful efficacy."[23] And insurance companies did not immediately lower their rates. Unhappily, the *Ville du Havre*, a vessel of the Compagnie Transatlantique that lacked the electric light, was struck in 1873 by a smaller vessel and sank.[24]

In New York, the steamship *St. Laurent* tied up at Pier 50, where it became a spectacle. A new vessel of the Compagnie Transatlantique, the *St. Laurent* was equipped with an Alliance lighting system, driven by a small steam engine. The system cost $3,000. An account in *Scientific American* indicated that the arc lamp could be used as a searchlight: "The light is displayed . . . through a Foucault lens, which can be turned by hand in any direction, placed on the bridge above the deck." In one playful example, the light was directed at a vessel in the middle of the Hudson River, bathing "her in a luminous halo, which enables the observer not only to see everything upon her deck, but also to fancy that he can detect the astonishment depicted upon the faces of the crew at being thus suddenly suffused with an illumination so powerful.' In reporting these goings-on, the *New York Times* noted that many maritime visitors admired the light, but that its steep price "will for a time prevent its general adoption on steamships."[25]

An Alliance system also debuted on a French dock, furnishing light for workers repairing ships. It was so bright that flags were seen on a vessel 700 yards away, and a diver 20 feet below the surface could read the scale on a ruler. Quoting a French source on this application, *Scientific American* stated: "It is now established that an electro-magnetic machine may be permanently fixed to light large workshops, submarine works, and narrow passages into harbors."[26]

In 1865, another marine application was undertaken: fishing at night by magneto light. Under the supervision of the engineer Ernest Bazin, an Alliance magneto was placed aboard the ship *Andalouse*. Near a pier on Belle Isle, the ship lowered an arc light into the sea. Fifteen hundred spectators watched from the pier as schools of fish swarmed to the brilliant light, easy prey for fishermen lowering their nets.[27] With magneto light Bazin also carried out underwater photography and searched for the Confederate steamer *Alabama*, which had been sunk by the U.S.S. *Kearsarge* off the coast of Cherbourg. I have not been able to ascertain whether any of Bazin's underwater applications were commercialized, but his exhibitions were good publicity for Alliance and for electric lighting in general.

It did not escape the notice of continental governments that magneto light might supersede or augment battery light in warfare, both in defensive and offensive operations. And so, in addition to the battery-powered lights that the French deployed in the defense of Paris during the Franco-Prussian War, the army used a large Alliance magneto on Montmartre to turn night into day on the Argenteuil plateau (figure 19.2; magneto at lower right).[28] When not scanning for German soldiers, the light drew Parisians in droves, and Bazin obligingly explained how it worked. Although the Montmartre light once foiled a Prussian thrust, Paris and the war were lost to a Germany that was united under Bismarck. Russia also bought a number of Alliance systems, apparently for military use.

St. Ignatius College in San Francisco (now the University of San Francisco) purchased an expensive Alliance system. The instigator was Father Joseph Neri, Professor of Natural Philosophy, who had ample funds with which to buy expensive apparatus. Father Neri first displayed an arc lamp on Pope's Day, 1871, from the College's office on Market Street, but I suspect that this light was battery-powered.[29] In 1874, however, he exhibited the Alliance system in the college's cupola. According to one account, "the light flashing across the sky first astonished, then delighted those who were on the streets at the time." This spectacle was achieved at a princely price "Machine, $2,500; regulator and mirror, $200; Fresnel lens, $500; four-horse-power engine $750;

Figure 19.2
A magneto light used in the Siege of Paris. Source: Guillemin and Thompson 1891: 889.

incidental expenses, $250. Total, $4,200"—that is, about six times the annual salary of a skilled laborer.[30] Father Neri's exhibitions fostered great interest in electricity among San Franciscans, but it confused some mariners because the light, vastly brighter than any lighthouse lamp in America, was not on the shore.

Sundry sales to wealthy individuals and government agencies sustained the Alliance Company throughout the 1870s and enabled it to continue remodeling the magneto.[31] In 1872 Alliance was selling for £320 a four-disk ac machine, driven by a 2.5-horsepower steam engine, that ran at 350 rpm.[32] Around 1880 Alliance was reorganized, rising again as de Méritens, which continued to refine and sell to the French lighthouse organization magnetos descended from the Nollet-Van Malderen design. (A lighthouse required two or even four magnetos, each with five to seven disks.) By this time, however, the modest market for hefty magnetos contracted even more in the face of competition from dynamos, development of which began in earnest in the 1860s.[33]

Although its applications were tightly circumscribed, the magneto-powered electric light's commercial success was a milestone. In the early 1860s, lighthouse and other applications announced to all in the know that a machine—admittedly cumbersome and costly—could replace the hated battery for electric lighting. It was equally obvious, however, that if electric lights were to reach the vast commercial, industrial, governmental, and military markets whose pent-up demand for bright lights had been signaled by trials with batteries and magnetos, generators would have to be redesigned to meet new performance requirements. The multi-ton magneto was a technology too massive, expensive, and inefficient to satisfy the lighting needs of so many varied consumers. In addition to lighting applications, inventors were stimulated to design new generators by the prospect of replacing the tens of thousands of big batteries used in electrometallurgy and telegraphy. Perhaps great riches might accrue to the person whose generators could penetrate these large, long-established markets.

Although experimenters had been churning out magneto designs since the 1830s, the technical adequacy of the magneto-powered light gave inventors a new urgency and a new focus. Now it was clear which performance characteristics of large generators had to be improved.[34] The flurry of new designs that followed the first electrically lit lighthouses was, I suggest, the result of purposeful calculation—indeed, raw opportunism—that required no new physics, much less a heroic inventor. Not surprisingly, during the mid 1860s an appropriate technology—the dynamo—was invented independently many times.

In improving the magneto's performance characteristics, inventors could exploit several design principles first embodied in earlier one-off machines. In addition, they had to devise some novel design principles that drew upon electrical *and* mechanical expertise. Although the developmental distance between magneto and dynamo was not insubstantial, by the mid 1870s dynamo design had almost become routine engineering. Moreover, many manufacturers in Western countries, some made wealthy by the telegraph, had the skills and the resources to commercialize dynamos—and did.[35]

One way to reduce a generator's size and weight was to replace bulky permanent magnets with electromagnets. Decades earlier, Joseph Henry and William Sturgeon had shown that, pound for pound, an electromagnet exhibited greater force than a permanent magnet. In 1845 Charles Wheatstone had experimented with a generator that used battery-powered electromagnets, but his hybrid technology attracted little interest.

In 1864 the Englishman Henry Wilde, a self-educated electrical inventor and manufacturer of telegraph magnetos, took another tack.[36] He mounted a small magneto atop a much larger one, and outfitted the latter with electromagnets; both were propelled with belts driven by a steam engine (figure 19.3). The small magneto fed power to the large one's electromagnets, which created a strong magnetic field that boosted the generator's power output.[37] Because the small magneto consumed power, some people feared that it would be inefficient, but this understandable concern proved to

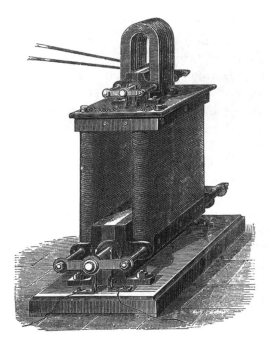

Figure 19.3
Wilde's early dynamo. Source: Alglave and Boulard 1884, figure 150.

be unfounded.[38] Wilde brought his dynamos to market in 1866, offering them in several varieties with armatures designed for electroplating or arc lighting. By 1877, he had sold 105 machines, more than 90 percent of which were used for electro-metallurgy.[39] Important electroplating customers included Elkington in Birmingham and Christofle in Paris.[40] In New York, the publisher Frank Leslie used a large Wilde system to electrotype the printing plates for his *Illustrated Newspaper*.[41]

In short order, experimenters in several countries, including Henry Wilde, figured out how to dispense with the external magneto. To achieve this economy, one could exploit an occasionally troublesome property of electromagnets: even after the current is turned off, the soft-iron core retains a small amount of magnetism. This residual magnetism in the field electromagnets was nonetheless sufficient to induce a small current in the revolving armature. The trick, then, was to augment the residual magnetism by connecting the generator's output to the field electromagnets. As the field strength increased in response, so did the output power, until reaching a maximum when the armature was rotating at full speed. This design principle is known as "self-excitation."[42] Commenting on self-excitation, *Scientific American* marveled at the efficiency of such machines: "This is a great result in its practical bearings, particularly

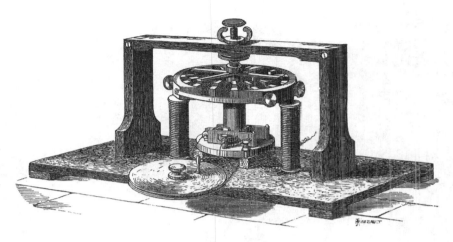

Figure 19.4
A Pacinotti dynamo. Source: Dredge 1882: 127.

on the advantageous conversion of mechanical power into light . . . It ought not to be long indeed, before night will become merely an optional indulgence."[43]

The dynamo also incorporated new armature designs. In a typical magneto, each coil on the armature was wound on a separate iron core. This arrangement was far from compact, especially with large numbers of coils ("windings"). Werner Siemens, founder of the German telegraph firm Siemens & Halske, and Antonio Pacinotti, an Italian physics professor, both devised compact armatures on the basis of a new design principle: winding independent coils on the same iron core (figure 19.4). With the addition of a rigid and durable axle, the coils and the core became the armature; there was no superfluous framework. Once that design principle was grasped, inventors created many variations by altering the armature's shape, number of windings, and so forth. Thus, the Siemens armature had a torpedo shape, whereas Pacinotti's was a ring. However, the form that was replicated most often in the years ahead resembled a barrel.[44]

An equally significant contribution to armature design came from Zénobe Gramme, a Belgian who had been trained as a carpenter and had made models for the Alliance Company.[45] After becoming interested in electrical things, Gramme went to Paris and worked for Heinrich Rühmkorff. There he began designing his own dynamos, at first mainly for electrometallurgy. In 1870 he patented a ring armature. What made Gramme's armatures especially noteworthy, as they became more barrel-shaped, was their many dozens of windings, which produced relatively constant direct current. Gramme first used his armatures in hand-cranked magnetos suitable for demonstrations, but was soon placing them in dynamos (figure 19.5). The Gramme dynamo was a piece of clever engineering that was widely praised, as in Alfred Niaudet-Breguet's

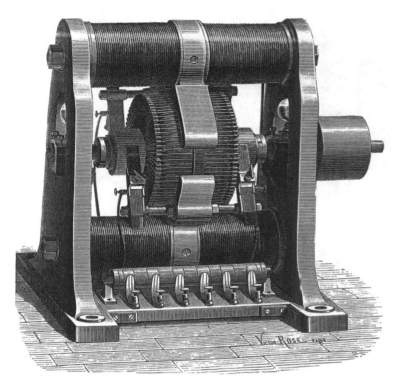

Figure 19.5
A Gramme dynamo. Source: Dredge 1882: 159.

statement in *The Telegraphic Journal*: ". . . the modifications recently introduced by M. Gramme are so original and so important that they constitute in themselves a work able . . . to revolutionise a host of industries."[46]

The principle of self-excitation and the operation of the Gramme armature both at first seemed somewhat mysterious.[47] However, these phenomena were explained almost immediately and shown to be in conformity with known laws, and so the dynamo received the blessings of scientific men. However, when Thomas Edison later announced his high-efficiency dynamo, scientists pounced on him for allegedly violating a venerable electrical principle.

The elimination of permanent magnets, with or without self-excitation, eventually marked the shift in terminology from "magneto" to "dynamo" (at first, "dynamo-electric machine," in Werner Siemens's phrase), although for a while "magneto-electric machine" designated all generators. Technologically, the transition was one of discrete changes that, in various combinations, resulted in many novel designs. Beginning in the early 1880s, the publication of a new genre of technical books, with page after page of recondite equations, showed that the design of dynamos (and motors) had become a formal engineering science, the province of university-trained specialists, many working in large corporations.[48]

The magneto's rapid evolution into the commercial dynamo testifies in this case to the strong pull of anticipated demand. Inventors, manufacturers, and investors all judged dynamos practical and believed that consumers would follow suit. However, if these forecasts were to be validated in the marketplace beyond purchases by wealthy people, companies, and government agencies, then the dynamo's price would have to descend far below that of equal-power magnetos. This, too, came about quickly because of the tremendous growth after mid-century of factories able to manufacture industrial products—from steam engines to machine tools—in large quantities. Economy of scale would soon make dynamos nearly ubiquitous in a process that was under way well before Edison's first light bulb glowed.

_⌒⌒⌒_

W. James King, an authority on the history of generators, credited Zénobe Gramme with being "the first to make the dynamo a success commercially."[49] Although that claim is not sustainable (Wilde and Siemens sold many dynamos before him), Gramme did establish a prosperous firm with the financial backing of one Count c'Ivernois. In late 1872, after a delay occasioned by the Franco-Prussian War, Gramme placed his first offerings on the market.[50]

The Gramme machine came in two models, one with low voltage for electrometallurgy and the other with higher voltage for arc lighting. The low-voltage machine weighed 750 kilograms (about 1,650 pounds), stood 1.3 meter high, and had a footprint of 0.8 square meter. Driven by one horsepower at 300 rpm, the dynamo was said to generate electricity at 40 percent efficiency. The high-voltage dynamo was even more impressive, producing four times as much light as a six-disk Alliance magneto. To achieve this output, the dynamo required 4 horsepower and operated with an efficiency of 50 percent. Weighing about a ton, it cost £400 in England, far less than Father Neri's enormous magneto.[51]

Gramme's dynamos were shown at the International Exhibition of 1873 in Vienna, capital of the Austro-Hungarian Empire. For many countries this trade show was a rehearsal for the extravagant Centennial Exhibition to be held 3 years hence in Philadelphia. As it turned out, a fateful "accident" made it possible for Gramme dynamos to exhibit more than their aptness to supply copious current. In its conventional telling, the story goes that, by chance, a worker connected the output wires from two dynamos together. When steam power was delivered to one dynamo, the second one's armature began to rotate rapidly: it had become a powerful motor. Gramme immediately exploited this startling occurrence by adding a new feature to the display: a dynamo, receiving current from another dynamo located nearly a mile away, was employed as a motor to drive a pump that fed a small waterfall.

One might reasonably doubt that this event was really an accident; after all, many electrical experimenters, from Moritz Jacobi onward, knew that motors and generators were reversible.[52] (In the motor mode, the commutator functions as a pole changer.) Yet, whether by happenstance or design, this arresting display made two momentous proclamations—beyond the obvious ability of one electric plant to supply current for both lights and motors. The first was that, by merely tweaking a dynamo's design, firms could now readily manufacture motors capable of heavy commercial and industrial work. The second was that it was possible to transmit power by electricity, perhaps over long distances. Thus, where mechanical power was cheap or conveniently produced, it could be converted by a dynamo into electricity, sent over wires, and turned back into mechanical power. Preliminary calculations indicated that these conversions could be made with efficiencies that might be deemed practical by investors and potential consumers.[53]

The possibility of using electricity to transmit power was recognized at a propitious time. As the pace of mechanized manufacture accelerated in many industries, engineers, industrialists, and military men became interested in power transmission.[54] In the United States there was a growing awareness of the vast power that might flow from Niagara Falls and other unexploited water resources. Many schemes were put forward for sending power over short and long distances, such as pipes containing compressed air or water, long shafts, and "wire ropes" (steel cables). In one ambitious example, the U.S. Army constructed a system to deliver water power by cable. Designed by Colonel D. W. Flagler, the project was located at the Rock Island Arsenal on the Mississippi River in Illinois. It included dams, a canal, and enormous turbine wheels. Driven by the turbines, a 15-foot pulley carried a continuous-loop cable that snaked its way overhead to the arsenal's shops nearly a half-mile away. Operational in 1879, the system drove the arsenal's machinery for two decades before being replaced by a hydroelectric plant.[55] In the decades ahead, electrical transmission of power would become a cornerstone of the modern world.

Suddenly the unfavorable calculus of electric power, which James Prescott Joule, Robert Hunt, and other scientific authorities had honed so finely, underwent a dramatic shift. At last there was a source of electricity vastly cheaper than batteries, and there were motors that decidedly were not philosophical toys. In the family of electrical technologies, the motor was no longer a pariah. Soon scientists, engineers, investors, manufacturers, and consumers would judge it to be eminently practical for many long-foreseen applications.[56]

In a promotional piece, an English manufacturer of electric motors set forth "the great value of electro-motors," listing four advantageous performance characteristics: it was portable, which made it possible "to bring the machine-driven tool to the work,

Figure 19.6
Electric plowing, 1879. Source: Hospitalier and Maier 1883: 443.

instead of, as at present, the work to the tool," it had a low weight-to-power ratio, it could be used at a distance from the prime mover, and it had the ability to propel machinery using high-speed direct drive, thereby avoiding losses from the mechanical transmission of power.[57]

In appreciation of the newfound potency of electric motors, people engaged in new trials (often quite varied) to replace other motive powers. At Marne, the French engineers Chrétien and Felix plowed a field with a plow pulled by Gramme motors (figure 19.6). Although it received attention in print media, this event was little more than a stunt that called attention to the capabilities of electric power and the ingenuity of Chrétien and Felix.[58] In a more serious application, Chrétien harnessed an electric motor to a crane that unloaded sugar barrels from boats at a claimed saving, relative to conventional practice, of 30 percent.[59]

⎯⌒⌒⌒⎯

The Centennial Exhibition of 1876 was an opportunity for Americans to flaunt—to themselves and to the world—their many industrial achievements, such as locomotives, tools and machines for agriculture and metalworking, finely crafted surgical instruments, and precision watches and surveying instruments. The centerpiece of the exhibition was a 1,000-horsepower Corliss steam engine. Driving the other machines in the Hall of Machinery through shafts and belts, this chugging behemoth epitomized the prime mover that helped to underwrite the industrial revolution. One of the fair's curiosities was the Statue of Liberty, represented only by the upraised arm and torch, which was on display to raise funds for the statue's erection in New York Harbor and which soon would hold an electric light. Another hall featured electrical exhibits, less monumental but still captivating, including an array of telegraphic apparatus, burglar alarms, electric clocks, and Alexander Graham Bell s telephone.

Gramme presented an entire electrical world writ small: an integrated system that included a steam-powered dynamo that furnished current for a motor (again driving a pump), electroplating, and the electric light. The latest Gramme dynamos were more lightweight, compact, and cheaper than the first models, yet achieved comparable power by operating at higher speeds (500–900 rpm). The new lighting dynamo, for example, took up scarcely more than a cubic meter.[60] With the widespread availability of steam power, these were machines that a factory or municipal building or ship could readily accommodate.

Thomas Edison, sharing a big booth with Western Union, exhibited a sample of his electrical inventions. The panel of distinguished judges, including William Thomson and the aged Joseph Henry, gave Edison awards for "his automatic telegraph system and his electric pen and autograph press."[61] Whether or not Edison himself visited Gramme's display, the American inventor would have soon known about the French electric light and power system. After all, by 1877 Gramme had sold 350 dynamos, and more than 1,000 by 1879.[62] Prices of the latest models, all of which could power arc lights, now ranged between £80 and £360.[63] Gramme's major investor, Count d'Ivernois, must have been pleased, for the anticipated markets had emerged.

Other manufacturing firms were contributing to the dynamo's success in the marketplace, forcing Carré, a major maker of carbon rods, to expand manufacturing capacity.[64] Edison, however, was not yet selling dynamos—or light bulbs. That would soon change. In August 1878, Edison began his experiments with electric lights; a few months later he bought Gramme and Siemens dynamos to test their performance characteristics.[65] The prolific inventor had now committed to creating an electric light and power system that could vanquish the avaricious gas interests and leapfrog arc lights. In the next several years this project would become an all-consuming enterprise.

Although the first arc-lit lighthouses in England and France were working well, beckoning ships to harbors or warning them away from hazards, the adoption of electric lights as a lighthouse illuminant was proceeding at a glacial pace.[1] In 1879 there only 10 arc-lit lighthouses in the world; at this technology's peak popularity in 1895, long after the advent of dynamos, the total was fewer than 30.[2] The distribution of electric lighthouses among countries was quite uneven: France and England together had about 20, a few had one or two, but most had none.

Why did a handful of countries, including the United States, adopt very few electric lights? Why did England and France install arc lights in some of their most important lighthouses? And why did the vast majority of lighthouse organizations judge the electric light, the brightest of all artificial illuminants, to be impractical for them?

In chapter 13, I contended that forecasting the behavior of consumers is the domain of social and behavioral—not physical—sciences. I now broaden that claim, seeking to show with the lighthouse case that crafting *explanations* of past decisions to adopt or not adopt a technology is also beyond the reach of physical science principles.

In explaining adoption decisions, the focus is on the performance characteristics of a new technology and its competitors in relation to the performance requirements of specific activities. Thus, our first task is to infer the relevant performance characteristics and the information about them that influenced consumers' decisions. The early trials of arc lights were amply reported in technical publications and so supplied lighthouse organizations—the consumers—with seemingly reliable data on performance characteristics. A lighthouse organization could compare the electric light with conventional oil lamps, weight the performance characteristics according to its own priorities, and then decide whether to adopt the new illuminant. By exploiting the information available to lighthouse organizations and inferring the latter's performance requirements from external factors (e.g., economic, political, and social), we can craft an explanation of past adoption decisions.

Table 20.1
A performance matrix for lighthouse illumination ca. 1860–1899, with + meaning "meets basic performance requirements" and – meaning "does not meet basic performance requirements." Source: Schiffer 2005b.

	Electric	Oil
Acquisition of components and installation of system		
Components commercially available	+	+
Can be installed in lighthouses anywhere	–	+
Easily installed in existing lighthouse structures	–	+
Affordable "first costs"	–	+
Existing expertise adequate for designing and installing system	–	+
Functions during use		
Yields whitest, brightest, most penetrating light	+	–
Produces sufficiently steady light	+	+
Long outages are avoidable	+	+
Does not cast confusing shadows	–	+
Avoids blinding mariners	–	+
Symbolizes special concern for safety of ships and sailors	+	–
Symbolizes a nation's wealth and political power	+	–
Symbolizes modernity	+	–
Symbolizes scientific or technological prowess	+	–
Operation, regular maintenance, repairs		
Operable with traditional staff of keepers	–	+
Operable without complete backup systems	–	+
Breakdowns easily repaired	–	+
Affordable operating expenses	–	+
Easy to administer	–	+

Performance characteristics can be conveniently organized by the major activities of a technology's life history, such as purchase and installation, use, and operation and maintenance. In a "performance matrix" we can juxtapose the competing illuminants, using + and – signs to indicate whether performance characteristics met basic requirements (table 20.1).[3] The contrast in the performance patterns of electric lights and oil lamps is striking.

Installation activities highlight some of the arc light's liabilities. A great deal of roofed space was needed to house the magnetos (and, later, dynamos), steam engines and their accessories, fuel and water, and supplies. These were formidable demands: a boiler consumed 500–2,000 pounds of fuel and more than 400 gallons of fresh water nightly. Also needed were a workshop stocked with tools and housing for additional workers, such as a mechanic or an engineer and a stoker.[4] These space requirements

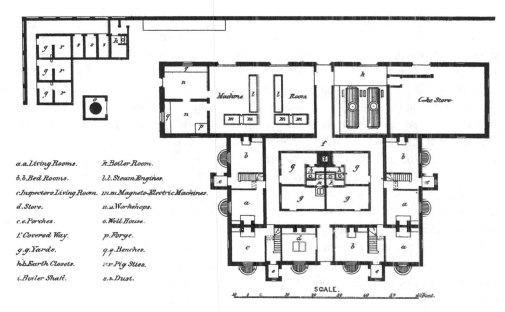

Figure 20.1
Floor plan of South Foreland structure. Source: Eliot 1874, plate II.

could not be met in all places, especially in lighthouses atop small rock outcrops at sea, and usually required new buildings. (See figure 20.1.)

To ensure that an electric lighthouse was never darkened for long by mechanical or electrical failure, a second arc light system was installed, and an oil lamp was retained as an emergency backup. Thus, the "first costs" of retrofitting a lighthouse with an arc light were considerable: In England, South Foreland cost $82,270 and Souter Point $75,740.[5]

In normal operation and maintenance activities, electric lights performed poorly. Failures were not common; however, if one did occur, the repairs might require shipment of the malfunctioning generator or regulator to a distant factory.[5] Oil lamps were eminently reliable, so long as the keeper trimmed the wicks, saw to the proper oil level, and kept the reflector or the lenses clean.

Before the mid 1870s, arc lights could flicker; occasionally the light went out, but it could be quickly restored. Changing worn-down carbons at intervals of 1–3 hours caused only the briefest interruption, because each light had two lamps. When the carbons in one lamp needed replacement, the second one was rapidly swung into position and turned on.

Faraday noted that, in comparison with an oil lamp, learning to operate an arc light was much more difficult, and the maintenance was far more complex. For example, keepers required special training far beyond that needed for oil lamps; administering a system having so many parts and regular inputs was taxing; and tenders (ships that periodically resupplied lighthouses) had to carry and unload more supplies. The electric light might have been a technical triumph, but it was an appreciable departure from the oil lamp's consummate simplicity.[7]

The operating expenses of electrical and oil systems differed, sometimes greatly. At La Hève, annual expenses were $3,215 for an electric light versus $2,829 for oil. The cost differential was even larger for South Foreland: about $4,000 per electric light, versus $2,000 for oil.[8] The only accounting method that gave electric lights an advantage was to calculate cost per unit of light. The lighthouse engineer Léonce Reynaud, considering the first French electric light on that basis, determined that its cost was one-fourth that of oil.[9] That sort of cost accounting was problematic because the arc lamp's total light output had to be used, whether needed or not.

Apparently, the vast majority of lighthouse organizations decided against electric lights by heavily weighting utilitarian, administrative, and financial factors. In view of these performance deficiencies, one might wonder why, after the early trials, any country installed an electric light. It turns out, however, that the electric light had certain advantageous performance characteristics during use that, in the reckoning of some countries, outweighed its huge handicaps—at least for a while.

Few disputed the claim that the arc light, because it was by far the brightest artificial illuminant, could penetrate furthest through haze and light fog.[10] It seemed ideally suited for important lights requiring the greatest illuminating power, but not for smaller lights and beacons. The electric light was relatively steady, in that—when properly regulated—it did not gradually depart from the optical focus, as some oil lamps did. In addition, the use of lenses and reflectors made it possible to properly direct most of an arc light's output. Nonetheless, a first-order Fresnel lens projected the light from a large oil lamp well enough so that, under very clear conditions, it might be seen from the horizon.

Mariners had mixed opinions on the electric light's merits. Some captains complained that the light was so bright it briefly impaired night vision, and that it cast confusing shadows.[11] Yet, on balance, mariners seem to have been positively disposed, if one can judge by selected testimonials and advocacy journalism.[12] Speaking for navigators who were familiar with the early English and French electric lights, *Mitchell's Steam-Shipping Journal* endorsed their general adoption because they could be seen better in hazy weather.[13] In his report on European lighthouses, George Elliot of the of the Army Corps of Engineers concluded: "Navigators acknowledge with pleasure the excellent service which the electric lights render them; the advantages of the system have been keenly appreciated, the range of the lights is sensibly increased, especially during somewhat foggy weather."[14] Later, an important mariner's guide proclaimed: "The ELECTRIC LIGHT IS THE MOST WONDERFUL of all the means now employed in Lighthouses" because of its "totally distinct character .. power, and colour."[15]

Beyond guiding ships, the arc-lit lighthouse served national interests by carrying out significant symbolic functions. Perhaps the most general meaning was that a country had wealth enough to purchase the most expensive new illuminant. An electric light also promoted the belief that a country put a high priority on maintaining the prosperity of shipping companies, whose business was growing every year, and on ensuring the welfare of all mariners. Electric lighthouses were also places where new, science-derived technologies could be conspicuously displayed. Indeed, both in England and in France, the American visitor Elliot sent by the Light-House Board, was shown every electric light.

I suggest that national pride and competitiveness contributed to the willingness of lighthouse organizations, especially in wealthy countries, to try out expensive illuminants. Although there had been experiments with other intense lights, the arc light had a special cachet as an *electrical* technology at a time when other electrical wonders, from telegraph to street lights to trolleys, were affecting, or promising to affect, daily life. Indeed, an enthusiasm for all things electrical, notably lights, suffused the consciousness of the West.[16] Clearly, the electric light's distinctive visual performance as the whitest, brightest light rendered it a potent symbol of a country's scientific and technological prowess. Not just a navigation aid, the arc-lit lighthouse was a beacon of modernity. Thus, a country's adoption of one or two arc lights could serve as a political technology, especially if the lighthouses were situated in prominent places and likely to be seen by mariners and voyagers from many nations.[17] The electric light advertised a country's commitment to safe maritime commerce and its expertise in science and technology, and also fostered local pride. A country that possessed only one electric lighthouse had at least a token of modernity. In any case, France and England—imperial powers, near neighbors, and perpetual competitors—went furthest in adopting arc lights.

The arguments in favor of adopting more electric lights in France were set forth by Émile Allard, an engineer and the director of the Central Service of Lighthouses and Buoys.[18] The present system was deficient, Allard maintained, because mariners were within sight of one or more lighthouses only under *ordinary* atmospheric conditions—the basis on which lighthouse spacing had been determined. But in some seasons, in some regions, extreme conditions put mariners at risk because oil-lit lighthouses were too far apart. Electric lights could ameliorate this problem. The three French electric lights, Allard emphasized, had "functioned with all the desired regularity."[19]

Allard pointed out that the English had "already established 6 electric lights on their coasts, and they appear to be soon installing many others."[20] French officials could not miss Allard's point: if France did not rapidly install more electric lights, England would take a decisive lead in lighthouse illumination. To guard against this indignity, Allard proposed a system of 42 electric lights, to be emplaced over a 12-year period in new and existing lighthouses, at a total capital outlay of 7 million francs. Once installed, the system would be economical, Allard insisted, because electric lamps had a lower operating cost per unit of light. Yet he did admit that annual operating expenses would actually rise by an estimated 300,000 francs.[21] The French Lighthouse Commission accepted Allard's proposal and began to implement it.

In addition to Allard's technical arguments, France's willingness to invest in electric lighthouses was conditioned by these lights' important symbolic functions, as indicated by their distribution along the French coast. Four lighthouses (Dunkerque, Calais, Gris-Nez, and La Canche) were tightly clustered in the extreme north, where, at the Strait of Dover, the English Channel is narrowest. This was one of the busiest shipping lanes in the world, a location where—probably not by coincidence—the English electric lights were also clustered. This placement ensured that nearly every ship heading to London (and to ports in northern Europe) from the Atlantic and Mediterranean would have been within range of one or more French arc lights, at least in good weather (figure 20.2). What is more, electrifying these lighthouses had a high priority, as all were operating before 1886.[22] While these lights promoted safe transit through a narrow and often stormy and fogged-in passage, they also advertised French scientific and technological eminence, a meaning that, I suspect, would have been apparent to any passing mariner. In the wake of the disastrous Franco-Prussian War, the electric lighthouse was one way for France to assert symbolically that she was still a world power while underscoring her contributions to electrical science and technology at the same time that other countries were taking the lead. France also brightened her tarnished image symbolically by lavishly lighting Paris with electricity.[23]

In 1881 the English laid plans to establish 60 electric lighthouses.[24] Before proceeding, however, Trinity House studied the penetrating power of gas, oil, and electric lights, finding that the arc light had poorer penetrating power in haze and fogs than oil or gas lights—a difference that, of course, did not matter, insofar as the electric light was much brighter.[25] It was concluded that "for ordinary necessities of Lighthouse illumination, mineral oil is the most suitable and economical illuminant, and that for salient headlands, important landfalls, and places where a very powerful light is required, electricity offers the greatest advantages."[26] This conclusion was not, however, followed by more than token adoptions of new electric lights, even though England and Wales together added 200 new coastal lights between 1887 and 1897. An English lighthouse engineer expressed the British consensus that "the electric light in its present condition is not, save in a few cases, to be too strongly recommended

Figure 20.2
The Planier electric light, near Marseille. Source: Alglave and Boulard 1884, figure 228.

for lighthouse service," owing to its high installation and maintenance costs.[27] None-theless, in 1886 an arc-lit lighthouse was established on the Isle of May, in Scotland, and in 1888 the last English arc lights were installed at St. Catherine's Point, on the south tip of the Isle of Wight.[28] Perhaps the new British electric lights, and the reten-tion of several old ones despite high operating costs, were half-hearted attempts to keep pace with France. Yet the French program, which had been partly predicated on the belief that the British were about to take the lead in electrifying lighthouses, was scaled back drastically in the mid 1880s. This retrenchment was understandable given the added responsibilities of lighting the coasts of France's North African colonies.[29]

_⌇⌇⌇_

Shortly after joining the Light-House Board as its Chairman and scientific member after Joseph Henry's death, Henry Morton reported on his promising tests of American, French, and German electric lighting equipment.[30] By mid 1882 the Light-House Board, aware of the growing enthusiasm for electric lights in France and England, had approved as an experiment the construction of its first arc-lit lighthouse. Several factors led to the placement of this lighthouse in New York City. Local interests had long been pushing the Board to solve navigation problems posed by rocks in an area of the East River known as Hell Gate. Although much of the rock had been removed by electrical blasting in 1876, the channel was still too shallow for vessels of large draft.[31] Interested, post-Henry, in gaining experience with an electric lighthouse, the Light-House Board seized the opportunity to illuminate a vast area of the East River, enabling boats to navigate around the hazards at night. Hell Gate was also conve-niently located just a short boat ride away from the Board's central depot on Staten Island. This convergence of local needs and a seemingly ideal test case augured well for the project and smoothed the way for congressional appropriations.

Supplied with electricity by an American-made No. 8 Brush dynamo, the nine arc lamps and their surrounding lens and lantern were placed atop a 250-foot tubular iron tower—the tallest of its kind in the United States—at Hallet's Point (figure 20.3), on the Astoria side of the East River. Two additional structures were erected to house equipment and workers.[32] On October 20, 1884, members of the Light-House Board were brought by lighthouse tender to witness the inauguration of the imposing struc-ture. As had been anticipated, the Hell Gate Light brightened the East River for several miles in each direction. (This lighthouse, it should be emphasized, was not a beacon to distant ships, but an enormous floodlight.)

Although the Hell Gate light was a creditable achievement, some shippers com-plained that it was too bright. So loud were the objections that the Board solicited the opinions of steamship companies that used the channel regularly. In late October 1886, owing to a preponderance of negative responses, the Board extinguished the Hell Gate Light and sold the electrical equipment as surplus.[33]

Figure 20.3

An artist's conception of the electric light at Hallet's Point. Source: Report of the Light-House Board (U.S. Department of the Treasury, 1883).

The case of the Hell Gate light reminds us that a technology's consumers may include distinct groups of purchasers and users. Moreover, each group can consist of subgroups that might have different performance requirements, and so might render rather divergent practicality judgments. In this case, captains who sailed up and down the East River, passing into and out of the lighted zone, had the most complaints, whereas ferry operators, who remained in the lighted area, were pleased. Clearly, adoption decisions made by one group can redound positively and negatively on different user groups.[34]

Americans acquired valuable experience during the Hell Gate Light's brief life, but the project did not become a springboard for the wholesale adoption of electric lights. The Light-House Board offered no proposal comparable to those floated in England or France for converting principal lighthouses to electrical operation.

⎯⌒⌒⌒⎯

Surprisingly, the world's most flamboyant example of an electric lighthouse serving as political technology was lit up in the United States about a week after Hell Gate was darkened. The Statue of Liberty, originally named "Liberty Enlightening the World," was a gift from the people of France to symbolize and cement the relationship between the two "liberty-loving" peoples.[35] Designed by the sculptor Frédéric-Auguste Bartholdi, the statue was paid for by contributions from the French people. However, the massive pedestal and modifications to the site required contributions from American citizens and the U.S. government.

Erected in 1886, Liberty stands atop an old army post on Bedloe's (also Bedlow's) Island in New York Harbor, where previously no lighthouse had stood. Bartholdi designed the statue to carry a torch lighted by kerosene, but the private American Committee in charge of the Liberty project decided to emplace a series of arc lights in the torch that could project into the sky and far out to sea.[36] Before the federal government became involved in the project, the American Committee had arranged with E. H. Goff, president of the American Electric Manufacturing Company of New York, to install an arc lighting system in time for the statue's inauguration. As the appointed date drew near, a concerned federal government effectively took control and put D. P. Heap, Engineer Secretary of the Light-House Board, in charge of the lighting. Heap forecast that the company could finish the work on time, but that it would be a temporary arrangement. Goff's work was permitted to proceed. With much fanfare, including a dedication speech by President Grover Cleveland, a parade, and an extravagant fireworks display viewed by multitudes, Liberty's torch shone at last on November 1, 1886.[37] Because the contract between the American Committee and E. H. Goff provided only for the donation and installation of the lighting equipment, the light was turned off a week later for lack of operating funds. (The *New York Times* prominently reported the bickering among federal officials.[38]) To remedy the national embarrassment of a darkened Liberty, which in the 10 years since its arrival in America had become a deeply evocative national symbol, President Cleveland directed the Secretary of the Treasury to place the statue under the jurisdiction of the Light-House Board. And so on November 22 the statue was made an official lighthouse. Once again, the Light-House Service had an electric light to tend; its hard-won expertise had not been wasted. A few years later, Goff's electrical system was replaced with a more permanent one, and the arc lights were moved to the crown.[39]

The Statue of Liberty was a political technology that also happened to have a limited lighthouse function. It was not designed as a significant aid to navigation (Bartholdi had not consulted the Light-House Board), and it never became one. We may suspect, however, that its arc light helped the statue to accrue new meanings. In the wake of other recent triumphs of American electrical technology publicized in magazines and newspapers on both sides of the Atlantic, including Edison's incandescent lighting system and Bell's telephone, Liberty's light announced that the torch of electrical progress had recently passed to the United States.

The United States continued to experiment with electrical navigation aids during the following several decades, and even deployed a few more arc lights.

In 1888, incandescent electric lights were emplaced on wooden buoys in Gedney's Channel in New York Harbor near New Jersey's Sandy Hook peninsula. Despite the difficulties and costs of keeping these buoys functioning (changing a light bulb required four men and a boat), the experiment was deemed a success because navigation through the narrow but busy channel had been improved.[40] Nonetheless, the Light-House Board continued to use hydrocarbon-burning lamps in most other buoys.

In August 1889, while the Gedney's Channel experiment was under way, the same Edison dc dynamo that fed current to the buoys was tapped to power an incandescent lamp in the nearby Hook Beacon. This installation was touted as the "first regular light-house in which this method of illumination has been employed." It was suggested that this system could be extended to the Sandy Hook Main Light, which was accomplished in May 1896. The handsome Main Light, an 85-foot-tall octagonal structure that survived shelling during the War of Independence, is the oldest standing lighthouse in the United States. However, this light was not a harbinger of others: few incandescent lights entered American lighthouses in the next several decades.[41]

In 1898, an arc light of some significance was installed in the South Tower of the Navesink lighthouse.[42] Situated in the highlands of northern New Jersey overlooking the Sandy Hook peninsula, Navesink's twin towers beckoned distant ships to the New York Harbor, America's busiest. The South Tower was now the brightest light in the United States, perhaps in the world. By adopting this overpowering electric light for one of its most important and highly visible lighthouses, the United States conspicuously signaled its concern for the welfare of mariners, emphasized its eminence in electrical technology, and announced its ascendance as a world power.

In the 1880s and the 1890s, more than a half-dozen countries built lighthouses with electric-arc illumination, including Italy (on Tino, in the Gulf of La Spezia, 1885), Spain (Cabo Vilán, along the Galician coast, 1896), Australia (Macquarie Lighthouse, in Sydney Harbor, 1883), and Denmark (Hanstholm, 1898).[43] Apparently none of these lighthouses was part of a major electric-lighting program. Rather, as unique lighthouses in each country, adopted despite the electric arc's well-known performance deficiencies, they were more than aids to navigation. By advertising mastery of electricity, they served as an emblem of national pride and a proclamation of modernity. Fittingly, birth announcements appearing in major science and engineering journals—and even later histories—include the boast, perhaps warranted, that a country's new electric light was the world's most powerful.[44]

The lighthouse was a logical—indeed, obvious—market for purveyors of powerful artificial lights to target; after all, these highly specialized structures needed bright illuminants. More than that, if a manufacturer could convince a lighthouse organization of its light's practicality, then dozens or even hundreds of sales might follow. Thus, manufacturers of successive arc light technologies—Staite's battery system, Holmes and Alliance magnetos, and many dynamo systems—hawked their wares to lighthouse organizations. Scientific authorities in England and France agreed that the arc light was a suitable illuminant on technical grounds. However, the always cost-conscious Joseph Henry preferred to retain oil lamps and invest instead in acoustic technologies. Because of his position on the Light-House Board, this man of science had the power to make his views prevail. And so, in Henry's lifetime the United States did not install a single electric light. However, after his death (in 1878), the United States undertook its short-lived experiment at Hell Gate, later acquiring the Liberty light and just a few additional arc-lit lighthouses.

In most countries, lighthouse organizations concluded, as had Joseph Henry decades earlier, that the arc light was impractical on economic grounds. Indeed, when arc lighting was at last enjoying substantial commercial success, beginning in the 1870s with the advent of dynamos, the lighthouse market had barely been penetrated. No country, not even France, wholeheartedly adopted the electric illuminant; the costs of installation, operation, and maintenance were simply too daunting, even for wealthy countries that had placed a high value on the technology's symbolic potency.

21   Enter Edison

In 1847, the year of Thomas Edison's birth, the joint-stock company had already become the template for creating corporations pursuing private profit. Brought into existence by state charter, its contracts enforced by state power, the corporation was well adapted to raising big money, spreading risk among investors, and insulating stockholders from a company's debts. Thus, corporations in the United States helped private enterprise to develop, for example, gas lighting, railroads, telegraph systems, and, later, electric light and power systems. When Edison took on the challenge of electric lighting, he did so in the context of a recognizably modern corporate capitalism; he would enjoy the advantages—and suffer the disadvantages—of dependence on corporate organization.[1]

By the time Edison projected his vision of an electrical system that could provide light and power for the masses, he was a wealthy inventor, highly charismatic, almost an inexorable force. These personal qualities along with accelerating demand for electric lighting, helped Edison to obtain favorable practicality judgments from most pertinent players, including wealthy investors. So great was Edison's social power, not to mention his engineering and organizational acumen, that he was able to press on despite claims by scientific authorities that major components of his system could not work as claimed because they contravened accepted principles.

Although Edison had to contend from time to time with carping men of science, his greatest power struggles would be with his corporations' stockholders and with a competing corporation's inventor-entrepreneur, George Westinghouse. With shrewd patent purchases, acquisitions, and hires, Westinghouse had assembled an alternative electrical system—employing alternating current—whose practicality was rapidly affirmed by players in the corporate world and by consumers. And so Edison and Westinghouse, both accomplished inventors and prosperous businessmen, became bitter enemies as they strove to defend and expand the markets for their respective technologies. While pushing their products, these two men, like Morse and Henry before them, resorted to the kinds of dishonorable acts that seem to be fostered by corporate capitalism. In the face of marketplace gains by Westinghouse, Edison was defeated, entangled in skeins of finance beyond his control.

_⌒⌒⌒⌒_

Thomas Alva Edison (1847–1931), the youngest of seven children, was born in the small town of Milan, Ohio. His father Sam was a lumber and feed dealer, his mother Nancy a teacher. As Thomas approached school age, the family's fortunes took an unfavorable turn. In search of new opportunities, the Edisons moved to Port Huron, Michigan.

Responsibility for educating Thomas fell mainly to his mother, who tutored him at home. When the boy was 9 years old, she gave him an elementary text on physical science, which he read and reread, repeating all the experiments. Soon his passion was chemistry, the paragon of experimental science. Engaging the material world firsthand, teasing out its secrets through painstaking experiments—sometimes with theoretical guidance, sometimes without—was to become the defining style of Thomas Edison's work.

Sam Edison's fortunes continued to founder, and so Thomas, at age 12, went to work selling newspapers and sundries on the train that traveled daily from Port Huron to Detroit and back. Even at this tender age he displayed an entrepreneurial flair, selling fresh fruits and vegetables along the route. Edison also set up a chemistry lab at one end of the baggage-mail car and experimented in odd moments. Only once did he set the car on fire.

In April 1862, Edison saw an edition of the *Detroit Free Press* that headlined the Civil War battle at Shiloh, where 60,000 casualties had been suffered. Sensing commercial possibilities, he had the news sent ahead by telegraph in advance of the train's arrival at several stops. Crowds gathered at the depots to await the train and its precious newspapers. The young businessman had obtained 1,000 copies, which he dispensed to eager buyers at suitably inflated prices. Edison did not fail to note the crucial role that the telegraph had played in this enterprise. With his windfall profits, he learned telegraphy.

From 1863 to 1868, Edison covered much ground as an itinerant telegrapher. Usually his earnings were spent on books, chemicals, and apparatus. Experiments and tinkering became his real life, interrupted only by the need to eat and earn a living. During this period, his experiments focused on electromechanical devices for telegraphy. He was supported in this work by Western Union and other companies, and his first commercially important inventions included duplex and multiplex telegraph equipment.

Edison had found his calling as an inventor. From his early work he had learned that earnings from the sale of patents could be invested in new projects, some of which might also pan out, generating still more money to reinvest in invention. He was also aware, however, that grand projects might require the sustained support of financiers, especially to capitalize manufacturing corporations In 1876, already wealthy from his telegraph and stock ticker inventions, Edison established an "invention factory" at Menlo Park, New Jersey, and equipped it lavishly for electrical and chemical experiments. Here he assembled a team of able men who assisted in experiments and translated his ideas and sketches into hardware. It was this unique organization, the first industrial research laboratory, that would give Edison a leg up on other inventors who also were working on incandescent lighting.

Edison and his team tackled a multitude of projects, some trivial and some monumental. Invention was Edison's passion, but it was also his business. Thus, before committing to a major project, he would forecast its commercial potential: what was the likelihood that financiers, the press, manufacturers, and consumers would judge the new technology favorably? In view of the avid adoption of arc lights during the late 1870s and the deep antipathy toward the gaslight monopolies, Edison didn't have to be prescient to anticipate robust sales of a technically competent and affordable system of incandescent lighting.[2] Many inventors perceived these possibilities and set to work, but Edison was the first to commercialize a fully functional system.

On November 22, 1875, Edison made a surprising discovery. From the iron core of a vibrator electromagnet—it oscillated on and off, rapidly and automatically—he drew sparks with a nearby metal bar. Similar effects, already familiar to some telegraphers and experimenters, had provoked little interest. Edison, however, was curious about the cause. His further experiments revealed far more novel effects whose explanation became mired in controversy.[3]

By attaching one end of a wire to the metal bar and the other to a stove or gas pipe, Edison learned, he could obtain sparks by touching the stove or the pipe with another piece of metal. In further tests, he found that this apparent electricity lacked polarity and passed through insulators; moreover, it could not budge a galvanometer or an electrometer, give shocks, or induce chemical reactions, yet it produced a scintillating spark. *Scientific American* reported that "signals have been sent long distances, as from Mr. Edison's laboratory to his dwelling house, in another part of the city, the only connection being the common system of gas pipes. Mr. Edison states that signals have also been sent the distance of seventy-five miles on an open circuit, by attaching a conducting wire to the Western Union Telegraph line."[4]

But a more astounding effect was yet to come. Edison built a small black box containing two graphite rods, their points separated very slightly as in a carbon arc. When he held the box at some distance from the vibrating electromagnet or near an operating telegraph line, he could see through a small opening that sparks jumped between the points.[5]

To Edison, a man intimately familiar with electrical phenomena, this congeries of effects was baffling. It was beyond the scope of known principles. Convinced that he had discovered something new to science, he named it the "etheric force" and proffered that it differed from electricity but might be comparable to heat or light. The name also suggested that the new force was propagated as a wave through the ether, the omnipresent medium believed to carry light vibrations. Perhaps for this discovery he would join Volta, Oersted, and Faraday in the highest rank of electrical scientists.

Edison did not hesitate to publicize his findings to the press, and he had an especially close relationship with the editor of *Scientific American*. Immediately after the first newspaper accounts appeared, the *New York Times* ran a spoof of the purported discovery—and of Edison.[6] Undeterred, in a presentation on December 15, 1875, at the Polytechnic Club in New York, Edison exhibited the etheric force's effects, dashing any doubts about its empirical reality. Reports of Edison's experiments, including a drawing of the apparatus, appeared in *Scientific American* over the next few weeks. The most detailed and authoritative description was printed on Christmas day.[7] This account turned out to be a gift to men of science, some of whom taxed their ingenuity to explain away with existing principles the varied, yet highly reproducible effects that Edison had reported.

A persistent American critic of Edison was the electrical inventor William Sawyer. In two lengthy and convoluted letters to *Scientific American*, Sawyer contended that the etheric force was merely induction; to maintain this position he had to dismiss the reality of several claimed effects, which was tantamount to accusing Edison of sloppy work or dishonesty. After granting that the etheric force was not a "deception," Sawyer nonetheless "pronounce[d] the whole thing, both as concerns the public and in a scientific point of view, as one of the flimsiest of illusions."[8] Edison did not dignify Sawyer's attack with a reply.

Men with scientific credentials more impressive than Sawyer's soon entered the fray. George Beard, a respected physician, electrotherapist, and pioneer of psychosomatic medicine, performed experiments with Edison at Menlo Park, conclusively showing that the etheric force had no physiological effects—no shock, no tingling on the tongue, nothing.[9] In these respects it was utterly unlike electricity. Beard also helped to exhibit the etheric force at the Polytechnic Club, arguing that "we have here something radically different from what has before been observed by Science." In further support of Edison, Beard suggested that the evidence favored "the theory that this is a radiant force, somewhere between light and heat on the one hand and magnetism and electricity on the other, with some of the features of all these forces."[10]

Far less supportive was Edwin Houston, a teacher of natural philosophy at Philadelphia's Central High School. Collaborating with Elihu Thomson, with whom he later founded an electric lighting company that would compete with Edison's, Houston argued that the etheric force was simply induction, as Sawyer and others—including the editor of *Scientific American*—had already suggested.[1] Curiously, Houston also asserted priority for the discovery that he had just claimed was not a discovery, recounting a similar experiment he had published in 1871 using an induction coil. An editor's note appended to Houston's letter in the *Scientific American Supplement* affirmed his priority.[12]

The consensus of the detractors, at home and abroad, was that Edison's discovery—if there had been one—was merely that of previously undescribed effects of induction; by implication this was not very important, and it was certainly not a new force. Edison, however, was sure that neither electromagnetic nor electrostatic induction was responsible for the effects. In one of his rare published contributions to the debate, Edison acknowledged Houston's priority, yet reiterated that "the spark has been observed by electricians for many years, and attributed by them to inductive electricity; and all that I can lay claim to is that perhaps . . I was the first to discover that it was not due to electricity."[13]

Though not swayed by the naysayers' sometimes contorted arguments, Edison doubtless came to realize that an electrical inventor, no matter how expert, was unlikely to prevail in this theoretical dispute. After all, he had not properly published his findings in a scientific journal, nor had he put his vague etheric-force theory into concrete, testable form. Perhaps more significantly, Edison lacked the academic credentials and mathematical skills to be taken seriously in the community of theoretical physicists. Preoccupied with other projects in 1876, such as preparations for the Centennial Exhibition and moving to Menlo Park, and pressured by Western Union to devote more time to telegraph work, he set aside investigations on etheric force. The controversy died quietly, the skeptics presuming, in the absence of new pronouncements from Edison, that they had carried the day.

Perhaps the most plausible interpretation of Edison's etheric-force experiments is that he had created and detected electromagnetic waves, the same invisible waves that Guglielmo Marconi would commercialize at the end of the century in his wireless telegraph. Edison later expressed regret that he had given up this line of research: "I did not think of using the results of my experiments on 'etheric force' that I made in 1875. I have never been able to understand how I came to overlook them. If I had made use of my own work I should have had long-distance wireless telegraphy."[14] This was just one of many episodes that convinced Edison of the fallibility of scientific authorities.

Heinrich Hertz, a German physics professor and a student of Hermann Ludwig Ferdinand von Helmholtz, is credited with having shown the existence of radio-frequency electromagnetic radiation.[15] He achieved this recognition by confirming, through simple experiments in 1886–1888, an implication of James Clerk Maxwell's dense and equation-laden electromagnetic theory. The theory, first published in a lengthy paper in 1864, had made slow headway, even among academic physicists.[16] Maxwell posited that light, electricity, and magnetism were all forms of electromagnetic radiation propagated as waves in the ether. When Hertz did his experiments—more than ten years after Edison's work—Maxwell's theory was finally receiving favorable attention, and so provided a context for appreciating the discovery's transcendent importance. Because Hertz was a brilliant mathematical physicist whose results were in accord with and embellished Maxwell's far-reaching theory, he—not Edison or one of the other contenders—achieved credit for discovering radio waves.[17]

In the course of such controversies, Edison acquired a reputation in the academic world for hostility toward scientists—even toward science itself. In later years he would often be caricatured as an empiricist plodder who eschewed theoretical guidance. However, Edison did hire university-trained scientists for his team, and in the early 1880s he bankrolled the journal *Science*.[18] Not only did he consider himself to be a "scientific man," and not only was he regarded as such by journalists and other non-scientists; he was also America's most visible practitioner of scientific invention as advocated by *Scientific American*.[19] More significantly, his projects sometimes ventured into virgin territory, where experimenting by trial and error was a sensible approach that might yield results amenable to theoretical cogitation.[20] And sometimes Edison had a deeper understanding of the scientific principles at issue than his critics. His conflicts with some men of science led him to conclude that their pretensions often exceeded their expertise—a point he was fond of making.

By the mid 1870s, gaslights had been adopted far and wide in America, but they were a mixed blessing. There were constant complaints against the monopoly companies, their pricing policies, and the gaslights themselves. Once the pipes and the fixtures were installed, coal gas was a cheaper illuminant than candles or even oil lamps, but it had many performance deficiencies. Gaslights polluted indoor air, deposited soot on exposed surfaces, gave off heat, had an unnatural yellowish hue, and flickered. And then there were the dangers of explosion, asphyxiation, and fire. Moreover, in a factory or on a public street, gas entailed the nightly ritual of lighting lamps individually—unless they had electric igniters. With consumers all too aware of gaslight's downside, advocates of electric lights could easily take aim at the entrenched illuminant. Makers of dynamo-powered arc lighting systems launched the first credible volleys against gas, but many targets were spared because the arc light's performance characteristics prevented it from competing effectively in all applications.

The era of relatively cheap electricity, ushered in by the commercialization of dynamos, attracted scores of inventors and manufacturers. Although many early dynamos had been patented, including Wilde's and Gramme's, the technology's inherent design flexibility meant that patents could be easily evaded—that is, invented around—through minor modifications. Indeed, between 1873 and 1882 Britain awarded 153 dynamo patents to inventors from many countries.[21] Other components of arc lighting systems, particularly regulators, also admitted numerous patentable variations.[22] Because demand for electric lighting was expanding exponentially, many manufacturers brought to market entire arc lighting systems, usually built around the contributions of a few inventors—sometimes "in-house" inventors. Apparently, companies that were able to enlist competent mechanical engineers, to develop reliable production processes, and to reach the market early with a product regarded favorably by consumers were the companies most likely to prosper, regardless of their patent position.[23] In some cases a telegraph manufacturer was well situated to take advantage of the boom in electrical lighting—Bréguet, a maker of telegraph equipment, produced Gramme machines, and Siemens was a major telegraph manufacturer.

One of the most successful American manufacturers of dynamos and arc lighting systems was Charles Brush (1849–1929), a graduate of the University of Michigan.[24] Although interested at an early age in electrical generators, he studied mining engineering because there was not yet a comparable curriculum in electricity. After graduation in 1869, he returned to his home state, Ohio, and worked in Cleveland as a chemist. Next he joined up for a few years with an old friend, Charles Bingham, and together they marketed iron and iron ore. That career apparently did not capture Brush's fancy, for he began to think again, more seriously, about generating electricity. As it happened, another of Brush's boyhood friends, George Stockly, was Vice President and General Manager of Cleveland's Telegraph Supply Company. After Brush shared his ideas about using dynamos for electric lighting with him, Stockly offered to supply the aspiring inventor with materials and a place to work.

While vacationing at his parents' farm in 1876, Brush built his first dynamo, which was literally horse-powered. Sufficiently encouraged, he quit the iron business and began full-time work on electric lighting, having also developed his own arc lamp. The regulator was especially elegant and compact. Employing a solenoid and a gravity-fed carbon rod, it did away with clockwork. It also contained a second set of carbons that turned on automatically when needed, and so the lamp could burn continuously through the night. In addition, because Brush employed shunts in his series wiring of lamps, the failure of one lamp did not affect others on the same circuit. Brush's system was well suited for lighting factories as well as streets, but did potential customers regard it as practical?

Indeed they did. By mid 1879, Stockly claimed that 200 Brush lighting systems had been sold to commercial and industrial establishments.[25] A dramatic demonstration in Cleveland's Monumental Park also helped to spur demand for the lighting of public places, an application that had been slow to take hold in the United States. There Brush set up 12 arc lamps on posts high above the 100 or so gas lamps they were destined to replace. On April 29, 1879, thousands gathered to see what the electric light was all about, and they were not disappointed when the lights went on at 8 p.m. The response was swift: people shouted, a band played, and artillery nearby sounded a salute. According to *Scientific American*, this was "the first public lighting with the electric light of any city in the United States."[26]

As news spread about the capabilities of the Brush system for street lighting, orders poured in from around the country, many from electric companies recently formed by local investors keen to replace gaslights (figure 21.1). To increase production, Brush and the Telegraph Supply Company, which had financed his project, established the Brush Electric Company. They produced Brush dynamos in many sizes, rated by the horsepower needed to drive them. In the next few years, the new firm captured 80 percent of the arc lighting business in America, and even made inroads across the Atlantic in the face of fierce competition.[27] A British authority on electric lighting acknowledged in 1882 that the Brush system was excellent, noting that it was in

Figure 21.1
A Brush system lights a New York street. Source: Alglave and Boulard 1884, figure 47.

extensive use in London, especially in railroad stations and for street lighting and that it had "taken a leading position among the lighting systems of Europe."[28] By 1886, when a large Brush system was used to light the Statue of Liberty, Charles Brush was a wealthy man.

_⌒⌒⌒_

Although some scholars assert that Brush's installation in Cleveland's Monumental Park was "the first central station in the country," it was not the first.[29] In fact, it was not even a central station—if a central station is defined as a facility that generates electricity and sells it to a number of independent customers. The idea of a central station had occurred often in the past, for it promised an economy of scale in electricity generation so long as transmission costs did not rise disproportionately. The question of transmission costs was on many minds at the end of the 1870s.

One person tackling transmission costs was William Ayrton (1847–1908), a physics professor who had studied under William Thomson at the University of Glasgow. Ayrton, who manufactured electric motors and instruments in England, advocated transmitting electricity from central stations to various customers. However, Ayrton, drawing on basic electrical principles, argued that sending current over long distances would be uneconomical with the low-voltage generators then in use.[30] The output voltage of a dynamo could be raised by employing appropriately wound armatures and high speeds. To lower the high voltage at the user's end (and thus make more current available), Ayrton suggested the adoption of motor-generator pairs: a motor, powered by the central station's high voltage, would drive a low-voltage generator that provided local current for lights and so forth. Ayrton suggested that if the most efficient motors and generators were used, energy losses might be kept in the neighborhood of 30 percent.[31]

In a breathtaking prognostication, Ayrton claimed in 1879 that electrifying the town of Sheffield—for arc lights, power, and heat—would save the town's residents about £400,000 per year.[32] His analysis assumed that electricity would be "produced in large quantities at certain centers." It could be distributed like gas, through main and small feeder lines, as long as the dynamos were furnished with "automatic current regulators," of which several examples were already extant. This was a remarkably clear vision of a central station, an idea that had been floated but never implemented.[33]

The first American central station had a Brush dynamo and Brush lamps, but it was located in San Francisco.[34] In June 1879, 27-year-old George Roe, a Canadian transplant, incorporated the California Electric Light Company. At Fourth and Market Streets, he installed two Brush dynamos that powered 21 arc lamps. The company offered service at a weekly rate of $10 per lamp, but electricity was available only from sundown to midnight (except on Sundays and holidays). Light-hungry businesses signed on for the service, and in a matter of months Roe had to install more dynamos. In the decades ahead, other entrepreneurs in northern California founded their own electric companies. (Those companies and Roe's were eventually consolidated into Pacific Gas & Electric.)

Ayrton also had advice about the economical use of arc lamps for street lighting. Such applications had not been entirely successful, he maintained, because municipalities followed the gaslight model and placed many lamps on low posts. As an example of an arrangement that better exploited the arc light's brightness, he pointed out that the interior of London's Albert Hall had been lighted by one brilliant arc lamp high above the floor, which replaced many gas lamps at a lower level. The operating expense of this light, including labor and depreciation, was "only one-third of that of the former inferior gas lights."[35] Streetlights could be more economical and effective, Ayrton believed, if they followed the Albert Hall model.

The earliest American lighting system consistent with the Albert Hall model was established in Wabash, Indiana; it became the country's first municipally owned electric company.[36] Seeking to replace their gas street lights, Wabash officials asked Brush to demonstrate his system, for which the city offered to pay $100—and to buy the installation if they were satisfied. Confident that his arc lights would give a virtuoso performance and stimulate sales, Brush agreed. Four lamps of 3,000 candles were placed 200 feet above street level atop the dome of the new courthouse; they received current from a dynamo in the basement. The inauguration of the system, scheduled for March 31, 1880, was publicized beforehand and attracted more than 10,000 spectators, including officials from other towns. When the lights went on at 8 p.m., "the people, almost with bated breath, stood overwhelmed with awe, as if in the presence of the supernatural. The strange weird light, exceeded in power only by the sun, yet mild as moonlight, rendered the courthouse as light as midday."[37] The light was so bright that it bathed an entire square mile, virtually the entire town. Pleased with the performance, the city bought the system for $1,800 and bragged that "Wabash enjoys the distinction of being the only city in the world lighted by electricity."[38] In the weeks and months ahead, Wabash became a pilgrimage site, with visitors paying homage to the electric lighting miracle. Its practicality affirmed by various consumers, the Brush system continued to sell well.

_⁀⌒⌒⌒⌒‿_

While Charles Brush was perfecting his simplified regulator, Paul Jablochkoff (1847–1894), a Russian telegraph engineer working in France, came up with a more radical invention that eliminated the regulator altogether. The Jablochkoff "candle" consisted of two carbon rods set side by side and joined by a thin strip of insulating material, such as kaolin. The tips of the rods were connected by a conductive ribbon that primed the arc, which continued to radiate on its own. Under the withering heat of the arc, which was fed current by an ac dynamo, the carbons and the insulation wore down in tandem. Because the carbons were always at a constant distance from one another, there was no need for a regulator. The Jablochkoff candle was commercialized almost at once, but performance disadvantages limited its adoption.[39]

Although removal of the regulator reduced the size and the cost of individual lamps, the candle was pricier than the rods it replaced, wore down more quickly, was noisy, had fluctuating output, and could not be re-lighted.[40] Moreover, it required about the same amount of power as a regulated arc.[41] In addition, to compensate for the candle's lower light output, more of them were needed to illuminate a large space. The Jablochkoff candle was in some ways more convenient to use than a regulated lamp, but high ongoing expenses made it difficult to sell.[42]

Paris, which fancied itself the City of Lights, funded an experiment in replacing its gas lamps with Jablochkoff candles. The trial was conducted on a small portion of the Avenue de l'Opéra. Each lamp in this installation held four candles; when one burned out after about 90 minutes, a second one was lit automatically, and so on. The Municipal Council was so disappointed by the trial's outcome (the Jablochkoff candles were three times more expensive than gaslights) that it did not immediately extend the lighting company's contract. Nonetheless, Jablochkoff candles were adopted in factories, theaters, concert halls, drapery shops, in parts of the Louvre, and in other places where proprietors were willing to pay a premium for safety, convenience, or having the light *la plus moderne* (figure 21.2).[43]

Jablochkoff candles were also tried in England. In 1879, Gramme dynamos at Charing Cross supplied current for candles along the Thames River between the Westminster and Waterloo bridges. The installation soon included 18 miles of wire; also added were ten candles for the interior of Victoria Station. Even with the advantages of a larger system, however, the operating expenses were greater than those of gaslights.[44] Despite a flurry of early adoptions, one economic historian declared: "The expensive electric candle was soon displaced in most installations by more economical though more complicated arc lamps of the traditional type."[45]

The Jablochkoff candle stirred little consumer interest in the United States, probably because of its high operating costs and because American dynamo pioneers made dc machines for electric lights and electrometallurgy.[46] A case in point is Edward Weston (1850–1936), a savvy inventor of electrical and chemical technologies, many of which became commercial products (eventually he held 334 patents).[47] An Englishman trained as a physician, Weston moved to the United States in 1870. After a stint in a chemical factory, he entered the electroplating business. Like so many others, he turned to the dynamo as the salvation from batteries. He invented his first dynamo in 1873.

After the Centennial Exhibition, at which Weston had exhibited a dynamo in action, he was approached by Frederick Stevens of the firm Stevens, Roberts & Havell.

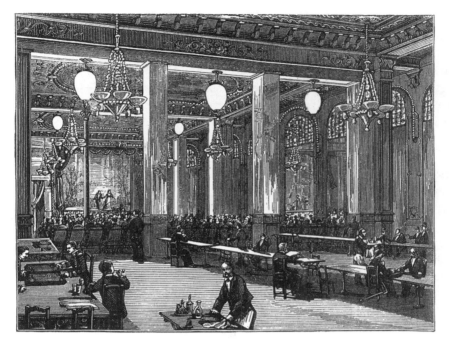

Figure 21.2
Jablochkoff candles light a music hall in Paris. Source: Alglave and Boulard 1884, figure 211.

Together they founded the Weston Dynamo Machine Company in Newark, New Jersey. The firm made dynamos for electroplating; it also made complete arc lighting systems. Prospering, the two-year-old company claimed in an 1879 ad that it had already sold more than 500 dynamos at prices from $125 to $500. Among its electroplating customers were Reed & Barton, the Rogers Cutlery Company, the Norwalk Lock Company, and the Domestic Sewing Machine Company.[48] By this time, nickel plating—a Weston specialty—was commonly used to protect iron on many machines and to create a handsome, silvery surface.

Other American dynamo companies, particularly Brush, Thompson-Houston, Maxim, and Wallace-Farmer, also did a brisk business. (See figures 21.3–21.6.) According to the historian Herbert Meyer, demand was so great that these companies could not keep up.[49] However, their lighting systems employed not Jablochkoff candles but regulated carbon arcs.

An appealing alternative to both the regulated carbon arc and the Jablochkoff candle was the incandescent lamp. Inventors had worked on this project since the time of

Figure 21.3
A Brush dynamo. Source: Dredge 1882, figure 194.

Figure 21.4
A Weston dynamo. Source: Dredge 1882, figure 214.

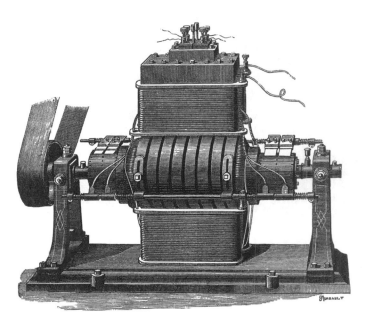

Figure 21.5
A Maxim dynamo. Source: Dredge 1882, figure 209.

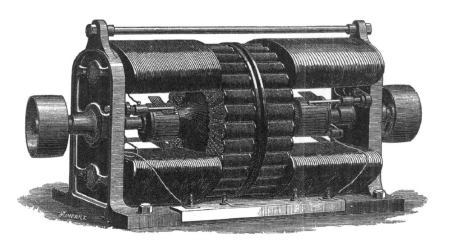

Figure 21.6
A Wallace-Farmer dynamo. Source: Dredge 1882, figure 165.

Humphry Davy, taking out more than a dozen patents, but no lamp was commercialized before the late 1870s.[50] Recall that an incandescent lamp worked by resistance heating: when current coursing through a conductor raised its temperature high enough, the material radiated white light. Decades of experiments had shown that platinum and carbon were promising materials for the conducting element. But even platinum melted at sustained white heat, and carbon, though radiating far more efficiently than platinum, oxidized in air. The incandescent lamps that preceded Edison's did establish some durable design features, including the use of an evacuated glass bulb, yet none of them produced much light, all were prodigious consumers of current, and most were short-lived. Thus, in failing to achieve the most basic performance requirements of an illuminant, these lamps were not brought to market. Evidently, the creation of a technically competent light bulb would entail a great developmental distance, perhaps along a path less traveled.

The commercialization of dynamos and arc lamps gave inventors a strong incentive to work hard on incandescent lamps, and they did.[51] The late 1870s were marked by premature claims of success and the founding of companies that speedily expired. One participant was William Sawyer (1850–1894), an Edison adversary in the etheric-force debates. In 1877s Sawyer—a notorious drunkard with telegraph experience and a creative streak—began work on a carbon lamp. Albon Man, a lawyer who lived to regret his association with the unpredictable inventor, financed and took part in his experiments, which yielded a series of patents. Several Sawyer companies were formed to exploit the lamp patents, but none had adequate capital to refine the invention, to bring it to market, or even to apply for foreign patents.[52]

Sawyer tried many materials, including charred paper, but achieved the most promising results with a thin carbon rod. This element, about half an inch long and made in varying diameters (e.g., $\frac{1}{16}$ inch), was held rigidly by a mount inside the glass bulb. To prevent oxidation, Sawyer filled the bulb with nitrogen. After witnessing an exhibition of a Sawyer lamp on February 20, 1879, *Scientific American* observed in a glowing account that "the light exhibited was soft, pure, and steady, and susceptible of perfect regulation. Any lamp in the circuit could be turned up or down, from a dull glow to brilliant incandescence without affecting the rest. . . . Comparison was made with gas light, and also with the voltaic arc, clearly demonstrating the superiority of light by electric incandescence for ordinary uses."[53] During this brief display, the shortcomings of Sawyer's low-resistance lamp did not become apparent. The rigidly held carbon rod would eventually crack from thermal shock (expanding when hot, contracting when cold), and it consumed great amounts of power. Desperate for funds, Sawyer invited Thomas Edison to join forces. Although Edison as yet had no functioning lamp, he could see that Sawyer's had fatal flaws, and doubtless he recalled Sawyer's contemptuous dismissal of his etheric force. Edison declined to collaborate, but his disdain for Sawyer and his lamps would come back to haunt him.

22 New Light

One of Thomas Edison's first inventions at Menlo Park was also his most novel. In 1877 a working model of a phonograph was constructed in accordance with Edison's drawings by John Kruesi, who ran the Menlo Park machine shop (figure 22.1). Edison went straightaway to New York, where he demonstrated the unprecedented device to Alfred Beach, editor of *Scientific American*. A large crowd gathered, including reporters, and they were amazed. Newspapers announced Edison's wizardry the next day. At Joseph Henry's behest, Edison exhibited the phonograph at the Smithsonian he also gave a demonstration at the National Academy of Sciences.[1]

Although commercial uses for the phonograph (e.g., recording dictation) were evident to Edison, he set aside these promising applications to tackle incandescent lighting. Nonetheless, lavish media coverage of the phonograph made Edison a public figure (President Rutherford B. Hayes had summoned him to the White House for a demonstration) and established his reputation as the Wizard of Menlo Park.

Robert Friedel, Paul Israel, and Bernard Finn point out in their book *Edison's Electric Light* that Edison skillfully incorporated the media into the invention process. Obviously he used newspapers and magazines to publicize his work, which kept Edison-the-Wizard in the public eye. Thus, when he or his lawyers came calling on potential backers for projects, no one had to ask "Who is Thomas Edison?" Less obviously, he made pronouncements, sometimes exaggerated or premature, about inventions in progress. These claims highlighted the laboratory's immediate research goals and put Edison's reputation at risk. Edison thrived with such an incentive, believing that the men in his laboratory could accomplish whatever task he set. Although Edison's claims often became self-fulfilling prophecies, his credibility might suffer if success did not arrive easily or on time.[2] Another risk of early engagements with the press was that his claims might provoke scientific authorities into responding with skeptical opinions in letters and interviews.

Edison's interest in incandescent lighting began in the summer of 1878, when, on the invitation of George Barker, a professor of physics at the University of

Figure 22.1
Edison and the world's first phonograph. Courtesy of Smithsonian Institution (negative 33553).

Pennsylvania, he accompanied an expedition to the Rockies to observe a total eclipse of the sun. Edison's role was to employ his "tasimeter" to measure the temperature of the sun's corona. Although the tasimeter proved useless, the trip gave him time to contemplate new projects. During discussions of electric lighting, Barker suggested that Edison visit William Wallace's factory in Ansonia, Connecticut, which was manufacturing the Wallace-Farmer dynamo and other equipment for arc lighting. Two weeks after his return from the West, Edison toured the Wallace facility and became enraptured with the vision of building a system capable of furnishing universal light and power.[3] Immediately he ordered two Wallace-Farmer dynamos.

As a latecomer to electric lighting, Edison was ideally situated to exploit previous work, especially on dynamos, and to identify other components whose performance characteristics had to be improved. Accordingly, the highest priority was to develop an incandescent lamp that "could be brought into private houses."[4] After only a few days, Edison declared success: "I have it now!" Claiming that his lamp was both brilliant and cheap, and predicted the eclipse of gas lighting.[5] Edison's announcement played well in the press, and gas companies' stocks slid on the news.

Investors from around the country approached Edison, inquiring about franchises to establish local electric companies.[6] However, Edison's first move was to launch a corporation for supporting the lighting project itself. Working with Grosvenor Lowrey, Edison's lawyer, a group of New York investors representing gas and telegraph interests as well as the investment banking firm of Drexel Morgan & Company founded the Edison Electric Light Company with $300,000 in capital. In exchange for his lighting patents, Edison received half of the company's stock and a $50,000 advance to continue his experiments.[7]

In the invention factory, however, it was rough going. Despite the submission of several caveats to the Patent Office, and despite boastful statements to the press, the Wizard had not yet conjured a functioning light bulb. Like others before him, he began with a platinum element, having early on rejected carbon. Because of his familiarity with electromechanical devices, Edison believed that he could build into a light bulb a small regulator that would shut off the current just before the element overheated. It was precisely this complicated device that was being patented when Edison trumpeted his breakthrough.

Comprehending that an entire lighting system required many components beyond the lamp itself, the Edison team was also working on a high-efficiency dynamo, motors large and small, meters, sockets, and other components. To outsiders the Edison system's most controversial component, apart from the lamp itself, was the dynamo, whose design was arrived at through hundreds of experiments undertaken with instruments such as a galvanometer and a dynamometer. Nicknamed "Long-Legged Mary Ann" because of its stretched-out field electromagnets (figure 22.2), the gangly dynamo occasioned snickers; moreover, its electrical characteristics and its claimed high efficiency drew severe criticism from men of science. Among the critics was Edward Weston, who cited an alleged violation of Ohm's Law.

Formulated in 1827 by Georg Ohm on the basis of experiments with battery circuits, Ohm's Law specifies a relationship between tension, current, and resistance.[8] (Today, Ohm's Law is represented by the equation $E = IR$ where E is tension in volts, I is current in amperes, and R is resistance in ohms.) According to conventional interpretations of Ohm's Law before 1880, the armature's internal resistance had to equal the external resistance of the dynamo's load.[9] Edison disagreed. Rather, he argued, the dynamo's efficiency would increase if one lowered the armature's resistance as much as possible, regardless of the load's resistance. Accordingly, Edison's armature had a resistance of about half an ohm, far lower than that of other dynamos. And his field

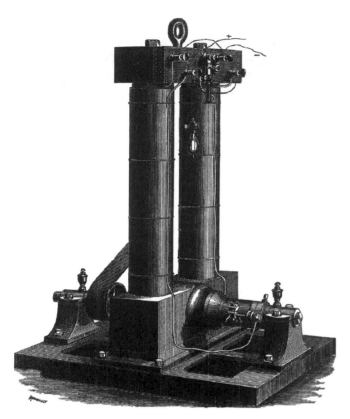

Figure 22.2
Edison's first commercial dynamo, "Long-Legged Mary Ann." Source: Dredge 1882, figure 255.

electromagnets had a relatively high resistance, which made more net power available to service the external load. Scientific authorities may have scoffed, but Edison's design principles were remarkably effective: Long-Legged Mary Ann had an efficiency of nearly 90 percent, the highest yet achieved.[10]

Developing the many components of the lighting system and building a machine shop large enough to manufacture a dynamo consumed cash at an alarming rate. While trying to mollify investors impatient with the project's slow pace, Grosvenor Lowrey negotiated a deal for more capital with Drexel Morgan & Company on the basis of foreign patent rights. J. Pierpont Morgan himself expressed confidence that Edison would eventually succeed.[11]

Other parts of Edison's system were also taking shape, but a working lamp remained elusive. For nearly a year after Edison's first announcement, his team tried to perfect the platinum bulb, but no amount of tinkering could get it entirely right. Moreover, platinum was a very dear metal, and Edison knew all too well that its expense threatened the economic viability of his vision of lighting the homes of ordinary people.[12] Near desperation, and dogged by the press and by anxious stockholders, Edison turned to carbon.

⎯⌒⌒⌒⎯

When American businesses, factories, and municipalities considered converting from gas to arc lighting, their decisions depended on assessments of many performance characteristics. Among these would have been the relative prices of gas and coal, the safety of the systems, the amount of light needed, the first costs of the equipment, the likely depreciation, the anticipated operating expenses, the expected reliability, and the value put on the symbolism of employing the most modern illuminant. Although cost data for European arc lights were being reported more often, such information—even when not contradictory—was an unreliable guide, because local factors in the United States and elsewhere varied greatly.[13]

In mid 1879, the *New York Times*—aware of the economic uncertainties—published an analysis based on American data. One general finding had been firmly established: "There remains no doubt from a collation of the 200 or more instances in this country in which . . . generators and lamps are in operation, that for large spaces the electric light can be supplied for from one-eighth to one-tenth the present cost of gas."[14] These figures did not take first costs into account, but they suggested that in a *large* factory already having steam power the purchase price of the electrical system could be amortized rapidly.

In view of the claimed economy and the apparent safety of electric lights, a growing number of American firms adopted them. The Riverside Worsted Mills, in Providence, Rhode Island, purchased a system with two Brush dynamos and 20 arc lamps, which replaced 230 gas burners. The electric light was said to be especially effective in the weaving room, where "the work requires an unusually strong illumination."[15] In Massachusetts, the Waltham Bleachery installed ten arc lights, which were fed current by two Wallace-Farmer dynamos. According to the Waltham Bleachery's cost analysis, the electrical system supplied far more light than the same expenditure on gas.[16] A mining company in Nevada put a huge Brush dynamo and three arc lamps on its placer claim so that the diggings could be worked at night.[17] The Wanamaker department store in Philadelphia used five Brush dynamos to light 20 arc lamps in its display windows, and found that pedestrians were drawn like moths to the brightly lit goods.[18]

European firms, especially French ones, were avid adopters of arc lights, but not every installation was applauded. At the Billingsgate Fish Market in London the arc lamps shed far too much light on the subject: "Soles that would have fetched a

Figure 22.3
The electric light in the Sautter & Lemonnier Factory. Source: Alglave and Boulard 1884, figure 232.

shilling a pair by gaslight looked dear at sixpence, while turbot fresh from the sea looked a week old." After just a few days, angry fishmongers forced the market's proprietors to restore the gaslights.[19]

Like their battery-powered and magneto-powered forebears, dynamo-powered arc lights found military applications almost immediately. In addition to lighting its own factory, which made Fresnel lenses (figure 22.3), the French firm of Sautter & Lemonnier sold the French government 40 Gramme-powered searchlights (then called "projectors") for use in defending forts and the coast. Costing 30,000 francs each, the lights had a useful range of about four miles.[20] Siemens's "dynamo-electric light apparatus" saw significant adoptions by the German Navy for similar purposes.[21] To direct the light, Siemens's shipboard installations used either parabolic reflectors or a projector with a Fresnel lens.

Several photographers in London and Paris used arc lamps in their studios, and so could accommodate requests for after-dinner portraits. With a hemispherical reflector of porcelain invented by P. H. Vander Weyde, the photographer could bathe customers in a gentle light lacking harsh shadows.[22]

In contrast to its solid endorsement of arc lights for large spaces, the *New York Times* declared with equal certainty that "bedrooms, parlors, and small spaces requiring light of only the power of a few candles, cannot be so illuminated." The performance characteristic of the arc light that made it most attractive for lighthouses and large spaces—incomparable brightness—was now its greatest liability. This wasn't just a matter of economics; it was also a matter of technical constraints.

To illustrate the technical argument with real numbers, let us take as an example a Brush No. 1 dynamo, the company's least powerful model. The No. 1 was driven by 1.5 horsepower and produced a total illumination of 1,000 candles. However, that immense amount of light was given off by just one lamp. Because a gas lamp for home use produced about 16 candles, 1,000 candles in a clerk's office or a bedroom would have been insufferably bright, even blinding—at a cost vastly greater than that of gaslight. Acceptable illumination in such rooms would have required dividing the dynamo's current among dozens of lamps, spread out among many offices or dwellings.

According to the *New York Times*, the electrical authorities Henry Morton and Moses Farmer claimed that "divisibility is an impracticable dream, contrary to well-ascertained doctrines of science," because electricity divided to such a degree could not raise a carbon arc or a metal element to white heat. Thus, the Brush No. 1 dynamo powered only one arc lamp because that was its effective limit. (Brush's largest dynamo of 1879 could support 18 lamps, but it required 13 horsepower.) Even some makers of electric lighting equipment agreed with this severe constraint: "Neither Mr. Brush nor Mr. Maxim believes in divisibility, or in the practical application of the electric lamp to the lighting of small spaces."[23]

In England, too, scientific authorities weighed in against the possibility of subdivision. In a paper published in 1879, William Preece, the Electrician General of the Post Office (which ran the national telegraph system), concluded after tedious mathematical analysis that "extensive subdivision" was "hopeless."[24] In framing his conclusions in terms of *all* electric lights, however, Preece had over-generalized his findings: the analysis had been based only on the electrical performance of arc lamps. Edison would prove that the laws of physics did not preclude the subdivision of current for an incandescent system.

In the meantime, several inventors found creative ways around the seemingly impenetrable barrier to subdivision. One idea was to use an optical system to divide the light (rather than divide the current). In 1879, Eusebius Molera and John Cebrián, Catalonian engineers residing in San Francisco, used an arc lamp to produce an intense beam of parallel rays that could be sent, with only small losses, to lamps in many rooms, or to a string of streetlights. The light was directed by a 24-inch Fresnel lens, routed through separate pipes, and guided to individual lamps by mirrors and prisms. At each lamp, the light was projected by means of prismatic reflectors and lenses.

Depending on the lenses and filters used, the intensity and color of the light could even be modified for specialized activities, such as a magic lantern show, photography, microscopy, or a medical examination. The inventors also claimed that their technology could be used with absolute safety in chemical factories and gunpowder works, and that it could produce 195 lights per horsepower at a cost one-twentieth that of gas.

The Molera-Cebrià system reached at least the prototype stage and was exhibited. According to one account of a 16-light demonstration, "the quality of the light was equal to pure diffused daylight" and it enabled a person to distinguish "several hundred shades of silk." Immediately before Edison's eventual breakthrough, *Scientific American* headlined the Molera-Cebrià system as the "practical divisibility of the electric light." I have found no evidence that this ingenious but complicated technology, whose first costs would have been stupendous, was ever adopted.[25] In any case, Edison's project soon made it superfluous.

_⌒⌒⌒_

The many months of toil that Edison's team had invested in the platinum bulb was not entirely wasted, for along the way they had perfected skills and technologies, such as sealing electrodes into a glass envelope, that would also serve in making carbon lamps. They also modified the conventional Sprengel pump until it achieved an ultra-high vacuum that removed extraneous gases from the platinum. A near-perfect vacuum also prevented oxidation of carbon.

On October 22, 1879, a cotton thread was carbonized at very high temperature (white heat) in an oxygen-free atmosphere. Contained in an evacuated bulb, this material put out a respectable amount of light, lasting more than 13 hours. The promising bulbs of late October stimulated a flurry of trials with other carbonized materials, and success followed success. Bristol board (a kind of paper) was tried, but bamboo was eventually preferred for the horseshoe-shaped filaments (figure 22.4). (This shape rendered the filament resistant to thermal shock.)

To publicize the breakthrough, and to show off the completed system, Edison hired a telegraph firm to string wires throughout Menlo Park and announced that there would be a full public display on New Year's Eve. In fact, through late December, Edison encouraged people to visit Menlo Park and watch the system take shape one lamp at a time. Reporters who came by heard Edison predict that a lamp would cost only 25 cents and would burn all day for 4 cents' worth of electricity; the same amount of electricity (about an eighth of a horsepower) would also run the motor of a sewing machine, or a small pump.[26] These published forecasts perhaps seemed like wild exaggerations; even so, reporters could see that the lamps were working continuously, day and night. By December 23, dozens of lamps were burning brightly in Edison's laboratory, shop, and office, and in six houses.[27]

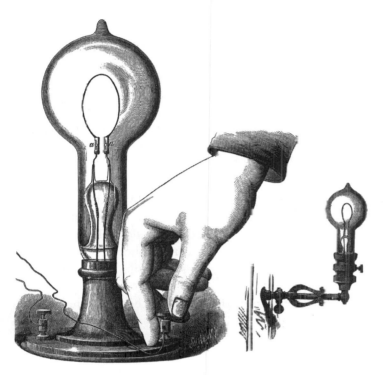

Figure 22.4
Edison's incandescent lamp. Adapted from *Scientific American* 19 (1880): 19.

Favorable coverage of Edison's lamp in the *New York Herald* elicited letters from scientific authorities who rendered negative judgments without having bothered to visit Menlo Park. Among the critics was the distinguished chemist Cyprien Tessié du Motay. Confining himself to matters ostensibly chemical, du Motay asserted that leakage of air into the bulb and disintegration of the filament would "show the impossibility of substituting this light, either in streets or in houses, for gas or even kerosene oil." Edison's lamps, du Motay continued, were "playthings . . . the experiments of a semi-practical prestidigitateur—something very amusing, perhaps."[28]

Cyprien Du Motay was not the only man to mock Edison's claims and cast aspersions on his honesty. William Sawyer, perhaps still smarting from Edison's recent rebuff, published a letter denying the Wizard credit for his earlier inventions and claiming that the phonograph was "of no earthly value." As for the new lamp, Sawyer discounted it on several grounds, one of them being that many experts—himself included—had been unable to make a similar paper element that worked. He went so far as to bluster that "all Mr. Edison's statements are erroneous." Sawyer also mentioned that his own carbon-rod lamp would be put on display in New York The letter ended with a challenge: if Edison's lamp could meet eight specific conditions, including maintaining a vacuum and running for 3 hours, Sawyer would pay him $100.[29] In the meantime, the *Chicago Tribune* reported in its Christmas Eve edition that the first lamps had been burning for 136 hours without a failure.[30] About two weeks later, Edison told a reporter: "I haven't time to bother with Mr. Sawyer or his challenges. I am always ready to explain matters connected with my light to any one who is really in earnest. But this man deals in absurdities." Edison added that Sawyer's own carbon-rod lamp was derivative.[31]

In commenting on Edison's achievement, Charles Brush managed to avoid blatantly *ad hominem* arguments, but he did malign the lamp. Edison, he claimed, had "taken up one of the oldest and least practicable forms of lighting by electricity, without apparently making any additions or improvements in it." Even if Edison had created a long-lasting lamp, Brush insisted, incandescent lighting would be less economical than arc lighting because of the filament's ravenous consumption of current. Brush even questioned whether Edison could keep his promise to exhibit the light on New Year's Eve.[32]

In an interview with the *New York Times*, Henry Morton expressed reservations about Edison's lamp and his electrical system. Declining an invitation to visit Menlo Park, he expressed doubt that Edison's system had achieved the performance requirements of practical incandescent lighting, including a long-lasting and inexpensive lamp, subdivision of the current, and economical transmission of electricity to users.[33]

Authorities who happened to be potential competitors of Edison in electric lighting had obvious economic incentives—e.g., free advertising—to offer their opinions to the press. But why did du Motay and Morton—who, I assume for the sake of argument, were genuinely disinterested parties—air their views to reporters? It would appear that scientific authorities, then and now, join in such public rituals to augment their prestige, call attention to (and reinforce) their positions as authorities, and exercise their often limited social power to influence others. I suggest that the penalty for erroneous views does not extend much beyond temporary embarrassment, because few readers will remember their stances anyway. More importantly, science is self-correcting over the long run, and so authorities enjoy wide latitude to express opinions that may turn out to be wrong. Thus, the popular media and scientific authorities enjoy a symbiotic relationship: the press arouses readers' interest by stirring up or feeding controversies about seemingly miraculous inventions, and scientific authorities receive ego-enhancing validation and (perhaps) influence other players. Morton and du Motay were merely participating in a peculiar eco-social system that thrives to this day.

Despite authoritative doubts, Edison's New Year's Eve show was a stunning success. According to the *New York Herald*, "extra trains were run from east and west, and notwithstanding the stormy weather, hundreds of persons availed themselves of the privilege. The laboratory was brilliantly illuminated with twenty-five lamps, the office and counting room with eight, and twenty others were distributed in the street leading to the depot and in some of the adjoining houses. The entire system was explained in detail by Edison and his assistants, and the light was subjected to a variety of tests."[34] Visitors were especially impressed by the ease with which motors and lamps could be interchanged on the power system. According to the *Herald*, everyone was convinced that Edison had shown the practicality of the electric light for household illumination. Newspapers around the country spread the news with cautious optimism, and shares of the Edison Electric Light Company that originally sold for $100 zoomed to over $3,000.[35] Two weeks later, *Scientific American* chimed in, detailing the lamp's elegant construction. Should these lamps last a decent interval, then Edison "has produced what the world has long waited for; that is, an economical and practical system of electric lighting adapted to the wants of the masses."[36] Even as the search for better filament materials continued in early 1880, lamps were already lasting, on average, nearly 800 hours.[37] Morton, du Motay, and other scientific authorities must have been at least a little chagrined, and a few months later du Motay died of "apoplexy."[38]

In Edison's view, the electric light would become the illuminant of choice for homes, businesses, and even factories. Edison believed that his system, in order to achieve this ambitious goal and to attract investors, would have to be cheaper than a gas lighting system. In his comparative cost estimates, incandescent lighting looked attractive.[39] Adding weight to these projections (which would turn out to be optimistic by an order of magnitude) was the remarkable achievement of subdividing the current. How had Edison accomplished this alleged scientific impossibility?

Arc lamps and the incandescent element in virtually every pre-Edison lamp had low resistance and thus consumed much power according to Ohm's Law (see above). For example, a Brush No. 1 dynamo having a modest efficiency of, let us say, 75 percent, would enabled a lone arc lamp to use about 700–800 watts. However, no electrical principle dictated that electric lamps had to be that power hungry, and no law of physics prevented a high-resistance lamp from radiating white light. Indeed, Edison showed that his carbon filament could reach incandescence, and that an 800-watt dynamo could power about a dozen of his lamps.[40] Edison used a high-resistance filament to subdivide the current. A brilliant move, this suggested that Edison understood the laws of electricity far more deeply than his detractors did.[41]

In a gloating editorial, *Science* struck back at Edison's critics: "We now extend our condolence to a certain class of professed scientific experts who have maintained, from first to last, the impracticability of Edison's well-devised plans. Never in the annals of scientific discovery has a grosser attempt been made to pervert the truth, and mislead public opinion."[42] Perhaps Edison, still subsidizing *Science*, had a hand in preparing this diatribe.

_⁓⁓⁓_

Edison's lighting of Menlo Park had been a wondrous spectacle. The next step was to refine the components, scale up the system, and electrify part of a major metropolitan area. Edison chose a location in New York's financial center that would become known as the Pearl Street District (because the central station was in a building on Pearl Street). This was an astute choice: a successful installation would demonstrate to capitalists and businessmen the practicality of generating electricity on a large scale. What is more, firms in this area could afford to splurge for the privilege of having the world's most modern lights.

In the meantime, consumers anticipated by neither Edison nor the capitalists came clamoring for "isolated plants" to light up ships, factories, and mansions. Henry Villard, who had seen the Menlo Park display, ordered a plant for the *Columbia*, a new steamship owned by the Oregon Railway and Navigation Company. Installed during the spring of 1880, the system worked well and did not foul the air in confined cabins as gaslights did.[43] Other isolated plants were placed in textile mills, where the risk of fire was so greatly diminished that lower insurance rates made electric lighting a practical proposition for mill owners. By early 1883, Edison had sold more than 250 isolated plants. Although these sales were profitable and pleased investors, Edison regarded them as a distraction and as not in keeping with his vision of electric light and power as an inexpensive utility available to all.[44]

To light the Pearl Street District, Edison and his companies developed many new technologies, including large dynamos, switches, fuses, junction boxes, and cables and tubes for subterranean installation. In addition, to improve the ability of the Edison system to compete with gas, it was necessary to work out cost-effective processes for manufacturing the components. Edison also had to obtain from city government a franchise for the privilege of tearing up the streets.[45] He could have placed the wires on poles far more easily and cheaply than underground, but he would have faced political resistance that had been stirred up by the criss-crossing of Manhattan with telegraph wires. He also feared that aboveground wires would be vulnerable to mischief.[46] For more than two years, Edison and his men were consumed with the challenges of building an electrical system on a scale never before conceived. Edison himself labored long hours in the streets of lower Manhattan, troubleshooting the system and supervising the laying of cable.

Establishing factories to manufacture light bulbs, dynamos, underground cables, and other components required a large amount of capital. Not only did Edison pour in his own money, and not only did he sell securities (including some shares in Edison

Electric Light Company); he also obtained investments from the members of his trusted laboratory team whom he had put in charge of the manufacturing plants. Managers of the Edison Electric Light Company, concerned about this disorderly growth as well as about the lavish spending on experiments, wanted to impose on Edison a bureaucratic structure similar those of other large corporations, including a requirement that Edison seek approval for each expenditure. Accustomed to a much looser organization that responded to his every request, Edison only reluctantly acquiesced to such impositions.[47] A few years later, however, Edison, threatening a proxy fight, retook control of the company and installed his own people in management positions.[48]

On September 4, 1882, the directors of the Edison Electric Light Company assembled in the offices of J. Pierpont Morgan, where Edison was to demonstrate the completed system. The dynamos were set in motion, and at 3 p.m. the Wizard turned on the lights. According to the *New York Herald*, "the luminous horseshoes did their work well." Answering the critics with uncharacteristic modesty, Edison merely claimed "I have accomplished all I promised."[49] Soon Edison would be selling an enormous dynamo that could power 1,000 lamps (figure 22.5).

The possibility that electric power might be employed in transportation was not lost on investors and entrepreneurs in many cities.[50] Siemens built the first trolley system, in Berlin in 1881. Three years later, E. M. Bentley and W. H. Knight inaugurated a

Figure 22.5
A 1,000-lamp Edison dynamo. Source: von Urbanitzky et al. 1890, figure 264. Note the directly coupled steam engine at left.

mile-long trolley line in Cleveland. However, the first major urban trolley in the United States was a 12-mile line in Richmond, Virginia, completed in 1887 by Frank Sprague, who had once worked for Edison. Although many people believed at the time that trolleys would be cheaper to operate than the horsecar systems they replaced, economy was often a secondary consideration. A city striving to attract businesses needed to have the most modern mode of transportation. Beginning in the late 1880s—decades after Charles Page showed that an electric motor could drive a large vehicle—there were 3,000 miles of trolley lines in American cities and towns.

_⁓⁓⁓_

During Edison's late-December demonstration at Menlo Park, William Sawyer went public with his own early trials of an incandescent lamp having a charred paper element.[51] Despite the admission that the element failed quickly, Sawyer's claims attracted the interest of Charles Cheever, a patent speculator, who urged Sawyer to secure a patent on the lamp. With $2 million in capital, the Eastern Electric Manufacturing Company was formed.

Sawyer's alcoholism was becoming ever more destructive. In April 1880, in a quarrel about Edison's lamp, Sawyer shot a neighbor. The Eastern Electric Manufacturing Company paid for his defense, but the case was hopeless and Sawyer was convicted. Before being sentenced, however, the 33-year-old inventor apparently drank himself to death.

Sawyer's inventions became the basis of a new firm, the Sawyer-Man Company. In 1884 this company and its lamp patents were acquired by Thomson-Houston, an electric lighting company whose principals had earlier cast doubt on Edison's etheric force. The Thomson-Houston purchase turned out to be fortunate. George Westinghouse was building a high-voltage ac light and power system to compete with Edison's low-voltage dc system.[52] However, Westinghouse's company—a highly profitable concern built upon his invention of the air brake for railroad cars—lacked its own incandescent lamp. Rather than pay Edison royalties, Westinghouse purchased the Sawyer patents from Thomson-Houston. Westinghouse also bought transformer patents and hired William Stanley Jr., who, exploiting earlier European efforts, perfected an efficient transformer that could be manufactured economically. And Westinghouse employed Nikola Tesla to work out the details of his own ac motor and polyphase system. By 1886, Westinghouse possessed all the basic components needed for an ac system of light and power.[53] However, because Sawyer's lamp was defective, Westinghouse ended up copying, without licenses, patented features of Edison's lamps.

With his collection of indifferent lamp patents, George Westinghouse could piously maintain that he owned all the technology needed for incandescent lighting.

Westinghouse commercialized his system in 1886, and it found eager buyers. One of its major selling points was that alternating current would allow long-distance transmission of power whereas Edison's dc system would not. However, as William Ayrton had noted in 1879, that wasn't strictly true: *high-voltage* direct current could be transmitted as easily as high-voltage alternating current. But an Edison central station, generating at most a few hundred volts, had a limited radius of action—no more than a mile—because greater distances required very thick and thus expensive copper wires.[54] Edison, familiar with the relevant electrical laws, was aware of this limitation. However, he believed that it could be overcome by building central stations at intervals of three or four blocks.[55] In contrast, a Westinghouse system sent current many miles, which invited profitable extensions, especially from downtowns to "horsecar suburbs" and (as hydropower) from rural rivers to cities. With its economy of scale, which permitted the construction of fewer and larger central stations, high voltage seemed the more practical system to investors.

Once Edison perceived Westinghouse's ac system as a serious threat, he picked up the gauntlet. So began the "war of the currents."[55] In truth, this was a war over control of an expanding industry believed to offer vast growth potential. Battles began during the late 1880s and raged before the public, with both principals granting interviews and authoring polemical articles in popular magazines.

Edison ostensibly objected to alternating current because its high voltage threatened human life.[57] To demonstrate the danger, Edison took advantage of contemporary controversies concerning capital punishment, especially those in New York State.[58] Opponents of capital punishment were especially unhappy with hanging, the most common method of state-sanctioned killing in America. They argued that hanging was inhumane: not only could it become a ghastly public spectacle, witnessed by jeering or cheering crowds, but the condemned person often survived the drop for 15 or 20 minutes, only to die of strangulation. As early as 1876, *Scientific American* had suggested that an induction coil with an 18-inch spark could do the job of execution swiftly.[59] Others, however, were opposed because the use of electricity might "degrade science by making it an accomplice in the executioner's task."[60]

In 1886, New York's legislature appointed a distinguished commission to advise on "the most humane and practical method" of execution.[61] The members conducted extensive research and sent a questionnaire to physicians, professors, judges, and wardens; even Thomas Edison received one. Of the 200 replies, 81 favored retaining hanging. Electricity, with 75 votes, was the favored alternative. The commission recommended that death by electricity be adopted for capital crimes. A bill was introduced to that effect, and it won support from the *New York Times*.[62]

Although the ability of electricity to kill humans and even large animals was already well known, Edison believed that experiments would reveal the best method for delivering current to the human body. He was certain, however, that the most efficient agent of execution would be his competitor's high-voltage ac system. On this matter, Edison—now revered as America's foremost expert on electric power—was regarded as the scientific authority, and his opinion was widely accepted.[63] Partisans of Westinghouse, however, took no pride in the prospect that their ac system might become the executioner's tool.

In Edison's Menlo Park compound, dogs, calves, and a horse were dispatched with electricity from Westinghouse dynamos generating from 500 to 1,000 volts. Among the witnesses were several members of a committee appointed by the superintendent of New York's state prisons to study electrical execution. The Edison camp claimed that the experiments showed "conclusively the utility and desirability of the alternating current as a means of producing sudden and painless death," and the committee agreed.[64] The calves, their meat still in good condition, were returned to the butcher who had supplied them.

In mid 1889, the State of New York outlawed hanging and decreed that all executions be carried out with electricity. Following Edison's advice, the state bought Westinghouse ac systems for its Auburn, Clinton, and Sing Sing prisons. The first electrical execution was carried out in a specially outfitted chair at Auburn in August 1890. An unrepentant William Kemmler, sentenced to death for killing his common-law wife, was rendered rigid and unconscious by the jolt. But when the current was turned off after 17 seconds, he began to breathe again, and the order was given to resume the electrocution.[65] Over the next three minutes, Kemmler was cooked to death, the stench causing one witness to faint.[66]

The Edison and Westinghouse camps issued contradictory interpretations of America's first official electrocution. Edison insisted that the doctors had bungled the execution by placing the electrodes improperly, yet he also contended that Kemmler had died from the first shock. Westinghouse, approached by a reporter for the *Times*, opined that it had been "a brutal affair" and that "they could have done better with an axe." A spokesman for the Westinghouse Company predicted that this would be the "last execution of the kind in this country."[67] But in the decades ahead, state authorities purchased ac-powered electric chairs for prisons throughout the United States. These adoptions could not have given Edison much satisfaction. Although he had prevailed as the scientific authority on this issue, his strategy of equating the name Westinghouse with sudden death would not earn him victory in the "war of the currents."

_⌒⌒⌒_

Because of Edison's prestige, his resources, and, above all, his tenacity, the "war of the currents" lasted about five years. Meanwhile, Westinghouse was gaining market share.

Local electric companies, with an eye toward the future, increasingly favored ac systems because they allowed larger service areas and long-distance transmission. Between 1886 and 1890, Westinghouse sold ac equipment for about 300 central stations. The triumph of the Westinghouse system was conspicuous at the Columbian Exposition of 1892–93, held in Chicago.[68] Determined to get the fair's lighting contract, Westinghouse undertook the project at a loss. Twelve enormous ac generators served current to 250,000 incandescent lamps, ensuring that everyone who visited or read about the fair would associate the name Westinghouse with progress in electric lighting.[69]

One wonders why Edison, a shrewd businessman and an industrious inventor, so defiantly resisted the demands of the market and the seemingly inevitable transition to alternating current. To the new generation of electrical engineers, alternating current looked very attractive. Even within the Edison organization there were ac advocates, but the "Old Man" (as he was affectionately called by his team) would not authorize the change. Several plausible explanations, by no means mutually exclusive, help to account for Edison's refusal to embrace the new electrical regime. One is that Edison and his financiers were protecting their large investment in dc technology. Clearly, though, Edison could have developed an ac system while continuing to sell dc systems where they could be deployed most effectively (e.g., in isolated plants for mansions, ships, hotels, factories, and trolleys). But Edison was mainly interested in selling equipment for central stations so that large numbers of ordinary people could summon the electrical genie at home. He was also a prideful and sometimes combative man who might have regarded a move to alternating current as an unthinkable capitulation to Westinghouse, that flagrant infringer on his patents.

Edison's most fervent objection to alternating current, repeated *ad nauseam* during the "war of the currents," was that its high voltage put human life at risk. One could view this position as mere posturing, affected in order to oppose Westinghouse, but there may be more to it. Had he employed dynamo-motor pairs, as proposed by Ayrton, Edison could have built a *high-voltage* dc system that, in principle, might have competed on an equal footing with ac systems.[70] Although he considered this possibility, even working up some plans, Edison did not commercialize it. This is understandable if one accepts his objections to high voltage at face value. Indeed, by rejecting high voltage so definitively and so publicly on the basis of its health hazards, Edison had effectively foreclosed his options. Bringing to market a high-voltage dc system would have damaged his credibility. In any event, because of their founder's refusal to bow to market forces, the Edison companies were losing the "war of the currents." Stockholders, including J. P. Morgan, were displeased.

Frustrated by Edison's stance toward ac technology, financiers took control of his companies, merged them with Thomson-Houston, which had extensive ac experience and a conventional bureaucratic organization, and in 1892 forged an electrical colossus: the General Electric Corporation.[71] Now they could build ac systems and compete with Westinghouse on all fronts.

_⌒⌒⌒_

Although Edison, eased out of the corporations founded on his inventions, lost the power struggles with Westinghouse and with the financiers, he remained an uncommonly wealthy inventor. Even so, he was no longer a player in the American electric power industry he had helped to found. The "Old Man"—still in his forties and quite energetic—moved on to new and challenging projects, financing many of them himself rather than relying on the capitalists he had come to loathe. Some of these projects—including making movies and records, milling iron ore and cement, and developing an alkaline storage battery for electric automobiles—would extend well into the twentieth century.[72] Electricity in the home turned out to be a luxury good whose slow adoption delayed for many decades the realization of Edison's vision of light and power democratized.[73]

Notes

The following abbreviations are used:

AEMC   *Annals of Electricity, Magnetism, and Chemistry*

AJSA   *American Journal of Science and Arts*

CT   *Chicago Tribune*

DP   Thomas Davenport Papers, Library of Congress

JFI   *Journal of the Franklin Institute*

GAPS   *Greenough's American Polytechnic Journal*

MAB   *Manufacturer and Builder*

MM   *Mechanics' Magazine*

MP   Samuel Morse Papers, Library of Congress

NMAH   National Museum of American History, Smithsonian Institution

NYT   *New York Times*

SA   *Scientific American*

SAS   *Scientific American, Supplement*

SIA   Smithsonian Institution Archives

SIAC   Smithsonian Institution, Archives Center

TE   *The Engineer*

USNA   United States National Archives

All NMAH catalog numbers refer to the electrical collections (E*).

Notes to Chapter 1

1. On the Pearl Street District and demonstration, see p 206 of Israel 1998 and chapter 7 of Friedel et al. 1986.

2. Quoted on p. 206 of Israel 1998.

3. I do not wish to imply that all technologies, even electrical technologies, have their origins in science or are merely applied science. However, natural philosophy *was* the starting point for many technologies treated in this book.

4. The corresponding definition of "practicable" is quite similar and equally general: "possible to practice or perform: capable of being put into practice, done, or accomplished: feasible . . . capable of being used: usable."

5. Carlson (1991: 7–8) argues that social and technical forces both shape technologies, and so neither can be neglected.

6. Staudenmaier (1985) surveys the historiography of technology through the pages of *Technology and Culture*. On social constructivism, see Bijker 1995; Bijker et al. 1987; Winner 1985, 1986; for a philosophical treatment, see Hacking 1999. Forman (2007) argues that historians of technology should take broad cultural presuppositions more seriously.

7. Sources on the anthropology of technology include the following: Appadurai 1986; Chaiklin and Lave 1993; Dobres 2000; Dobres and Hoffman 1999; Douglas and Isherwood 1996; Foucault 1973, 1977; Gell 1998; Hayden 1998; Hutchins 1995; Keller and Keller 1996; Kingery 1996; Lansing 1991; Latour and Woolgar 1979; Lave and Wenger 1991; Lemonnier 1992, 1993; Leroi-Gourhan 1993; McCracken 1988; Meskell 2004; Miller 1995a,b, 2005; O'Brien et al. 1994; Pfaffenberger 1992; Schiffer 1992, 1993, 2000, 2001a, 2002, 2004, 2005a, 2005b; Schiffer et al. 1994; Schiffer et al. 2003; Suchman 1983; Wallace 2003.

8. On the relationship between the historical and archaeological records, see Schiffer 1987.

9. These are, obviously, modern Western categories that do not necessary have direct equivalents in societies of other times and places.

10. Schiffer and Miller 1999.

11. Behavioral studies of technology and technological change include the following: LaMotta and Schiffer 2001; Schiffer 1992, 1993, 2000, 2001b, 2002, 2004, 2005a,b; Schiffer et al. 1994; Schiffer et al. 2003; Schiffer and Skibo 1987, 1997; Skibo and Schiffer 2001, 2008; Skibo et al. 1995; Walker and Schiffer 2006.

12. On performance characteristics, see Schiffer 2003, 2005b; Schiffer and Miller 1999: 16–20; Skibo and Schiffer 2001, 2008.

13. I follow Layton's (1971) definition of engineering science as the principles arising in application-oriented contexts. See also Schiffer and Skibo (1987) on "techno-science."

14. For further discussion of these three processes, an obvious revision of Schumpeter's duo of invention and innovation, see Schiffer 1996, 2001b. The entire sequence of activities in the life history of an artifact is termed a "behavioral chain" (Schiffer 1976).

15. Pacey (1992) argues that many inventions, from cathedrals to moon landers, are "idealistic" technologies because they do not make economic sense, their main functions being to instantiate

cultural values. Friedel (2007) makes the case for a "culture of improvement" that impels inventiveness.

16. On "developmental distance," see Schiffer 2005a.

17. Rosenberg (1970) emphasizes the importance of know-how and organization.

18. On user perspectives, see Edgerton 2007; Oudshoorn and Pinch 2003; Schiffer 1991, 1992; Schiffer and Skibo 1997.

19. According to Anderson (2006), the Internet enables extreme market segmentation, allowing niche products to reach consumers—products that might otherwise be unprofitable. Sometimes consumers salvage an apparently failed technology by finding applications not anticipated by its promoters.

20. Elsewhere (Schiffer 2007) I define the totality of technology-specific role players as an important unit of societal organization and term it a "cadena" (from the Spanish word for chain).

21. On how Edison, in discourses with other players, employed different "languages," see Bazerman 1999. Bazerman's brilliant analyses have appreciably influenced my approach.

22. For present purposes, I take "social power" to include political and economic power—any manifestation of the ability to allocate human and/or material resources. This definition is adapted from Nielsen 1995: 49.

23. Electrical technologies figure prominently in Mumford's (1963) "neotechnic" phase.

24. These changes are covered in the following works: Bernal 1965; Bruce 1987; Chandler 1977; Cowan 1997; Giedion 1969; Jacob 1997; Kirney 2004; Landes 1969; Licht 1995; Pursell 1995.

25. The present work is concerned with how the role of scientific authority is played by particular individuals in specific cases. I take the existence of the role and its recognition by players as givens. On the social construction of scientific authority, see chapter 1 of Biagioli 2006a.

26. On the role of scientific authority in twentieth-century society, see Pielke 2007; Redner 1987; Toumey 1991; Walters 1997.

27. Nineteenth-century electrical technologies are richly represented in patent specifications, in technical manuals, in journal articles, in inventors' magazines, and in museum specimens. For later time periods, I drew heavily on *Scientific American* (founded in 1845) as an information source. A weekly magazine published by a New York patent agency, *SA* lavishly covered electrical technologies in text and drawings, and also gave voice to the practicality assessments of many players, including inventors, patent attorneys, journalists, scientists, engineers, and sometimes consumers.

28. In the late twentieth century, the scientific authority's role expanded to include assessment of a technology's effects.

Notes to Chapter 2

1. For a survey of eighteenth-century electrical technologies, see Schiffer et al. 2003.

2. Franklin 1986: 130.

3. Galvani 1953.

4. On the Galvani-Volta controversy, see Pera 1992; for boiled-down versions, see Schiffer et al. 2003; Dibner 1964. On Volta's life and scientific activities, see Pancaldi 2003.

5. As an archaeologist, I am often asked about the so-called Baghdad battery, made about two millennia ago. This find consisted of a ceramic jar and a copper cylinder containing an iron rod; the metal components were cemented into the jar with asphalt, an insulating material. On the artifact's nature and probable function, see Keyser 1993.

6. Volta 1800. For an English translation, see Dibner 1964.

7. On Benjamin Thompson and his works, see Brown 1979.

8. Dray (2005) develops this argument at length.

9. See Carrier 1965.

10. On Davy's lectures and their audience, see Foote 1952.

11. For an early use of the term "cell," see Davy 1812: 152.

12. Davy 1808: 3. This source furnishes information about Davy's isolation of potassium and sodium.

13. Ibid.: 3.

14. Children 1809.

15. In a sentence that startles the modern reader with its prescience but was unnoticed in its time, Children (1809: 37) came close to formulating the law of electric power: "The absolute effect of a VOLTAIC apparatus, therefore, seems to be in the compound ratio of the number, and size of the plates: the intensity of the electricity being as the former, the quantity given out as the latter."

16. Source of quotation: Davy 1812: 152.

17. For the quotation and for details of the battery's construction, see ibid.: 152.

18. Ibid.: 151–152.

19. Singer 1814: 410.

20. Davy 1812: 152.

21. See, e.g., McMillan and Cooper 1910.

22. On the safety lamp, see Davy 1816; Jones and Tarkenter 1992: 16–17.

23. Davy 1816: 3.

24. Source of quotation: ibid.: 15.

Notes to Chapter 3

1. Dibner (1961: 12) mentions that several people had stumbled upon the effect before Oersted, but assigned it no significance.

2. Some information on Oersted's life is paraphrased from Williams 1974. The most useful source on Oersted's commitment to *Naturphilosophie* is Stauffer 1957.

3. Jelved et al. 1998: 378.

4. I follow Stauffer's (1953, 1957) reconstruction of the crucial experiment.

5. For this surmise I am indebted to Henry Noad (1857: 644).

6. For English translations of Oersted's published works, see Jelved et al. 1998. Dibner (1961) also reproduced the 1820 paper in the original Latin and in English translation.

7. Williams 1970: 143. Other biographical details on Ampère are from this source.

8. On Ampère's early discoveries in electrodynamics, see Williams 1983.

9. Ampère 1820 is probably his most accessible early work. It is unencumbered by equations.

10. Ampère 1826: 195.

11. Ampère coined the term "galvanomètre" (1820: 67) and distinguished it from electrometers (which indicated tension). On the history of galvanometers, see chapter 14 of Keithley 1999.

12. Biographical information on Faraday is condensed from Williams 1971. See also Williams 1965; Hirshfeld 2006.

13. Information on this device is from Faraday 1952a: 797–798.

14. Faraday 1952b: 807.

15. Davis 1838: 69.

16. Among those attributing invention of the electric (or electromagnetic) motor to Faraday are Baird (1994), Bruno (1997: 316), Gooding (1990), Hacking (1983), Lindley (2004: 73), and Williams (1971). Williams (1971: 533) even asserts that Faraday's rotating device was "the first conversion of electrical into mechanical energy." Greenslade (1936: 26) incorrectly asserts that it "led to the various modern forms of electric motors."

17. Franklin 1996: 38; Ferguson 1778: 24–33 and plate 2.

18. Cardwell (1992) debunks the claim that Faraday invented the electric motor, but on different grounds than mine. Gee (1991) concludes that Faraday's rotating device "does not constitute an

electric motor," yet allows that "the tangible evidence of continuous rotary motion from electromagnetism did inspire others to develop the idea."

19. Biographical information on Sturgeon is from Finn 1976 and Farrar 1974. On the history of the electromagnet, see King 1962a.

20. Sturgeon (1826: 357–358) is explicit about the performance requirements of electromagnets used in demonstrations.

21. Sturgeon 1850: 36.

22. Ibid.: 106.

23. Sturgeon 1826: 360.

24. Watkins's 1828 text on electromagnetism illustrates Sturgeon's display devices, including many that rotated (but none is a motor by my criteria), which were "constructed and sold by Watkins and Hill" (p. 69). For an illustration of the fabled horseshoe electromagnet, see plate 1, figure 9.

Notes to Chapter 4

1. Henry 1935: 9.

2. The best sources on Henry's life are Moyer's 1997 biography and the introductory chapters in the eleven-volume *Papers of Joseph Henry*, published by the Smithsonian Institution. Also useful are Reingold 1972; Crowther 1937; Jahns 1961. For a survey of Henry's scientific contributions, see Taylor 1879a.

3. Jahns 1961: 65.

4. Unless otherwise noted, information on Henry's experiments with electromagnets is from Henry 1831a.

5. Ibid.: 400. Much later, in a court deposition, Henry stated that Sturgeon had "produced the first electromagnetic magnet" (Henry, in Smithsonian Board of Regents 1858: 108).

6. For an illustration of the galvanic multiplier, see Schweigger 1821, plate 1, figure 3. On early galvanometers, see Committee on Electrical Standards 1913; Keithley 1999.

7. Henry 1831a: 405.

8. Henry to Silliman, January 15, 1831, in Reingold et al. 1972: 320.

9. Henry's (1831a: 407) exact statement was ". . . it is evident, that a much greater degree of magnetism can be developed in soft iron by a galvanic current, than in steel by the ordinary method of touching." This generalization was not overturned until the late twentieth century, with the commercialization of permanent magnets containing rare-earth elements.

10. I have relied on accounts given by Moll (1830, 1831).

11. Moll repeated Henry's experiments, but his most powerful electromagnet lifted only 254 pounds (Moll to Faraday, June 7, 1831; Williams et al. 1971: 1197–1198).

12. Henry to Silliman, December 10, 1830 (Reingold et al. 1972: 301–302).

13. Silliman to Henry, December 17, 1830 (Reingold et al. 1972: 302–303).

14. This paper, Henry's first on electromagnetism, appeared in the *Transactions of the Albany Institute* and is unlikely to have been seen by many Europeans.

15. Quotation is from Henry to Silliman, December 28, 1830 (Reingold et al. 1972: 316).

16. Reingold et al. 1972: 348, note 4.

17. Pixii 1835: 23. Daniel Davis, an American instrument maker in Boston, sold smaller Henry-style electromagnets for 50 cents to $3 (Davis 1838: 69).

18. Pope 1891a: 2.

19. Information on this project is from the following letters in Reingold et al. 1972: Henry to Silliman, December 8, 1830, pp. 316–317; Silliman to Henry, January 6, 1831, pp. 318–319; Henry to Silliman, January 15, 1831, pp. 319–320; Silliman to Henry, January 25, 1831, pp. 321–322; Silliman to Henry, March 12, 1831, p. 330; Henry to Silliman, March 28, 1831, pp. 331–333; Silliman to Henry, April 5, 1831, pp. 333–334; Henry to John Henry, July 9, 1831, pp. 348–349; Henry to Parker Cleaveland, November 16, 1831, pp. 375–376.

20. Henry and Ten Eyck 1831.

21. NMAH catalog no. 181,343.

22. The Bowdoin project is described in Henry to Parker Cleaveland, May 8, 1832 (Reingold et al. 1972: 420–426; the quotations are, respectively, on pp. 421 and 426). This is the most detailed account of how Henry manufactured large electromagnets.

23. Information on this project is from Penfield and Tait to Henry, May 30, 1831 (Reingold et al. 1972: 340–341) and from Pope 1891a.

24. Source of quotation: Reingold et al. 1972: 341, note 1.

25. Davenport 1851: 5. Later in 1831, Henry built a second magnet for another iron mining firm, also to magnetize pieces of steel in an iron-ore separator. See Henry to Rogers, November 4, 1831 (Reingold et al. 1972: 364–372).

26. Henry 1831b.

27. Palmer, a London instrument maker, sold a Henry-style teeter-totter without crediting the American; it used horseshoe magnets instead of bar magnets. See Palmer 1840: 45.

28. Henry 1831b: 340.

29. Source: "The Statutes of the Albany Academy" (Reingold et al. 1972: 235). I follow Reingold et al. (p. 232) in assuming that Henry authored this definition.

30. Kahn (2005) points out that inventive activities were much more democratic in the United States than in Europe.

31. Quotations in this paragraph are from Henry to John Henry, July 9, 1831 (Reingold et al. 1972: 348, 349).

Notes to Chapter 5

1. On early, non-electrical telegraphs, see Aschoff 1984; Beauchamp 2001. My paragraphs on electrostatic telegraphs include information taken almost verbatim from Schiffer et al. 2003: 249–254.

2. Cited by Fahie (1884: 68–71), who also reproduced the letter.

3. On Chappe's system, see ibid.: 93–95.

4. Ibid.: 101–108; Noad 1857: 748.

5. Ibid.: 127.

6. On early galvanic telegraphs, see Fahie 1884; King 1962c.

7. Unless otherwise noted, information on Sömmerring's telegraph is abstracted from Fahie 1884: 227–243.

8. Sömmerring (1811: 479) used that event to refute a critic who had questioned his telegraph's ability to transmit very far.

9. Fahie 1884: 235.

10. Ampère 1820: 73.

11. On Barlow's contributions, see Sharlin 1970.

12. Source of quotations: Barlow 1825: 105.

13. To promote ease of understanding, I have remodeled Barlow's law; in the original formulation, current strength is expressed as a function of the square root of the distance.

14. Henry 1831a: 404.

15. In Ohm's publication of 1827, the main idea—that there is a direct relationship between voltage and resistance, holding constant current—is lost amidst a forest of abstruse equations. As Caneva (1974) points out, Ohm sought acceptance by mathematical physicists, and that accounts for his mode of presentation. As a result, Ohm's law entered textbooks very slowly, and only in later decades was it used explicitly in circuit design. Henry learned about Ohm's law sometime between December 1834 and 1837 (Reingold 1972: 278).

16. Henry 1831a: 404.

17. Smithsonian Board of Regents 1858: 94.

18. Source of Henry quotation: ibid.: 94.

19. For Henry's description and illustration, see ibid.: 104. See also Moyer 1997 69–70.

20. Henry, in Smithsonian Board of Regents 1858: 106.

21. Source of Henry quotation: ibid.: 111.

22. Henry also set up a similar apparatus at Princeton in 1835, including a line between his home and the campus; the latter used a single wire with a ground (Nickerson 1885 17–19). This line may have been used for some kind of communication.

23. Henry to MacLean (Reingold et al. 1972: 435–437). Source of quotations: ibid.: 436–437.

24. Silliman to MacLean, September 23, 1832 (Reingold et al. 1972: 456).

Notes to Chapter 6

1. Henry 1832: 403.

2. In discussing the discovery of electromagnetic induction, I rely on Faraday (1952c: 265–284). Williams (1965) supplied a theoretical structure for the major experiments.

3. On the basis of his iron ring with two coils, Faraday often receives credit for inventing the "transformer"—a term introduced only decades later. However, that interpretation is unsustainable because Faraday used it only to demonstrate an inductive effect, and most others treated it similarly. For example, Newman's (1835: 21) catalog contains the following listing: "FARADAY'S compound helix, for exhibiting electric currents, produced by induction from magnetism [and] Double helix, for exhibiting the effects of induction upon a neighbouring wire upon making and breaking contact with the galvanic battery." Moreover, Faraday did not suggest that his apparatus could be used to raise or lower intensity or quantity, which it could not have done without an additional device to convert a battery's dc into ac or at least pulsating dc; yet he did mention the single coil (in self-induction) as being "one of the very few modes we have at command of converting quantity into intensity as respects electricity in currents" (Faraday 1834: 351). However, an anonymous experimenter believed that the "practical importance" of Faraday's ring had been overlooked. He suggested that, by using a crank-driven pole changer to apply alternating current to one coil, he could conveniently generate a high-intensity current in the other. Thus, the ring could replace the magneto; see MM 21 (1834) 113–116, 191, 273–275.

4. Henry 1832, 1835. When the battery is disconnected, the magnetic field collapses, momentarily inducing a current in the coil.

5. Faraday 1952c: 267. In the diary, his conclusion is "Hence effect evident but transient" (Martin 1932: 368).

6. Faraday 1952c: 277. However, he does not mention this goal in the diary.

7. Ibid.: 279. In the diary he merely states, without the slightest hint of elation, that "whilst the plate continued to revolve the Galvanometer needle was permanently deflected" (Martin 1932: 385).

8. Faraday 1952c: 281, note 1.

9. Source of quotations: ibid.: 284.

10. Williams 1965: 196. This claim is also erroneous because electromagnetic generators of the 1830s were magnetos; the term "dynamo" was not applied to generators until about 1870 or later. (See chapter 19 below.)

11. On early electromagnetic generators, see Bowers 1982; Gee 1993; King 1962b.

12. The Pixii firm published a pamphlet that described and advertised these machines (Pixii 1833?). There are also listings in later catalogs (Pixii et Fils 1835: 23–24, 1845: 31–32). Both machines were described briefly in *Silliman's Journal* (Hachette 1833).

13. The small Pixii machine was first described and illustrated in Hachette 1832.

14. Thus, in later magnetos having a stationary magnet and moving coils, when the coil sweeps by one pole of the horseshoe magnet, the induced current flows in one direction; when it goes by the other pole, current flows in the other direction.

15. My description of the large Pixii magneto is based on my study of the machine in the National Museum of American History (catalog no. 323,353); although it bears the date 1832, it may have been a reproduction. The NMAH also possesses a smaller Pixii machine (catalog no. 244,882). For an illustration, see Hodgins and Magoun 1932: 223. See also Le Roux 1868: 11.

16. At every half-revolution of the magnet, the cam moved an insulated rocker arm, which pressed brass pins alternately on the front or back of crossing copper strips, thereby reversing the current and creating a unidirectional flow.

17. Pixii 1833?: 4, my literal translation. The mention of "weather" alludes to the electrostatic generator's sensitivity to changes in ambient humidity.

18. The first person to publish a design for this kind of magneto was William Ritchie, an English cleric, whose spartan machine apparently lacked technological descendants. (See Ritchie 1833.) Curiously, on December 18, 1832, two weeks after Saxton had made his first sketch, Ritchie paid him a visit. Saxton "proposed to show him the plan I had contrived to get the full effort of a magnet in producing the spark but he declined seeing it and said he was at something of the kind himself" (SIA, Record Unit 7056, Joseph Saxton Papers, Notebook 2, unpaginated).

19. For a brief biography of Saxton, see Frazier 1975. Henry (1935) furnishes more details on Saxton's life, but he is not a reliable source on the priority controversies that erupted over the invention of magnetos. These happenings are recounted in detail by Morus (1998), on whose account I rely except where I note otherwise.

20. On the Adelaide Gallery, see Morus 1998: 75–83.

21. In 1832, in a set of letters forwarded by Bache and published in *Silliman's Journal* (Bache 1832), Saxton briefly described this device, and Lukens and Say, friends of Saxton in Philadelphia, reported their modifications of it. See also Emmet 1833.

22. King (1962b: 345) sometimes states that a magneto was more "efficient" than an earlier one, but provides no quantitative data to support such claims until discussing the machines of the 1860s (ibid.: 369), when generator inventors themselves began to measure efficiency.

23. For a photograph of an early Saxton magneto, see the frontispiece in Henry 1995. For illustrations of a different early Saxton machine, see King 1962b: 351; MM 21 (1834): 95. For a description and an illustration of a later version of a Saxton machine, see Saxton 1836. Watkins and Hill (1836: 62) also sold "Saxton's Magneto-electric Inductive Machine . . . the best arrangement hitherto devised for illustrating the electrical, chemical, physiological, and mechanical powers evolved by magnetism . . . £12 4s."

24. Some drawings show three coils (see previous note). Saxton (1836: 362) also illustrates four coils.

25. SIA, Record Unit 7056, Joseph Saxton Papers, Notebook 3, unpaginated.

26. Saxton 1836: 364.

27. Watkins (1835: 239) mentions the ownership of the Pixii machine, but it is not clear which model was tested. Wealthy collectors had long purchased pricey electrical apparatus. See chapter 5 in Schiffer et al. 2003.

28. Quoted on p. 89 of Morus 1993.

29. A Saxton magneto is curated by the American Philosophical Society.

30. Sturgeon 1834. This brief note supplies no information on the commutator itself.

31. Sturgeon 1835: 231.

32. Sturgeon 1850: 251.

33. For a schematic illustration of a Saxton-style machine with Sturgeon's early commutator, see ibid., plate 12. Some illustrations of early Pixii magnetos show a similar commutator.

34. This quotation is from a reprint, of uncertain provenance, titled "Description of a Magnetical Electrical Machine. Invented by E. M. Clarke." Bound with Sturgeon 1830 in the Dibner Library at NMAH (pp. 89–95), it most likely dates to 1837 or later.

35. For an overview of nineteenth-century electromedicine, see Rowbottom and Susskind 1984. On eighteenth-century electromedicine, see Schiffer et al. 2003: 133–160.

36. Clarke 1836. Elsewhere (1835: 169) Clarke mentions that "for some time past [he had] been engaged in the manufacture of the new magnetic electrical machines."

37. This design change may have been made to "increase the current" (King 1962b: 349; cf. Sturgeon 1835: 234); it also made the machine more compact and portable.

38. Saxton 1836: 360.

39. Ibid.: 365.

40. Saxton could not have made this four-coil device before January 1835, when in a notebook he recorded his first understanding that coils having different lengths of wire are appropriate for achieving different effects (January 1, 1835, SIA, Record Unit 7056, Joseph Saxton Papers, Notebook 5, unpaginated). By that date, electrical experimenters in America, including Henry (1832) and Emmet (1833), had published the finding that electricity of higher tension was generated in coils with many turns. Henry (1835) reiterated and augmented this finding with reference to self-induction.

41. Clarke 1837.

42. Clarke (1836) employed these terms when referring to the variant armatures.

43. In England, Newman (1836) offered a magneto, but it was neither illustrated nor described. Davis (1838) of Boston sold a $25 machine almost identical to Saxton's 1833 model (illustrated on p. 13 of Davis 1842); however, for $50 one could also buy a magneto designed by Charles Page, an American experimenter, that shared certain design features with those of Sturgeon and Clarke (illustrated on p. 15 of Davis 1842). Davis sold several other magnetos, including a portable Page machine for electromedicine (Davis 1842: 16).

44. Merton 1968.

45. Sturgeon 1850: 274.

Notes to Chapter 7

1. Henry to Rogers, November 4, 1831 (Reingold et al. 1972: 367).

2. On the changing role of patents in Western societies, see Biagioli 2006b.

3. Henry's views on patents were expressed in a letter to Rogers dated November 4, 1831 (Reingold et al. 1972: 369–372). See, especially, the footnotes to this letter.

4. For an extensive discussion of this point, see Molella 1976.

5. On the pervasiveness of this ideology in the nineteenth and twentieth centuries, see Forman 2007.

6. Unless otherwise noted, biographical information on Davenport is from Davenport 1929 or Davenport 1851. These and other sources (e.g., Davenport 1900; Kent 1928; Pope 1891a,b) sometimes differ on the order of events of the mid 1830s; this is not surprising given that these accounts were based on recollections written down decades later. I present a reasonably coherent narrative but cannot reconcile the dating discrepancies. Entries for Davenport also appear in *National Cyclopedia of American Biography* 3: 339 (New York, 1893) and in *Dictionary of American Biography* 3: 87–88 (New York, 1959).

7. This story is based on Oliver's recollection (Davenport 1929: 52–54; Kent 1928: 217–218). Thomas Davenport's (1851) spare account does not mention Oliver, but says he went to Crown Point in December 1833 to buy some iron and maybe see the electromagnet. Impressed with the latter, he bought it instead of the iron.

8. Davenport 1851: 5.

9. On the history, technologies, and uses of steam engines, see Hunter 1985.

10. Information about steamboats in this paragraph is from Redfield 1833.

11. Ibid.: 317. This accident occurred in 1832 or 1833.

12. Davenport 1851: 6.

13. Ibid.: 6–7.

14. King (1962a) and Gee (1991) give histories of early electromagnetic motors; Michalowicz (1948) focuses on the inventions of Davenport, Jacobi, and Page. A letter requesting back issues is evidence that Davenport was familiar with *Silliman's Journal* before he constructed his rotary motor (Davenport to Silliman, August 4, 1833, Yale University Library, Manuscripts and Archives, Silliman Family Papers, Box 19, Folder 24).

15. On the history of this shop, which still stands, see Thomas D. Visser, "Smalley-Davenport Shop, Forestdate, Vermont: Birthplace of the Electric Motor in 1834" (www.uvm.edu/~hispres/SD/hist.html#1).

16. Davenport 1851: 8. In that same year, T. Edmondson (1834) of Baltimore created a small rotary electromagnetic device, but he did not suggest that it could drive machinery.

17. Davenport 1851: 7.

18. Ibid.: 8.

19. Ibid.: 8.

20. Ibid.: 9.

21. H. S. Davenport, in Pope 1891b: 96.

22. Davenport 1929: 82.

23. Van Rensselaer to Henry, June 29, 1835 (Reingold et al. 1975: 416)

24. Quotations in this paragraph are from Davenport 1851: 14.

25. Henry to Silliman, September 10, 1835 (Reingold et al. 1975: 446). In a later letter to Bache, Henry wrote that "the discovery of a new truth is much more difficult and important than any one of its applications taken singly" (quoted on p. 251 of Moyer 1997).

26. Davenport 1851: 14.

27. Ibid.: 20.

28. The patent specifications were quickly reproduced and the invention discussed in scientific and technical journals, e.g., "Specification of a Patent for the Application of Electro Magnetism to the Propelling of Machinery," *JFI* 24 (1837): 340–343; 'Speculations Respecting Electro Magnetic Propelling Machinery," *JFI* 25 (1838): 35–36; "Davenport and Cook's Electro-Magnetic Engine," *MM* 27 (1837): 159; "Notice of the Electro-Magnetic Machine of Mr. Thomas Davenport," *MM* 27 (1837): 204–207; Davenport's Electro-Magnetic Engine," *MM* 27 (1837): 404–405.

29. Davenport's English patent is no. 7386 (June 6, 1837).

30. NMAH catalog no. 252,644.

31. The brass pieces may have been a modification made many years later. Not only does the solder appear relatively fresh, but the pieces are mounted ineffectually—rotated 90° from where they should be. Proper placement of copper wires, flattened or not, could have obviated the twisting problem, but would have remained this motor's Achilles' heel.

32. On the 1877 Patent Office fire and efforts to restore the surviving models, see Robertson 2006: 67–70.

33. NMAH catalog no. 181,825. Pope (1891a) reports that Davenport made at least two other model trains, one of which he dates to 1837, the other to 1836 or 1837. Willard Davenport (1900) claimed that Davenport and Cook, after their arrival in New York, "constructed a larger circular railway, fourteen feet in diameter, having a train of cars attached to the engine, upon which a little child could ride."

34. On Henry's European trip, see Reingold et al. 1979: xiii–xxv, 172–266.

35. Bache, quoted on p. xvii of Reingold et al. 1981.

36. Henry 1981: 79.

Notes to Chapter 8

1. Daniell 1836a,b. Mertens (1998) notes that the term "constant battery" had been used in earlier years.

2. Daniell 1836b: 125.

3. In showing that electricity, regardless of how it is generated, differs only in quantity and intensity, Faraday first enunciated this law: "the *chemical power, like the magnetic force . . . , is in direct proportion to the absolute quantity of electricity* which passes" (Faraday 1952d: 318). Later he calculated that oxidizing 3.5 oz. of zinc in a voltaic cell could produce enough current to decompose almost one ounce of water, yielding 2,400 cubic inches of hydrogen (Faraday 1952f: 413). Faraday (1952e: 361–362) introduced a host of new terms—still used today—applicable to electrochemical phenomena, including electrode, anode, cathode, electrolyte, electrolyze, electrolytic, anion, and cation. On his contributions to electrochemistry, see Robertson and Hartley 1933.

4. Daniell 1837.

5. Ibid.: 160.

6. Henry, "European Diary," April 27, 1837 (Reingold et al. 1979: 329).

7. For surveys of nineteenth-century batteries, see Niaudet-Breguet 1880; von Urbanizky et al. 1890.

8. Smee (1841: xxvi) coined the term "electro-metallurgy," whose meaning today encompasses more than electrodeposition (see McMillan and Cooper 1910: 1); I employ the original mid-century meaning.

9. Faraday 1952e: 395–396.

10. "On the Decomposition of metallic salts by the Voltaic pile, and on the state of chlorides, iodides, &c. in solution," *Journal of the Royal Institution* 1: 377–378. This account of Matteuci's experiments was perhaps written by Faraday.

11. Faraday 1952e: 414.

12. Here "rare" means expensive and difficult to come by; some of these liberated elements are abundant on Earth. At the end of the century, relatively cheap electricity from hydroelectric plants and steam-powered dynamos became the foundation of an enormous electrochemical industry (McMillan and Cooper 1910; Trescott 1981).

13. On early developments in electrometallurgy, see Mertens 1998; Smith 1974; Pavlova 1968.

14. Smee (1841: xix–xx) mentions several people who observed this effect, including Daniell himself.

15. Smith 1974. On Jacobi's work, whose contributions were comparable to Spencer's, see Pavlova 1968; Jacobi 1841a. Jacobi named the process "galvanoplasty." Mr. Lettsom and C. J. Jordan also have fair claims to having made the invention at this time (Hunt and Rudler 1878: 209; see also C.W. letter, *MM* 37 (1842): 85–87.

16. Spencer 1840: viii.

17. I rely mainly on Spencer 1840.

18. Smith (1974) points out that Golding Bird reported in 1837 his use of such a cell to produce beautiful crystals of several metals from their nitrate or chloride solutions. On Crosse's experiments, see Crosse 1857.

19. Spencer 1840: 32.

20. Ibid.: 32.

21. Ibid.: 34.

22. Ibid.: 34.

23. Ibid.: vi.

24. Ibid.: iii.

25. Smee 1841: xvi. He added that "The extended use of galvanism for manufactures requires the utmost encouragement, and the improvements must not be shackled by patents" (ibid.: 142).

26. Pavlova 1968: 28–29.

27. The patent (no. 8447, March 25, 1840) is reported in Great Britain, Patent Office 1859: 39–40. On the Scheele connection, see Smith 1974: 47.

28. Doubtless many important processes were not patented but rather were held as trade secrets.

29. Watt 1860: 2.

30. Wylde's (1881–82, volume 2: 263–330) discussion of electrometallurgy makes clear that the industry involved management of complex chemical systems.

31. Byrne and Spon 1874: 1380.

32. According to a letter from M. Vergnes, "the operator frequently suffers very much from taking the articles from the bath, his hands becoming impregnated with the poison, causing them to inflame very much, burst open, and discharge an acrid humor" ("Advice to Electrotypers," *SA* 11, 1856: 251).

33. Walker (1841: 22, 25) was one of the first to discuss and illustrate the use of a separate tank for electroplating.

34. Wilcox's statement appears after the title page of Davis's 1848 catalog, which occurs at the end of Channing 1849. (The copy I examined is in the NMAH Library.)

35. Byrne and Spon 1874: 1382.

36. It is possible, perhaps likely, that Wilcox was referring to the entire book (Channing 1849), not just the catalog appended to it, as having been electrotyped.

37. "Back Numbers Electrotyped," *SA* 1 (1859, n.s.): 89; "The Electric Art Applied to Printing," *SA* 1 (1859, n.s.): 177; on *Harper's Magazine* and the U S. Coast Survey, see *Annual Report of the Smithsonian Institution for 1858* (Washington, D.C., 1859, p. 412–414).

38. New York City Mercantile and Manufacturers 1857: 103; *Wilson's Business Directory of New York City* 27 (1874): 215.

39. On non-electrolytic processes of plating, see Pavlova 1963: 14–16. On the safety issue in using mercury, see Byrne and Spon 1874: 1381; Mangin 1866: 223; Pavlova 1963: 14.

40. This point is made by Trescott (1981: 17).

41. Davis 1848: 8; Palmer 1840: 40, 43.

42. Unpaginated advertisement, *Journal of the Society of Arts* 9 (1861).

43. Gee 1885. This source also furnishes a detailed description of battery maintenance, including amalgamating zinc plates with mercury (p. 151–155).

44. On the use of electroplating in preparing and duplicating daguerreotypes, see Hunt 1854: 14; Snelling 1849: 57–80; Watt 1860: 32–33; "To Reproduce Photographic Impressions," *SA* 5 (1849): 80.

45. "Engraving by Light and Electricity," *SA* 11 (1856): 363.

46. Byrne and Spon 1874: 1381.

47. Some of these applications are mentioned by Hunt and Rudler (1878: 213–214), by Smee (1841), and in "Electro-Metallurgy" (*SA* 24, 1871, n.s.: 135)

48. Smee 1841: 98.

49. Pavlova 1968: 41, 46.

50. Smith 1974: 40.

51. Sturgeon 1850: vi.

52. Unless otherwise noted, information about Woolrich's inventions and their applications is from King 1962b: 350–353.

53. "Woolrich's patent process of Magneto-Plating," *MM* 38 (1843): 146–149. An image of this machine can be seen on the cover of the issue of *MM* dated February 25, 1843.

54. King 1962b: 350. The exact dimensions of the 1844 magneto are as follows: height 1.760 m; width 1.730 m, depth 0.780 m (Jack Kirby, personal communication, December 8, 2005).

55. On the size of the Elkington firm, see "Electrotype," *SA* 8 (1853): 402.

56. The all-metal version of Woolrich's second magneto bears the imprint of "W. CANNING & Co BIRMINGHAM" on its base (King 1962b: 354, fig. 30).

57. Pavlova (1968: 57–59) mentions the use of magnetos in Russia.

58. On the Beardslee magneto for electroplating, see "Machine Electricity—Electrotyping—Telegraphing—Electric Light," *SA* 4 (1861 n.s.): 276–277; "Improved Magneto-Electric Battery," *SA* 5 (1861 n.s.): 353–354; "Beardslee's Magneto-Electric Machine," *SA* 5 (1861 n.s.): 376. The U.S. patent is no. 26,558, the commutator is no. 26,557; both from December 27, 1859.

59. "Beardslee's Magneto-Electric Machine," *SA* 5 (1861 n.s.): 376.

60. "Miscellaneous Summary," *SA* 11 (1864 n.s.): 323. A Beardslee magneto also powered portable military telegraphs during the Civil War (Beardslee Magneto-Electric Co. 1863).

61. Pavlova (1968: 58) claimed that the magneto's pulsating current produced inferior products. If that were generally the case, it is doubtful that Elkington would have continued using the machines for more than a decade. The bulk of evidence indicates that the plate was of good quality. P. H. Vander Weyde commented that magnetos were no longer being used in electrometallurgy "as an ordinary steam engine gives much more battery power than the largest electroplating establishment can make use of; perhaps also for reason of the difficulty of

subdividing a powerful current properly between the different electroplating troughs, for which purpose different smaller zinc batteries are more easily adjusted" (letter, *SA* 25, 1871, n.s.: 36).

62. "Mr. Elkington stated that they had never been induced to abandon the voltaic battery, which they employed in their manufactory, finding it more economical than the magneto-electrical machine;" from "Electric Batteries—Useful Discovery of Plating," *SA* 5 (1849): 54. On the other hand, Hunt and Rudler (1878: 216) state that "plating and gilding are successfully, and, in point of economy, advantageously, carried on at Birmingham, in more than one manufactory, by means of magneto-electricity."

63. The use of batteries is mentioned in the following electrometallurgy textbooks: Gore 1884; Napier 1876; Watt 1860. Batteries continued to be used well into the twentieth century because line voltage was too high for electroplating and alternating current, increasingly adopted beginning in the mid 1890s, was unsuitable. However, two technological fixes were available: (1) in dc systems, one could charge a number of storage batteries in series to reduce the voltage, and (2) in ac systems, a motor-generator set could transform alternating current to low-voltage direct current (McMillan and Cooper 1910: 76).

Notes to Chapter 9

1. Barlow 1825: 105.

2. The "cascade model" (Schiffer 2005a) shows that life-history processes, such as manufacture, use, and maintenance, are incubators of invention cascades during development of a complex technological system; Morse's electromagnetic telegraph is the case study.

3. Among the many nineteenth-century books on the telegraph are Culley 1878; Davis 1851; Noad 1857; Pope 1869; Preece and Sivewright 1891; Sabine 1869; Schellen 1850; Shaffner 1859. Recent histories include Beauchamp 2001; Dawson 1976; Hubbard 1965; Israel 1992; King 1962c; Marland 1964; Standage 1999. Another useful source is Shiers 1977.

4. On Morse as artist, see Larkin 1954; Mabee 1969; Staiti 1989.

5. Morse 1973: 2.

6. Unless otherwise noted, information in this chapter and in chapter 12 is from Silverman's book.

7. Silverman (2003) spells the elder Morse's name "Jedediah," but it appears as "Jedidiah" in Library of Congress catalog entries.

8. Ibid.: 6.

9. Ibid.: 21.

10. On Morse and other itinerant portrait painters, see Lipton 1981.

11. Morse, quoted on p. 48 of Silverman 2003.

12. Ibid.

13. The oil lamps in the Capitol were not replaced by gas until 1847 ("The Capitol at Washington Illuminated," *SA* 3, 1847: 88).

14. Morse, quoted on p. 66 of Silverman 2003.

15. Ibid.: 85.

16. Morse and Shaffner 1855: 400–401.

17. Silverman 2003: 104.

18. Morse and Shaffner 1855: 401.

19. For excerpts or copies of the testimonials, see Vail 1845; Morse 1973; Morse 1855.

20. Hindle (1981) argues for the importance of Morse's artistic training and spatial thinking in the invention of the telegraph.

21. The date is given by Morse in a later chronology, "Dates of Invention of my Telegraph," MP, Reel 6, p. 98301.

22. In many accounts, this triangular frame is referred to as a "pendulum," but this is misleading because the motion was perpendicular to the plane defined by its three sides.

23. The motions were not actually *identical* because the device riding on the port-rule merely had to move up and down enough to open and close the circuit.

24. Morse and Shaffner 1855: 405. Morse's prototype sender and receiver are in the NMAH (catalog nos. 181,249 and 181,250).

25. Ibid.: 405. Leonard Gale affirmed in court testimony that Morse came up with the relay idea "early in the spring of the year 1837" (Gale, quoted on p. 73 of Kendall 1854). Several others, including Wheatstone and Henry, independently invented the relay (Henry, in Smithsonian Board of Regents 1858: 111–112).

26. Gale, quoted on p. 73 of Kendall 1854.

27. The account of Gale's earliest contributions to the design of Morse's telegraph is from Gale 1875. Morse and Shaffner (1855) make no mention of Gale's Henry-inspired contributions, for by this late date Morse was deeply involved in litigation, defending his claims to originality on virtually every issue.

28. "Telegraph Signs," U.S. patent no. 1,647, issued to Morse on June 20. 1840.

29. Morse, quoted on p. 74 of Vail 1845.

30. "Articles of Agreement," September 23, 1837, MP, Reel 6, pp. 98407–98411. On this agreement in Morse's hand is a note stating that Vail's share was reduced to three-sixteenths. The date of this note—March 1838—coincided with the conclusion of a more encompassing agreement involving Morse, Smith, Gale, and Vail.

31. Morse to Sidney Morse, January 13, 1838. MP, Reel 6, p. 98435.

32. I have inferred the transmission rate from Silverman's (2003: 166) discussion of improvements made just before the Franklin Institute demonstration.

33. Hamilton 1838: 107–108.

34. Ibid.: 108.

35. "Telegraphs," *AJSA* 32 (July 1837): 201.

36. Vail 1914: 12.

37. U.S. House Commerce Committee Report of April 6, 1838, quoted on p. 77 of Vail 1845.

38. Silverman (2003: 170–171) notes that Smith promised Morse that he would resign his seat in the House to avoid a conflict of interest with his telegraph work. In fact, Smith served out his term, still casting votes in early 1839, but was not reelected (source: http://bioguide.congress. gov/scripts/biodisplay.pl?index=S000531).

39. "Smith would acquire a one-quarter interest in the American patent and a five-sixteenth interest in any foreign patent. Vale agreed to a two-sixteenth interest in both patents, Gale to a one-sixteenth" (Silverman 2003: 171). Obviously, this agreement superseded the earlier one between Morse and Vail.

40. "Articles of Agreement," March 1838, MP, Reel 6, p. 98435.

41. Silverman 2003: 171.

42. Ibid.: 178.

Notes to Chapter 10

1. Numbers in this paragraph are from Davenport 1929 and Paine 1838.

2. Silliman 1837: 3.

3. Ibid.: 2.

4. Paine 1838: 36.

5. *AEMC* 2 (1838): 158–159, 257–264, 284–285, 347–350. That article contains the patent specification.

6. Silliman 1837: 6.

7. Ibid.: 7.

8. Ibid.: 7. Later that year, Silliman published a brief update on the motor project: "We are informed that they have constructed a seven-inch wheel, with two tiers of magnets in the revolving part, or four crosses, which will be applied to a turning lathe, and will raise over one hundred lbs. from the floor," *AJSA* 32 (1837): 399.

9. The authorship of this pamphlet is uncertain, but I cite it as Electro-Magnetic Association 1837. The cost of printing the pamphlet was $197.77; apparently, some copies were sold (Paine 1838: 37).

10. Electro-Magnetic Association 1837: 11.

11. Ibid.: 12.

12. Patterson 1975: 521.

13. Electro-Magnetic Association 1837: 26.

14. Ibid.: 28.

15. Ibid.: 30–31.

16. Ibid.: 34. A "roasting jack" was a spring-operated spit.

17. Ibid.: 40.

18. Ibid.: 27.

19. Paid advertisements also appeared in a number of newspapers toward the end of 1837 (Paine 1838: 37).

20. Electro-Magnetic Association 1837: 38.

21. Quotations in this paragraph are from *AEMC* 2 (1838): 285.

22. Jones (1837: 342–343), after seeing the motor displayed, reported that Davenport had articulated this goal; it was also reported in *AJSA* 33 (1838): 193 and in *AEMC* 2 (1838 : 350

23. Amon Shwing (?) to Davenport (by way of *The Sentinel*), February 20, 1837. DP, Box 1, Folder 3.

24. Redwood to Cook, June 29, 1837. DP, Box 1, Folder 6.

25. Robert Farmer to Davenport, May 1, 1837. DP, Box 1, Folder 5.

26. Lane to Davenport and Cook, October 19, 1837, DP, Box 1, Folder 8.

27. On the performance characteristics of steam engines, large and small, see Hunter 1985; Hunter and Bryant 1991.

28. Hudson to Cook, August 3, 1837, DP, Box 1, Folder 7.

29. Jones 1837.

30. Hare 1840: 67. His witnessing of Williams's demonstration is mentioned on p 38 of Electro-Magnetic Association 1837.

31. *AJSA* 33 (1838): 194.

32. Page 1839a: 107.

33. Quotations in this paragraph are from Henry to Bache, August 9, 1838 (Reingold et al. 1981: 97).

34. Henry to Bache, August 9, 1838 (Reingold et al. 1981: 97). The following year, Henry again vented to Bache about the prevalence of "Quacks and Jimcrackers" in New York City, though he did not specifically mention the Davenport enterprise (Henry to Bache, October 28, 1839; see Reingold et al. 1981: 290).

35. Henry also might have been envious because Davenport had received so much credit for inventing an electromagnetic engine—an invention underpinned by the principle embodied in his teeter-totter (Henry 1831b).

36. Quotations in this paragraph are from p. 13 of Griglietta 1838.

37. Allusions to the sale of exhibition rights occur in an undated letter from Edwin Williams to Ransom Cook. DP, Box 1, Folder 11, and in Davenport to Cook, June 11 (?), 1837. DP, Box 1, Folder 6.

38. One source states that Davenport had achieved the goal of operating a Napier printing press by early 1838 (Paine 1838: 3); this was probably accomplished with a rotary motor but was not widely publicized.

39. Pope 1891b: 96.

40. Davenport, quoted in *SA* (6, 1850: 32). He contended in the same source that the printing press had also been driven with a rotary motor. (See note 38 above.)

41. These numbers are from pp. 33–34 of Paine 1838.

42. Ibid.: 10.

43. Morse to Cook, June 15, 1840, MP, Reel 7, p. 98758.

44. Davenport (1929: 128–131) furnished these accounts. Pope (1891b: 94) found a witness who had seen the telegraph in Davenport's New York workshop in 1837. A register that Davenport used when exhibiting electromagnetic devices is in the Smithsonian (NMAH catalog no. 315,892).

45. Davenport 1900: 72.

46. Ibid.: 80.

Notes to Chapter 11

1. Source of *Royal George* story: Slight 1849.

2. Source: http://www.memorials.inportsmouth.co.uk/churches/st_marys/royal_george.htm.

3. Pasley to Faraday, January 12 and 14, 1839 (Williams et al. 1971: 329–333). Stotherd (1872: 209–210) states that Pasley's experiments with electrical blasting began in 1812.

4. Hare 1832, 1834. On nineteenth-century electrical blasting, see Abel 1884; Hunt and Rudler 1878; Stotherd 1872.

5. Hare 1832: 141.

6. Abel (1884: 108) reported that "the first practical application of the voltaic battery in this direction was made . . . by French military engineers."

7. Slight 1849: 97.

8. In the United States, early uses of electrical blasting to remove submarine obstacles to shipping took place in New Orleans ("Blowing up Wrecks by Electricity," SA 5, 1849: 131) and at Diamond Reef, near Governor's Island, just below the tip of Manhattan in New York Harbor ("Submarine Explosion," SA 7, 1851: 7; "Submarine Blasting," SA 11, 1856: 386).

9. Hare 1834: 358. Material on Schilling in this paragraph is condensed and paraphrased from Lundeberg 1974.

10. For a nuanced treatment of the relationship between the military and technological development, beginning in the nineteenth century, see Smith 1985.

11. Lundeberg 1974.

12. A contact mine explodes when struck by a moving ship.

13. Ibid.: 8–9.

14. Ibid.: 15

15. Quotations in the paragraph are from Colt to Tyler, June 19, 1841 (Lundeberg 1974: 59–60).

16. The Americans were unaware that, under Jacobi's leadership of a secret program, Russia was far ahead in developing electrically detonated mines (Lundeberg 1974: 19).

17. Source of quotation: ibid.: 26.

18. Source of quotation: ibid.: 29.

19. Source of quotation: ibid.: 30.

20. Ibid.: 32.

21. Ibid.: 35, in legend to figure 30.

22. Ibid.: 37.

23. Ibid.: 39.

24. Ibid.: 45–46.

25. Source of quotation: ibid.: 46.

26. Source of quotation: ibid.: 49–50.

27. Source of quotation: ibid.: 52.

28. Unless otherwise noted, information on the *Cairo* is from Bearss 1980.

29. After the loss of the *Cairo*, Rear Admiral David Porter acknowledged that "these torpedoes have proved so harmless heretofore (not one exploding out of the many hundreds that have been placed by the rebels) that officers have not felt that respect for them to which they are entitled" (ibid.: 101).

30. Ibid.: 95.

31. Source of quotation: ibid.: 99.

32. "Vessel Lost—Men All Saved," *CT*, December 19, 1862; "The Iron-Clad Gunboat Cairo Destroyed by a Torpedo," *NYT*, December 19, 1862. The *Times*'s account was a brief note; it was followed a few days later by a story that detailed the mine's construction ("The War in the Southwest," *NYT*, December 25, 1862).

33. Scharf's (1887: 752–753) inconsistent account implies electrical detonation of the mine, but his paragraph is ambiguous.

34. Bearss 1980: 97. Bearss's major source is Scharf 1887.

35. Wideman 1993. Perry (1965) had earlier described the mechanical detonator of the mine that destroyed the *Cairo*.

36. For descriptions and images of Confederate torpedoes, see Barnes 1869; Scharf 1887. Hicks (2002) tells the story of the *Hunley*, a Confederate submarine that used a spar torpedo to destroy the *Housatonic*, which was blockading the harbor at Charleston, South Carolina. The *Hunley* itself sank, and all aboard perished.

37. Scharf 1887: 768.

38. Ibid.: 750.

39. On Maury's life, see Jahns 1961; Williams 1963.

40. On Maury's system, see Maury 1915; Navy Department, Naval History Division 1971.

41. On post-Civil War applications of electrically detonated mines, see Abel 1884; Sleeman 1880.

42. Barnes 1869: 161.

43. Merrill 1874a: 3. An outline of the history of Goat Island, the location of the Naval Torpedo Station, can be found at www.nuwc.navy.mil/hq/history/0002.html. See also "Interesting Torpedo Trials at Newport, R.I.," *SA* 33 (1875): 195–196.

44. For discussion and illustrations of the Siemens-Halske and Smith generators, see Schellen 1884: 149–154. On the Wheatstone machine, see Merrill 1874b: 13–16. On Beardslee's machine, see Barnes 1869: 170–178; Beardslee Magnetic Electric Co. 1863. Rühmkorff induction coils were

also tried (Hunt and Rudler 1878: 197–198; Wheatstone and Abel 1861). Spon (1874) lavished much attention on electrical detonation in a general overview of boring and blasting technologies.

45. Abel 1984: 140.

46. Background information on the Hoosac Tunnel is from Harrison 1891 and from "The Hoosac Tunnel," *Scribner's*, December 1870: 143–159.

47. Coyne (1995) places the Hoosac Tunnel in the context of regional competition for the western trade, and highlights the arguments of proponents and opponents.

48. Quoted in "The Tunnel," *Boston Daily*, November 28, 1873.

49. Some sources report the project's total cost as $14 million, others as $21 million.

50. On the dangers of hanging fire, see, e g., Abel 1884: 108; "Improved Electric Fuse," *MAB* 1 (1869): 116

51. "The Hoosac Tunnel," *SA* 22 (1870 n.s.): 106–107, quotation on p. 106.

52. "Hoosac Tunnel," *Scribner's Monthly*, December 1870, pp. 143–159, quotation on p. 150. A firsthand account of the blasting process, published in the *North Adams Transcript*, was reprinted as "Blasting with Nitro-Glycerin at the Hoosac Tunnel," *SA* 20 (1869 n.s.): 27–28.

53. Merrill (1874b) described and illustrated frictional generators used for detonation of mines and torpedoes, and also specified performance requirements for generators used in offensive torpedoes (pp. 3–4).

54. On this ceremony, see pp. 18–19 of Harrison 1891.

55. "The Hoosac Tunnel," *NYT*, November 28, 1873. Other examples of newspaper coverage: "The Tunnel," *Boston Daily Globe*, November 28, 1873; "Details of the Opening," *Boston Daily Globe*, November 28, 1873; "The Hoosac Tunnel," *CT*, November 14, 1873.

56. Nye (1994) does not mention the Hoosac tunnel, but he might have.

57. See Coyne 1995: 20.

58. Post (2003: 69) notes that "the success—the very existence—of many transportation technologies has depended on the way that people with their hands on the levers of political power have 'manipulated' those levers, typically by providing overt or hidden subsidies, sometimes by finding ways to shoulder capital or developmental costs."

59. "Improved Electric Fuse," *MAB* 1 (1869): 116.

60. Merrill 1874a: 3.

61. Abel 1884: 146–148. A similar system was also used in the West; see "The Great Submarine Blast in the Harbor of San Francisco," *SA* 22 (1870): 304.

62. In New York one could buy fuses from the Bishop Gutta Percha Company, in Boston from the George E. Lincoln Company. See, respectively, "Improved Electric Fuse," *MAB* 1 (1869): 116; advertisement, *SA* 26 (1872 n.s.): 189.

63. Abel 1884: 107.

64. See, e.g., Stauffer 1906.

Notes to Chapter 12

1. These two letters—Morse to Henry, April 24, 1839 and Henry to Morse, May 6, 1839—were published in Morse 1855. On Henry's contrasting views on the practicality of the electric motor and the telegraph, see Molella 1976.

2. Morse's queries and Henry's answers: May 1839, MP, Reel 7, p. 98672.

3. "This separation of the proprietors of the Telegraph so that no consultation can be had is I fear doing an injury to us all. I cannot move a step without such consultation." Morse to Vail, May 14, 1839, MP, Reel 7, p. 98663.

4. The offer was tendered in a letter from Cooke to Morse, January 17, 1840, MP, Reel 7, p. 98742. An account of the successful operation of the English telegraph appeared in "The Electro Magnetic Telegraph of the Great Western Railway," *JFI* 25 (1840): 271–272.

5. Henry to Morse, February 24, 1842 (Vail 1845: 87–88).

6. Morse to F. O. J. Smith, July 16, 1842, MP, Reel 7, p. 98933–98934.

7. Silverman 2003: 215. Author of this quotation is unspecified.

8. Morse to Sidney Morse, December 18, 1842, MP, Reel 7, p. 98975.

9. House Commerce Committee report, quoted on p. 219 of Silverman 2003.

10. Source of quotation: ibid.: 219.

11. Morse to Sidney Morse, January 20, 1843, MP, Reel 7, p. 98998.

12. That same day Morse complained to Vail that "for two years I have labored all my time, and at my own expense, without assistance from the other proprietors . . . to forward our enterprize [*sic*]." Morse to Vail, February 23, 1843, MP, Reel 7, p. 99015.

13. Some details in this paragraph are from "History of the Telegraph: Difficulties and Success of an Inventor," *SA* 11 (1855): 19. Although parts of this story may be apocryphal, "An Act to test the practicality of establishing a system of electro-magnetic telegraphs by the United States" did not become law until March 3, 1843—the last day of the 27th Congress (Statutes at Large, 27th Congress, 3rd Session, chapter 84, pp. 618–619).

14. Silverman 2003: 222.

15. Morse to Sidney Morse, December 30, 1843, MP, Reel 8, p. 99324.

16. Unaddressed Morse note, June 25, 1844, MP, Reel 8, p. 99697.

17. This is inferred from surnames on laborer receipts, including Flaherty, Patrick, and McMahon (MP, Reel 8, pp. 99523–99525). Laborers earned from $3 per week to $1 per day.

18. For a detailed description of the register (which was awarded U.S. Patent 4,453), see Vail 1845: 26.

19. Numbers 23: 23. The register tape of this message is in the Smithsonian (NMAH catalog no. 1,028).

20. "Morse's Magnetic Telegraph," *Baltimore American*, May 31, 1844.

21. "The Electro-Magnetic Telegraph," *Utica Daily Gazette*, June 5, 1844.

22. At the scheduled end of the project (February 15, 1845), Morse expected to have a balance of around $500. Morse to George Bibb (Secretary of the Treasury), January 28, 1845, MP, Reel 9, p. 100063.

23. Silverman (2003: 241) reproduces a relevant portion of this report.

24. Vail 1845: 59.

25. Wilkes to Morse, June 13, 1844 (Vail 1845: 59–60).

26. "Application of the Electro-Magnetic Telegraph to the Determination of Longitude," *MM* 41 (1844): 111.

27. Silverman 2003: 250.

28. On the history of the Patent Office building, see Robertson 2006. On the history of the American patent system, see Cooper 1991; Post 1976a.

29. On use of the Page magneto, see Post 1976a.

30. On the Page-Davis collaboration, see Sherman 1988 Page's electromagnetic devices first appeared in Davis's 1838 catalog (Davis 1838). In later catalogs (e.g., Davis 1848) he also offered Morse apparatus.

31. Page's receipt for $99.50 (MP, Reel 8, pp. 99674–99675).

32. On the compound magneto, see Vail 1845: 145–149. The Smithsonian Institution eventually acquired the telegraph magneto, but it was destroyed in the fire of 1865 (Post 1976a: 69).

33. At least 5 years earlier, Page (1839b: 252) had envisioned the construction of compound magnetos: "The avenue, then, to an indefinite power, is too obvious to escape notice. Increase the number of pairs of magnets, extend the series of armatures upon the same shaft, or in any way in which they may be brought to bear on the same terminal pole."

34. Page, quoted on p. 69 of Post 1976a.

35. On the experiments with the Beardsley magneto, see "Telegraphing by Magneto-Electric Machines," *SA* 11 (1864): 340.

36. "Application of Dynamo-Electric Machines to Telegraphy," *SA* 42 (1880): 63–64, "The Future of Electricity," *SA* 42 (1880): 64–65.

37. On attempts to market the telegraph to the sovereigns of Japan, Egypt, and the Ottoman Empire, see Bektas 2001. In such contexts, "the electric telegraph was not merely a technological artefact of wonder but also a political symbol that represented what was often called American 'inventive genius' and technological power" (p. 202).

38. The histories of these companies are too complex to recount here. See Reid 1879; Thompson 1972.

39. Du Boff 1984: 573.

40. "Telegraphs—Europe and United States," *SA* 20 (1869): 183.

41. Du Boff 1984: 572.

42. "By 1849–1852, profit rates on 'the majority' of the lines constructed were variously described as 'enormous' or 'vigorous' by contemporary observers." (Du Boff 1980: 475)

43. Henry to Spencer F. Baird, October 19, 1862 (Rothenberg et al. 2004: 283–284).

44. Unspecified author, quoted in Du Boff 1980. See also Friedlander 1995.

45. Du Boff 1980: 462.

46. Scharlott's (1986, 2004) studies of Cincinnati demonstrate that the telegraph had a considerable effect on the city's commercial activities.

47. Du Boff 1980, 1984.

48. Du Boff 1980: 462.

49. "History of the Telegraph: Difficulties and Success of an Inventor," *SA* 11 (1855): 19; "Railways of the United States," *SA* 10 (1855): 146.

50. "Progress of the Telegraph," *SA* 27 (1872): 146.

51. "Samuel F. B. Morse," *SA* 23 (1870): 357. These numbers are, at best, informed guesses.

52. "Telegraphs—Europe and United States," *SA* 20 (1869): 183. Du Boff (1984: 573), however, reports the average cost of a telegram in the United States in 1868 was $1.05; by 1877 it was $0.39.

53. Morse, quoted on p. 309 of Silverman 2003.

54. Henry, hired by a railroad to assess House's telegraph, judged it "workable and, moreover, clear of patent infringement" (Moyer 1997: 246).

55. Henry, in his Supreme Court deposition, made clear his irritation about this matter (Henry, in Smithsonian Board of Regents 1858: 115).

56. Henry, in Smithsonian Board of Regents 1858: 113.

57. Morse 1855: 9.

58. In a letter to Henry, Gale reiterated his role in apprising Morse, on the basis of Henry's 1831 paper, of the need to use a battery of higher intensity and a coil with more turns to obtain a greater transmission distance. Gale to Henry, April 7, 1856 (Rothenberg et al. 2002: 347–348).

59. Taylor 1879b.

60. He had to share this sum with his agent in France, F. O. J. Smith, and with Vail's widow.

61. Standage (1999) argues that the telegraph was a Victorian Internet.

62. Silverman 2003: 322.

63. On the effects of the telegraph on diplomacy, see Headrick 1981; Hugill 1999; Nickles 2003.

64. Advertisements in *The Telegrapher* and other trade journals indicate the variety of artifacts and materials needed to establish and maintain a telegraph system.

65. Kinsey (2004) describes this process in detail for the nineteenth-century carriage trade in America. Israel (1992) shows how changes in the manufacturers of telegraph components affected invention processes. On the early electrical manufacturers, see MacLaren 1943; Passer 1953. A general work on changes in American manufacturing technology is Hounshell 1984.

66. For a social history of the telephone, see Fischer 1992. Bowers (1982: 39) stresses that telegraphy created the manufacturing infrastructure for later electrical technologies.

67. Of course there were women telegraphers, especially in later years.

68. McMahon (1984: 7–8) makes a similar point.

69. The arguments in this paragraph derive substantially from Chandler 1977.

70. The telegraph was capital-intensive in relation to electrometallurgy, the other widely commercialized electrical technology of that era. However, telegraph capital costs were insignificant in comparison with those of the railroad, which could cost many thousands of dollars per mile. See, e.g., "Western Railroads," *SA* 7 (1851): 64.

71. On the importance of electrical measurement in telegraphy, see Clark 1868; Gooday 2004. On the history of the electrical engineering profession in America, see McMahon 1984. Hunt (1997) attributes the creation of formal electrical units and measuring instruments to the particular requirements of *cable* telegraphy.

72. These units, along with the farad (for capacitance) and the coulomb (for electric charge), were given formal definition in 1881 (Committee on Electrical Standards 1913). Appropriately, the unit of inductance is called the henry. On the social construction of electrical units, see Gooday 2004; Morus 1988.

73. Hunt (1994) emphasizes the contributions of telegraphers. See also Keithley 1999.

Notes to Chapter 13

1. Sturgeon 1836.

2. Jacobi 1841b.

3. On the *Galvani*, see Post 1974.

4. On early electric motors, see Bowers 1982; Martin and Wetzler 1891; Du Moncel et al. 1883; Gee 1991; Michalowicz 1948. On post-1875 electric motors and their uses, see Ayrton and Perry 1883; Urquhart 1882.

5. Sturgeon 1839: 437.

6. Bijker 1995: 84–88

7. Post (1976a) mentions the changing personal relationship between Henry and Page.

8. The induction coil was the first transformer brought to market. In 1839, Henry, who had not yet met Page, established some of the basic principles of transformer design. He noted, for example, that *"an intensity current can induce one of quantity*, and, by the preceding experiments, the converse has also been shown, that *a quantity current can induce one of intensity"* (Henry 1839: 312). This paper was widely reprinted in both American and European journals (for a listing, see Taylor 1879a: 361), but it had little immediate influence. During the nineteenth century, the term "induction coil"—increasingly employed after about 1860—was also sometimes applied to a single coil that, through self-induction, could generate momentarily a current of high tension. On experiments with coiled devices of early design, including induction coils, see Cavicchi 2006.

9. Page 1839b; *AEMC* 3 (1839): 478–486.

10. Another early version of the induction coil was invented by Nicholas J. Callan, a professor of natural philosophy at Maynooth College, Ireland. However, Page in 1838 was seemingly unfamiliar with Callan's work, which had been published in 1836 and 1837 (e.g., Callan 1837). In his history of the induction coil, Page (1867: 14–19) discussed Callan's invention, but maintained his own claim of priority.

11. Pantalony et al. (2005: 157–159) illustrate what may be a prototype of Page's induction coil. It is in the instrument collection of Dartmouth College.

12. Davis 1838: 71, 1848: 38, 1857: no pagination; in the 1857 catalog the electrotome had been completely redesigned.

13. On the role of induction coils in nineteenth-century electromedicine, see Rowbottom and Susskind 1984.

14. Although Page had invented the essential parts of an induction coil, the spark on his 1838 model was short. A dozen years later, he made a coil yielding 8-inch sparks (Page 1867: 42).

15. On the Apps coil, see Pepper 1869; Routledge 1879.

16. Contributors to the modern induction coil were Edward S. Ritchie, a Boston instrument maker (Warner 1994); Jonathan N. Hearder, an English experimenter (Cavicchi 2006); and Heinrich Rühmkorff, a Paris instrument maker (Finn 1975). Armand Fizeau added the condenser. On the large Ritchie coil, see Wahl 1871. Ritchie made a similar coil for Dartmouth College at a cost of $700, "Induction Coil of Unusual Size," *JFI* 90 (1870): 7.

17. "The Applications of Electricity," *TE* 18 (1864): 253.

18. Ibid.: 253; "Medals for Inventions in France," *SA* 12 (1865): 148; "Submarine Photography," *NYT*, July 8, 1873.

19. "Luminous Electrical Tubes. *SA* 27 (1872): 66.

20. See, e.g., Huggins 1864.

21. See, e.g., Wright 1897.

22. See Warner 1994.

23. On this controversy, see Post 1976a,c. Page is not mentioned in Du Moncel's 1867 history of the induction coil.

24. The patent model survives in the Smithsonian (NMAH catalog no. 309,254).

25. Page (1839a) presented these arguments.

26. See Page 1839a.

27. Jacobi 1837: 412.

28. See, e.g., Page 1845a,b.

29. Assuming that battery current is held constant, the variables influencing electromagnet strength include size and shape of the iron core and its constituents (e.g., solid, laminated in sheets, wires; horseshoe, disk, cylinder); composition of the insulation and its manner of application; and diameter, length, and cross-section of the wire and how it was wound. The contest was announced in *Annals of Electricity, Magnetism, and Chemistry* (6, 1841: 168).

30. Fox 1969: 77. Other useful sources on Joule: Cardwell 1976; Cardwell 1989; Steffens 1979. Joule's scientific papers are collected in Joule 1884.

31. Joule 1838: 123.

32. Joule 1839a: 134.

33. Ibid.: 135.

34. Curiously, Joule, who would have been familiar with the articles on Davenport's motor and patent in *Sturgeon's Annals*, claimed that one reason for publishing new motor designs was to prevent others from patenting them first, people "who seem to regard this most interesting subject merely in light of a pecuniary speculation" (ibid.: 203).

35. Joule 1840a: 481.

36. Joule 1840b: 192.

37. Ibid.: 188.

38. Joule 1842: 220–221. An earlier version of this paper had been presented at the Victoria Gallery in early 1841 (Fox 1969: 83). The efficiency measure implicit in this statement—how much work a motor could do per pound of zinc consumed in the battery—was also advocated by Jacobi (1841b: 159).

39. Page 1845a: 131.

40. Page (ibid.: 134–135) believed that motors of this design obviated the problems created by self-induction. He also reported a somewhat larger, more powerful motor that contained two iron rods, with four coils acting on each (Page 1845b).

41. Post 1976a: 89.

42. Page 1845a: 132.

43. In 1848, with his friend John J. Greenough (editor of the *Polytechnic Journal*) as a front man, Page sought a British patent for his motors. A patent (no. 13,613) was issued for "Electro-Dynamic Axial Engines" on May 3, 1851 (Great Britain, Commissioners of Patents 1859: 220–222). The draft specifications, correspondence between Page and Greenough, and related materials can be found in the SIAC (Charles Grafton Page Papers, 1844–1870).

44. King 1962a: 268. Hunt (1860: 98), in his entry on "Electro-Motive Engines," stated that Froment's use of electric motors "appears to have been abandoned, on account of the great cost of the battery power."

45. Post 1976a: 82. See also Post 1972.

46. Benton, quoted on p. 84 of Post 1976a. Post (ibid.: 87–88) notes that the zinc mines in Benton's home state of Missouri stood to profit from greater use of batteries.

47. Ari Davis was referred to in one article ("Electro-Magnetism as a Motive Power," *SA* 7, 1851: 64) as an "electric engineer," an early use of that term.

48. Page 1973a: 411.

49. According to the *National Intelligencer*, "Henry said he had witnessed with great interest Dr. Page's experiments before the Smithsonian Institution" and that "Dr. Page had produced by far the most powerful electro-magnetic engine ever made" (quoted on p. 92 of Post 1976a).

50. Quotations in this paragraph are from Page 1973a: 412.

51. Ibid.: 413.

52. On congressional opposition to granting additional funds to Page, see Post 1976a: 93–96.

53. "Money Paid by Government for Inventions," *SA* 8 (1853): 221. On federal support for developing technology in the nineteenth century, see Hill 1960.

54. The motor's construction is nowhere presented in detail, because Page feared that others might patent the design. I have fashioned a composite description from the following sources: Post 1976a: 96; "Electro-Magnetism as a Motive Power," *SA* 7 (1851): 64, 68; "The First Locomotive That Ever Made a Successful Trip With Galvanic Power," *GAPJ* 4 (1854): 257–264.

55. The Grove cell was said to yield abundant current at a higher tension than, say Daniell's cell. (See von Urbanitzky et al. 1890: 112.) When the volt was first defined as an electrical unit, the Daniell cell was taken to be the standard at one volt.

56. Another report, either approved or authored by Page, stated that six cells were already broken at this time. "The First Locomotive That Ever Made a Successful Trip with Galvanic Power," *GAPJ* 4 (1854): 258.

57. Page 1973b: 425.

58. According to one observer, "the wood-work of this whole apparatus, was constructed by a common house-carpenter, who had never seen a carriage built before; the truck was a miserable cast-off affair, unworthy [of] the position it was made to occupy, and the driving wheels were ill proportioned, and worse balanced" ("The First Locomotive That Ever Made a Successful Trip With Galvanic Power," *GAPJ* 4, 1854: 258).

59. According to another report, the trip was interrupted 15 times and the battery at the end of trip had fewer than half its cells functioning (ibid.: 259–260).

60. At 19 miles per hour, I calculate—assuming no gearing—that the engines were making 106 strokes per minute. The driving wheels were 5 feet in diameter.

61. Page 1973b: 425. Presumably this followed from his supposed principle that the "axial force . . . is as the square of the quantity of galvanic current" (ibid.: 425). As for the battery's woes, he suggested that "the defect of the cells is easily remedied" and that better insulation of the coils might solve the arcing problem (ibid.: 424).

62. Ibid.: 425.

63. Ewbank 1973: 446–447. (Ewbank's essay originally appeared in a Patent Office publication.)

64. Hunt 1851: 308.

65. Synopses of the paper and the ensuing discussions were published in "Institution of Civil Engineers," *TE* 3 (1857): 325, 364–365.

66. Ibid.: 364.

67. Ibid.: 364. Sprague (1875: 343), who also believed on economic grounds that "all electro-magnetic engines fail," nonetheless provided a table showing that "Man's power" and electro-magnetic power were essentially equivalent in cost per unit of mechanical energy.

68. See Hunt's (1860: 100) entry on "Electro-Motive Engines."

69. "Liebig on Electro Magnetism as a Motive Power," *SA* 6 (1851): 35; "Liebig's Opinion of the Value of Electro-Magnetism as a Moving Power," *MM* 60 (1851): 295–296.

70. Thomas Allan (1858), a motor inventor, offered a vigorous defense of electric power. Surprisingly, even at this late date, *Mechanics' Magazine* was not entirely convinced by the arguments of Joule and Hunt: "Mr. Joule professes to understand the *modus operandi* of Nature in producing electro-magnetic force from zinc and acid. But is not this presumption on his part? . . . we think it an undue stretch of scientific principles to announce what amount of work an electro-magnetic engine will do, before we have even found out what is the best construction to give to an electro-magnet." "Electro-Magnetism as Motive Power," *MM* 68 (1858): 31.

71. "Electro-Magnetism as a Motive Power," *SA* 7 (1851): 91.

72. "The New Light and Motive Power," *SA* 8 (1853): 298. At this time, Watson was seeking to sell a battery-powered light for use in English lighthouses (James 1999).

73. "Electro-Magnetism as a Motive Power," *TE* 2 (1856): 367.

74. "Science and Arts, Improvements &c," *SA* 7 (1851): 11.

75. "Electro-Magnetism as a Motive Power," *SA* 9 (1851): 68.

76. Example: "We now know that nothing is to be hoped from electro-magnetism as a motive power" ("The Impossible in Constructive Science," *SA* 19, 1868: 18). A few years later, the editors stated—also referring to the use of motors to drive machinery—that "scientific men . . . have long ago given up the problem as incapable of a practical solution" ("Recent Progress in Electromagnetism," *SA* 29, 1873: 352).

77. On sewing machine applications, see "Sewing Machine Driven by Electricity," *SA* 21 (1869): 311; "Electromagnetic Motor for Sewing Machines," *SA* 24 (1871): 390. On operating vents and valves, see "Electricity and Some of Its Practical Applications," *SA* 4 (n.s. 1861): 53. On better batteries, see "A Chance for Inventors," *SA* 31 (1874): 128.

78. On the favorable performance characteristics of electric motors, see Allan 1858; Vergnes 1857.

79. For evidence, see Hunter and Bryant 1991.

80. An early example, made in Philadelphia in sizes from 1/32 to 1/2 horsepower, was the battery-powered "Bastet Magnetic Engine" (*SA* 35, 1876: 13).

81. Vander Weyde 1889: 234.

82. Henry, quoted on p. 58 of Molella et al. 1980.

83. Henry, quoted on p. 1275 of Molella 1976.

84. Engineers, of course, are routinely required *by their employers* to take costs (capital and operating) into account when designing a product or technological system. However, in such cases the

consumer's stated preference for economy is a "quasi-technical constraint" on design that is specified explicitly in advance (see Vincenti 1995); i.e., it is a performance requirement.

85. For examples of the adoption of prestige or "idealistic" technologies, see Friedel 2007; Hayden 1998; Pacey 1992.

86. MacLaren 1943: 92.

87. A plugger pounded gold leaf into drilled-out cavities; on battery-powered dentist's tools, see White 1995; Franklin Institute 1885. Urquhart (1882) emphasized that battery-powered motors were practical for applications such as routers, stone engravers, dentists' drills, sewing machines, lifts, hoists, fans, clockmaking tools, and ventilators. On the post-1880 adoption of electric motors, see Bell 1897.

Notes to Chapter 14

1. On the growth of the U.S. Patent Office, see Post 1976a,b.

2. The former number is from "Progress of Patents," *SA* 29 (1873, n.s.): 272; the latter is from Post 1976a: 53.

3. "Inventors—the Scientific American," *SA* 8 (1853): 322.

4. Post 1976b: 24. On the interplay of factors affecting Patent Office policy, see Post 1976a,b; see also Kahn 2005.

5. The term "scientific inventions" occurs as early as 1850 in an editorial, "Waste of Ingenuity," *SA* 5 (1850): 171.

6. In practice, however, patent applications—even successful ones—could obscure essential details, perhaps to slow down competitors who would have to replicate the invention through trial and error.

7. Quotations in this paragraph are from *SA* 22 (1870, n.s.): 95.

8. Henry, quoted on p. 333 of Molella and Reingold 1973.

9. During the period 1879–1883, *Scientific American* published 87 articles on "American Industry" that profiled specific firms and described and illustrated their manufacturing processes. For an index, see Pursell 1976. Hounshell (1980a) points out that these articles were in fact paid advertisements.

10. "Science—Its Truths and Falsehoods," *SA* 7 (1852): 373.

11. "What Remains for Inventors," *SA* 21 (1869, n.s.): 407–408.

12. "An Invention Wanted—Chance for Electrical Engineers," *SA* 7 (1852): 213.

13. "A Good Chance for Inventors," *SA* 13 (1858): 325; "Premium for Steam-Plow," *Scientific American* 14 (1859): 286; "Ivory—A Substitute Wanted," *SA* 12 (1857): 261.

14. "Inventions Wanted," *SA* 12 (1856): 20.

15. *Scientific American* 2 (1847): 337.

16. "Wonderful Announcement," *SA* 4 (1848): 101.

17. Ibid.

18. Paine, letter, *SA* 4 (1848): 117.

19. Quotations in this paragraph are from G.C.T.'s letter (*SA* 4, 1849: 347).

20. Letter signed by "Carburetted Hydrogen," *SA* 5 (1850): 158.

21. Paine, letter, *SA* 5 (1849): 85.

22. Letter signed by R, *SA* 5 (1850): 317.

23. Paine, letter, *SA* 5 (1849): 85.

24. Paine, letter, *SA* 5 (1850): 203.

25. On Henry's handling of "visionary theorizers" (whom he hoped to turn to serious study, unless they were humbugs like Paine), see Molella 1984.

26. Henry, quoted on p. 160 of Reingold 1991.

27. In a letter to the *New York Tribune* (June 13, 1850), a writer, having talked to Henry, cited the latter's objections to Paine's supposed discovery (see Rothenberg and Dorman 1998: 64, note 8). Two decades later, writing to George A. Clark, Henry stated undiplomatically that "Paine was a charlatan who claimed to have made an impossible discovery," November 12, 1869, SIA, RU-33, Official Correspondence Outgoing.

28. Henry to Henry R. Schoolcraft, July 9, 1850, SIA, RU-7001, Joseph Henry Collection, Letters Outgoing, 1847–1850.

29. Letter signed by Carburetted Hydrogen, *SA* 5 (1850): 158.

30. Quotation and information in this paragraph are from "Report of the Scientific Committee to Investigate Paine's Light," *SA* 5 (1850)332.

31. "Paine's Electric Light," *SA* 5 (1850): 341.

32. Paine, letter, *SA* 5 (1850): 355.

33. Quotations and information in this paragraph are from Mathiot's letter and the appended comment, published as "The Electric Light—Mr. Paine's Discovery Corroborated by Experiment," *SA* 5 (1850): 371. Curiously, Mathiot subsequently authored, under the pseudonym "Volta," a multi-issue treatise on batteries and electrotyping: *SA* 6 (1850): 3, 11, 19, 27, 35, 43, 51, 59, 67, 75, 83, 91, 99, 107; Volta's identity was revealed in "Electrotypes," *SA* 8 (1853): 402.

34. Paine, letter, *SA* 6 (1850): 22.

35. "Light From Water," *SA* 6 (1850): 21.

36. Wright's article was reprinted as "Mr. Paine's Light," in *SA* 6 (1850): 123, to which was appended an editorial comment.

37. "Paine's Light—Evidence of an Eyewitness," reprint from the *Boston Transcript* in *MM* 54 (1851): 114–115.

38. On the importance Henry attached to specialized knowledge for evaluating claims, see Reingold 1991: 160–161.

39. Quotations and information in this paragraph are from "Expose of Paine's Light," *SA* 6 (1851): 154.

40. P.M.H., letter, *SA* 6 (1851): 172.

41. Mathiot, letter, *SA* 6 (1851): 205.

42. J. B. Blake, "Practical Remarks on Illuminating Gas," *SA* 6 (1852): 310.

43. "Paine's Electric Light—The Patent," *SA* 6 (1851): 248.

44. "Paine's Light," *SA* 6 (1851): 260.

45. "False Lights," *SA* 6 (1851): 301.

46. "The Atmospheric Lamp! Another New Light from Paine's Laboratory," *SA* 6 (1851): 268.

47. Henry W. Adams to Ewbank, June 2 and 21, 1851, USNA, RG-241, Patent Office, Patented Files, Box 147, patent no. 9,119.

48. Paine to Ewbank, January 7, 1852, ibid.

49. Ewbank to Henry, April 28, 1852, ibid.

50. Henry to Ewbank, May 11, 1852, ibid.

51. Ewbank to Henry, May 26, 1852, ibid.

52. Henry to Ewbank, May 29, 1852, ibid.

53. Henry to Ewbank, June 1, 1852, ibid.

54. The draft letter survives: Leonard Gale to Paine, June 2, 1852, ibid.

55. On the threat to petition Congress, see Paine to Ewbank, April 3, 1852, ibid.

56. Report of experiments, June 22, 23, and 26, 1852, ibid.

57. On political pressures on the Patent Office to liberalize its policies—i.e., grant patents to a higher percentage of applicants—in view of the latitude for judgment afforded by the 1836 patent law, see Post 1976a,b.

58. This was first suggested in a letter in February 1851: "A Mr. Mansfield, constructed, last year, an arrangement for simply forcing *atmospheric air* through an exceedingly volatile hydro-carbon, known by the name of Benzole. He succeeded in producing, by this means, a brilliant gas

light. . . . Perhaps Mr. Paine slily [*sic*] uses some such liquid instead of turpentine, or in connection with it." J.T., "Hydrogen—Benzole," *SA* 6 (1851): 178. See also, "Expose of Paine's Light," *SA* 7 (1851): 19.

59. In reference to benzole as an illuminant, Wylde (1881–82, volume 1: 653) wrote that "As an addition to alcohol it affords the necessary amount of carbon to give a bright white light, and may be used instead of turpentine in the lamp just described . . . if pure hydrogen be passed through it, the gas will burn with a brilliant flame. Even atmospheric air may be so 'naphthalised' by it as to burn like coal-gas."

60. Paine, letter, *SA* 10 (1855): 286.

61. Patents 102,856, 103,228, 103,229, 103,230, and 103,231. (These patents were tracked down by Robert C. Post.) By this time, Charles Page had passed away and was no longer a patent examiner. Apparently the patent office had become lax in scrutinizing electromagnetic inventions.

62. "Novel Electro-Magnetic Engine," *NYT*, June 20, 1869.

63. Quotation and information in this paragraph are from "Paine's Electro-Motor," *SA* 24 (1871, n.s.): 374; an earlier account is "Payne's [*sic*] Electro-Magnetic Motor," *SA* 24 (1871): 167.

64. Paine, letter, *SA* 24 (1871, n.s.): 404. For quantitative claims, see Paine, letter, *SA* 25 (1871, n.s.): 36.

65. Vander Weyde letters, *SA* 25 (1871, n.s.): 36–37, 51–52; the quotation is on p. 36. On Vander Weyde's interesting careers, see his "Reminiscences of an Active Life," in *MAB*, which begin in volume 25, no. 2 (1893): 25–26; his electrical work is mentioned in volume 25, no. 4 (1893): 88.

66. Paine, letter, *SA* 25 (1871, n.s.): 52.

67. Rowland, letter, *SA* 25 (1871, n.s.): 21. This visit, which took place a few months before Rowland's July, letter, was probably to Paine's New Jersey office. For biographical information on Rowland, see "Henry Augustus Rowland," *The National Cyclopædia of American Biography* 11: 25–26 (1967, University Microfilms). Vander Weyde also visited Paine and witnessed a demonstration of the motor. He, too, inferred that the saw was connected to another power. To test his conjecture, he secretly disconnected the battery, yet the saw continued to run (Vander Weyde 1886). Henry Morton (1895a) furnished additional details about the fraudulent motor.

68. B.D., letter, *SA* 25 (1871, n.s.): 116.

69. Smith first airs his suspicions: *SA* 25 (1871, n.s.): 100–101; Paine's replies to Smith: *SA* 25 (1871, n.s.): 101, 132. Smith's challenge to Paine: *SA* 25 (1871, n.s.): 180.

70. Vander Weyde, letter, *SA* 25 (1871: n.s.): 212.

71. Paine, letter, *SA* 25 (1871, n.s.): 244.

72. Morton 1895b: 205.

73. Paine, letter, *SA* 6 (1851): 246.

74. Rothenberg and Dorman 1998: 254.

Notes to Chapter 15

1. Schiffer 2008.

2. Koestler 1964.

3. Davis (1842, 1848). The Morse-Vail apparatus appears in the Appendix to the 1842 catalog, p. 24; by 1848 this device was thoroughly integrated.

4. On the American patent-management system during this period, see Cooper 1991; Kahn 2005; Post 1976a.

5. Often I was not able to learn if the inventions were commercialized and adopted, but I believe many reached the prototype stage.

6. "A New Electric Annunciator," *SA* 11 (1864): 315.

7. J. P. Joule, "Self-Acting Apparatus for Steering Ships," *SA* 12 (1865 n.s.): 194–195.

8. J. A. Ballard, "Guiding and Controlling the Movements of Torpedo Boats," *MM* 24 (1870 n.s.): 153–154.

9. "Sounding the Sea by Electro-Magnetism," *MM* 34 (1841): 320.

10. "The Atlantic Submarine Telegraph," *MM* 67 (1857): 127–130, 128.

11. Reported by Du Moncel (1859: 521–522).

12. "Electrical Loom," *MM* 60 (1854): 272–273.

13. "Electro-Magnetic Engraving Machine," *MM* 60 (1854): 539–540.

14. "Scientific Detection of Burglars," *MM* 68 (1858): 345. In 1847, John Rutter patented a similar electromagnetic burglar alarm, using a mercury switch (Great Britain, Commissioners of Patents 1859: 90–91).

15. Ibid.: 122–123.

16. On the Varleys' invention, see Great Britain Commissioners of Patents 1871: 306, "Sternberg's Electro-Magnetic Regulator for Dampers and Valves," *SA* 23 (1870 n.s.): 126; Du Moncel 1859: 532–534.

17. "New Galvanic Clock," *MM* 40 (1844): 325–326.

18. *SA* 18 (1868): 193; *SA* 33 (1875): 62.

19. Bain 1852.

20. "Greenwich Time," *SA* 17 (1867 n.s.): 90.

21. "A Time Ball in New York City," *SA* 36 (1877): 257.

22. Christopher and Christopher 1996: 149. Houdin's estate in Saint Gervais was filled with electromagnetic contrivances; see "Curious Applications of Electricity," *SA* 18 (1869 n.s.): 178.

23. Jahns 1961: 212–213.

24. "A.B." (Alexander Bain), "Music by Electricity," *MM* 68 (1858): 544.

25. "Pneumato-Electric Organ," *SA* 14 (1866 n.s.): 285; "An Electric Organ," *SA* 17 (1867 n.s.): 310; "Electric Actions for Organs," *SA* 20 (1869 n.s.): 347; New Applications of Electricity to Organ Building," *SA* 34 (1876): 117; "The Largest Organ in the World," *SA* 41 (1879): 385).

26. "Telegraph in America," *SA* 6 (1850, n.s.): 81.

27. "The Electric Telegraph," *SA* 23 (1870): 196–197.

28. "The Domestic Telegraph," *SA* 27 (1872): 39.

29. "Kite Tails and Telegraph Wires," *SA* 31 (1874): 113.

30. See "William Francis Channing," *The National Cyclopædia* 23: 284–285.

31. King 1883: 15.

32. See "Moses Gerrish Farmer," *Dictionary of American Biography* 3: 279–281.

33. Channing (1852) gives a technical description of the Boston fire alarm telegraph with some possible variations. See also "The Municipal Fire Telegraph," *SA* 7 (1852, n.s.): 219, 227.

34. Channing 1855.

35. King 1883: 18.

36. A system, probably invented independently, was installed by Siemens in Berlin in 1851 (Channing 1852: 59); see also "Electric Fire Telegraphs on the Continent," *TE* 14 (1862): 185. For descriptions of a later New York system, see "The Working of the New York Fire Department," *SA* 24 (1871): 213; "The New York Fire Alarm Telegraph," *SA* 26 (1872): 177.

37. "Railroad Collisions," *SA* 10 (1854): 67.

38. On the importance of the telegraph to railroad operation, see Chandler 1977; Salsbury 1988; Thompson 1972.

39. "Railroads in the United States," *SA* 22 (1870): 42.

40. Langdon 1877: 45. From this work I have derived a description of the early block signaling systems.

41. Salsbury 1988: 46. He also notes that the U.S. time zones were established by the railroads in 1883.

42. On the influence of everyday technical problems on Einstein's thinking, see Galison 2003.

43. "Improved Switch Signal and Alarm," *SA* 16 (1867): 277.

44. "Grand Central Depot Signal System," *SA* 33 (1875): 399, 402.

45. Carlson 1991: 7.

46. Bain, who patented his copying telegraph in England in 1843, exhibited it in New York City during 1848. "New Telegraph Line," *SA* 4 (1848) 29. On the early copying telegraphs, see Prescott 1888.

47. Information on Caselli's pantelegraph is from Prescott 1888: 744–756, Société d'Histoire de la Poste et de France Télécom en Alsace 1992 and from "Caselli's Pantélégraphe," www.hffax. de/html/hauptteil_caselli.htm.

48. Prescott 1888: 744–756. Caselli's pantélégraphe was awarded U.S. patents 20 698 (1858) and 37,563 (1863).

49. "Caselli's Pantélégraphe."

Notes to Chapter 16

1. This section derives largely from Schiffer 2005b.

2. For overviews of U.S. lighthouse organization, see Johnson 1890; Noble 1997; Weiss 1926.

3. For histories of Trinity House, see Elliot 1874; Hague and Christie 1975; MacLeod 1969.

4. On French lighthouses, see Reynaud 1871. Fresnel's numbering system was based on the focal length of the lens. The correlation between a Fresnel number and brightness was imperfect, but first-order lights generally were the brightest and used in the most important lighthouses.

5. MacLeod 1969.

6. On lighthouse science, see Allard 1876; Douglass 1886; Henry 1887; Stevenson 1871; Tyndall 1874.

7. "Our Artificial Light." *SA* 14 (1858): 125.

8. On the composition of carbon rods, see Bright 1949: 24–26.

9. On early batteries, see Niaudet-Bréguet 1880. On regulators, see King 1962b. Hospitalier and Maier (1883, volume 1) list 33 arc lamps, most of them French.

10. On feedback mechanisms, including arc light regulators, see Mayr 1971.

11. Bright (1949: 25–26) lists some battery-light inventors.

12. Du Moncel (1880) mentions early applications of arc lights. In England, the unit of light was the candle, in France it was the Carcel lamp; one Carcel lamp equaled about 7.5 candles (Holmes 1884: 86).

13. Grove 1849: 211.

14. Great Britain, Commissioner of Patents 1859: 108–109.

15. Information about Staite's electric light and its reception is from Staite 1882. On Staite's patents for arc light inventions in the years 1846–1849, see Dredge 1882.

16. Staite, quoted in "More About Staite's Electric Light," *SA* 3 (1848): 219.

17. *Manchester Courier*, October 18, 1851, quoted on p. 9 of Staite 1882.

18. *Chemical Times*, December 2, 1848, quoted on p. 6 of Staite 1882.

19. "Progress of Lighting by Electricity—Mr. Staite's New Patent," *MM* 46 (1847): 621–622, p. 621.

20. "New Electrical Light," *SA* 4 (1848): 92.

21. Source of quotations: "Lighting by Electricity," *SA* 2 (1847): 416.

22. W. E. Staite, letter dated July 5, 1847, *MM* 47 (1847): 44.

23. *London Times*, November 2, 1848, quoted on p. 5 of Staite 1882.

24. *Sunderland Herald*, June 7, 1850, quoted on pp. 6–7 of Staite 1882.

25. Wilkins to Staite, July 22, 1850 (p. 8 of Staite 1882).

26. On Faraday's involvement with Trinity House, for which he received £200 per year, see James 1999, 2000. A second battery light, offered by Dr. Joseph Watson, was tested under Faraday's supervision and was also rejected (James 1999).

27. Faraday to T. T. W. Watson, December 20, 1852 (p. 674 of Williams 1971b).

28. "More about Staite's Electric Light," *SA* 3 (1848): 219.

29. "False Light," *SA* 6 (1851): 301.

30. The French experiments are mentioned in M'Gauley 1867.

31. This is a close paraphrase from p. 478 of Du Moncel 1859.

32. Grove 1849: 211.

33. Hunt 1860: 89.

34. See, e.g., Hayden 1998. "Expensive" in non-market societies is usually computed on the basis of total labor invested in acquiring materials, manufacturing, and distributing a good.

35. "Electric Light for Daguerotypes [*sic*]," *SA* 4 (1849): 308.

36. Advertisement, *NYT*, June 1, 1857.

37. *NYT*, May 29, 1857.

38. Snelling 1849: 78.

39. "Improvement in the Photographic Art," *SA* 6 (1851): 356.

40. Gernsheim and Gernsheim 1969: 429.

41. "The Magic Lantern," *SA* 22 (1870, n.s.): 13.

42. The "limelight," invented by the American chemist Robert Hare, was reinvented in England, where it was known as the Drummond light.

43. "The Magic Lantern," *SA* 22 (1870, n.s.): 13.

44. Gage 1908.

45. Alglave and Boulard 1884: 388.

46. Hearder 1865: 7.

47. See Gorman 1977; Alglave and Boulard 1884.

48. For additional examples of the early use of battery-powered carbon arcs in public spectacles, see Zöllner 1865.

49. "Lighting Streets by Electricity," *SA* 11 (1855): 22.

50. "Electric Illumination," *SA* 12 (1857): 284.

51. These examples are from Alglave and Boulard 1884: 408–411. On Duboscq's employment, see Hospitalier and Maier 1883, volume 1, p. 420.

52. Quotations in this paragraph are from "France," *NYT*, June 25, 1867.

53. Information in this paragraph is from "Affairs in France," *NYT*, August 30, 1868. Use of the electric light in French national festivals had begun by 1858; see *NYT*, September 1, 1858.

54. "The Great Boston Organ," *NYT*, November 6, 1863.

55. "The Carnival at Washington," *NYT*, February 21, 1871, p. 1.

56. On the docks, see *NYT*, September 14, 1854. On the mansion, see *NYT* November 16, 1854.

57. Higgs 1879a: 229–230.

58. "Electric Head-Light for Locomotives," *MAB* 7 (1875): 16.

59. "Electric Illumination," *SA* 14 (1858): 4.

Notes to Chapter 17

1. Morse to Spencer, August 10, 1843, MP Reel 7, p. 99117.

2. "The Telegraph," *SA* 7 (1852): 330. Even earlier, it was remarked that "the great question of uniting England, or rather Ireland and America [by submarine cable], has been the theme of conversation in various circles, and sage opinions expressed in regard to its feasibility or

possibility;" from "Sub-Marine Telegraph Under the Atlantic," *SA* 6 (1850): 35. Also in 1850, the American civil engineer John Roebling claimed that a cable across the Atlantic would be "entirely practicable" ("Transatlantic Telegraph," *SA* 5, 1850: 260).

3. Early works on the Atlantic cable project: Briggs and Maverick 1858; Bright 1908; Field 1866, 1893. Later works: Dibner 1959 (which includes stunning but often unattributed images); Carter 1968; Coates and Finn 1979; Finn 1973; Gordon 2003; Hearn 2004; Lindley 2004; Silverman 2003.

4. For a brief discussion of this cable, see Dibner 1959: 5.

5. This cable was engineered by John Brett. On the earliest submarine cables, see Brett 1857.

6. Gutta percha was the preferred insulation for cables well into the twentieth century (de Guili 1932: 4–6).

7. Mechanical considerations are discussed in "The Atlantic Telegraph," *NYT*, August 28, 1857. For other discussions, see "Letters to the Editor," *TE* 7 (1859): 84, 278, 356.

8. Carter 1968 is a biography of Field.

9. Unless otherwise noted, information in this section is from Carter 1968, Dibner 1959, Field 1866, or Williams 1963.

10. Henry to Dorothea L. Dix, August 6, 1858 (Rothenberg et al. 2004: 54–56). This letter also furnishes reasons why Henry "never had faith in the enterprise" (p. 55), an opinion he apparently kept to himself. On Henry's tangential involvement in cable discussions and experiments, see Henry to Bache, November 30 and December 22, 1859 (Rothenberg et al. 2004: 124–126, 130–133).

11. Williams 1963 is a biography of Maury. See also Jahns 1961; Hearn 2002.

12. "Ocean Telegraphs and Surveys," *SA* 11 (1856): 357; "Sounding the Ocean," *SA* 12 (1856): 110; Williams 1963: 235–247.

13. Dibner 1959: 9.

14. Ibid.: 12.

15. Werner Siemens had reported this effect in 1850 and attributed it to the (inductive) electrostatic charging of the cable (Hunt 1991: 5–6).

16. On Faraday's analysis, I have relied on Faraday 1854 and Hunt 1991. Instead of Faraday's term "specific inductive capacity," I employ "electrostatic capacity."

17. Thomson, largely because of his work on the cable project, was made Lord Kelvin, Baron of Largs, in 1892. Biographies include Lindley 2004; Sharlin 1979; Smith and Wise 1989.

18. For the mathematical analysis, see Thomson 1855. On his additional discussions of cable design, see Thomson 1856a, 1856b, 1884; for a later summary, see Thomson 1884.

19. Hunt's (1996) nuanced discussion of the conflicts between Thomson and Whitehouse backgrounds their implications for cable design.

20. These experiments are described in "The Atlantic Telegraph," NYT, August 27, 1857. On earlier Whitehouse experiments, see Whitehouse 1857; "On Ocean Telegraph Cables," SA 11 (1855): 118.

21. On his reanalysis, see Thomson 1884: 92–102. Later experiments by Varley (1862), a close collaborator of Thomson's, added support for the law of squares.

22. Window's (1857: 44) estimate was based on Latimer Clark's experiments.

23. His actual claim is 17 letters a minute (Thomson 1884: 101). The cost figures are on the same page.

24. For a description of this experiment, see "The Atlantic Telegraph," NYT, August 27, 1857.

25. Morse to Field, October 3, 1856 (Dibner 1959: 14).

26. Ibid.: 16.

27. Ibid.: 16.

28. "Electric Telegraph Across the Atlantic," TE 2 (1856): 633.

29. For a list of important personnel and directors, see "The Atlantic Telegraph," NYT, August 27, 1857.

30. Silverman 2003: 351–355.

31. Seward, quoted on p. 18 of Dibner 1959.

32. "The Transatlantic Telegraph Bill," SA 12 (1857): 211.

33. Dibner (1959: 17) describes the 1857 cable. See also "The Atlantic Submarine Telegraph," SA 12 (1857): 216.

34. The cable was so flexible "that it may be tied in a knot around a man's arm" with ease ("The Atlantic Submarine Telegraph," SA 12, 1857: 216).

35. Hunt (1996) notes that even Whitehouse complained that insufficient testing had been done beforehand.

36. Thus, regarding Thomson's law of squares, "Nature recognizes the existence of no such law, "The Atlantic Telegraph," NYT, August 27, 1857.

37. In a prescient piece originally published in the London Times, critics assailed the design of the paying-out machinery, asserting that it could easily sever the slender cable. "A Gloomy View of the Success of the Enterprise," NYT, August 6, 1857.

38. Information in this paragraph is from "The Atlantic Cable," NYT, August 27, 1857. The cable itself had consumed 73.4 percent of the money spent up to that time.

39. Unless otherwise noted, information in this section is from Dibner 1959: 19–25.

40. Ibid.: 22.

41. "The Atlantic Telegraph," *NYT*, September 11, 1857.

42. "Mishap to the Atlantic Telegraph—The Cable is Broken," *NYT*, August 27, 1857. For Bright's account of this event, see his letter to the Directors of the Atlantic Telegraph Company, in "The Atlantic Telegraph," *NYT*, September 4, 1857; another assessment is "Failure of the Atlantic Telegraph—Its Cause and Prospects," *SA* 13 (1857): 29.

43. De Cogan (1985) describes the storage conditions, which were inimical to the preservation of gutta percha.

44. Field, quoted on p. 24 of Dibner 1959.

45. Ibid.: 25; "The Atlantic Telegraph," *NYT*, August 28, 1857. Briggs and Maverick (1858: 56–91) describe how the cable was manufactured and tested for electrical and mechanical performance.

46. For details of the mirror galvanometer's construction and operation, see Dibner 1959: 27–28; Keithley 1999: 160–162.

47. "Failure of the Second Attempt to Lay the Atlantic Cable," *NYT*, July 21, 1858. There are many contradictory accounts of the second expedition, but all agree on the cable's eventual fate; I follow Dibner 1959: 29–30.

48. Ibid.: 30–32.

49. On the second try of 1858, see ibid.: 31–36.

50. Field to the Associated Press, in "Telegraphic News," *CT*, August 6, 1858.

51. Dibner 1959: 37.

52. Ibid.: 37–38.

53. Advertisement for book (Briggs and Maverick), *NYT*, August 14, 1858; souvenir cable, *NYT*, September 20, 1858.

54. For the messages, see "The Atlantic Telegraph," *SA* 13 (1858): 406. For a description of the sending and receiving apparatus, see Whitehouse 1858.

55. News of celebrations and so forth appeared on the front pages of the *NYT* on August 18 and 19 and September 1, 1858.

56. Henry, quoted on p. 39 of Dibner 1959.

57. I have simplified this part of the story. For a richer account, see de Cogan 1985.

58. On transmission rates, see Dibner 1959: 39, 44.

59. "Probable Rupture of the Atlantic Cable," *NYT*, September 22, 1858. De Cogan (1985) provides a message count, noting that some were sent until about October 20.

60. De Cogan (1985) claims that Whitehouse's use of high voltage was essentially irrelevant because defects in the cable's construction and in its subsequent handling and storage (for specifics, see Dibner 1959: 44–45) would have rendered it unusable anyway. On Whitehouse's interpretation of the cable's failure, see his letter, *NYT*, September 22, 1858. On Thomson's delineation of the location where the insulation failed, see "Atlantic Telegraph," *NYT*, September 30, 1858. For an abridged version of Whitehouse's reply to the directors, see Dibner 1959: 91–93. Hunt (1996) believes that Whitehouse was the convenient fall guy, who could be jettisoned from the project without serious repercussions because he lacked support from either scientists or telegraph engineers.

61. Commission report, quoted on p. 5 of "Sir James Anderson's account of Marine Cables Laid," *The Telegraphic Journal and Electrical Review* 1 (1878): 5–9. This article, which contains an exhaustive list of cables laid between 1850 and 1872, noted that "all light cables have been short-lived, and that all heavy cables have continued working, often under most adverse conditions," p. 7.

62. "Prospects of a New Atlantic Telegraph," *NYT*, March 31, 1862, p. 4.

63. On Shaffner's project, see "The Atlantic Telegraph," *CT*, April 28, 1862; Bright 1860–61: 70–79.

64. Dibner 1959: 45.

65. These numbers are from Dibner 1959: 48. Using them, I calculated the transmission rate as follows: 300 days × 16 hours × 60 minutes = 288,000 minutes per year; £413,000/288,000 minutes = £1.434 per minute = 344 pence per minute; 344 pence per minute/30 pence per word =11.5 words per minute.

66. Field (1893: 251–254) notes Thomson's influence on the design of the new cable and also reports the latter's performance characteristics. Information in this paragraph is from "The Atlantic Cable," *CT*, July 10, 1866.

67. The ship's vital statistics vary among sources; I have relied on Dibner's numbers (1959: 53, note **) and on "The Steamship Great Eastern," *SA* 12 (1857): 264; see also Routledge 1879: 87–91. Prophetically, the *London Artizan* suggested in 1856 that the *Great Eastern* would be ideal for laying the Atlantic cable; see "The Atlantic Telegraph Cable," *SA* 12 (1856): 57.

68. Dibner 1959: 53.

69. Information in this paragraph is from ibid.: 57.

70. Ibid.: 58.

71. Ibid.: 60–63.

72. Ibid.: 63–67.

73. For details of the financial dealings, see ibid.: 70. The cost of the new cable was estimated at £500,000 (ibid.: 69).

74. On outfitting the 1866 expedition, see "The Atlantic Cable," *NYT*, July 30, 1866.

75. Field, quoted on p. 78 of Dibner 1959.

76. For the £2.5 million figure, see ibid.: 85.

77. "The Atlantic Cable," *CT*, August 1, 1866.

78. "A Guinea a Word," *NYT*, August 1, 1866. An official statement of the tariff: "A Corrected Statement of the Tariff for Dispatches to Europe," *NYT*, August 2, 1866.

79. "Telegraphic Messages for Europe—First Day's Business, *NYT*, August 2, 1866.

80. "The Atlantic Telegraph," *CT*, August 3, 1866. By 1875, signaling had reached an average rate of 17 words per minute: "Telegraphy," *SA* 33 (1875): 199.

81. Foster 1866 is a useful source on the electrical-engineering science embodied in the cable's operation as it was understood in 1866.

82. "The Atlantic Cable," *CT*, September 21, 1866.

83. "Rates Reduced on Despatches by Atlantic Cable," *CT*, October 26, 1866; "Ocean Telegraphy—The Foreign Connections of New York City and the Existing Rates of Charges," *SA* 32 (1875): 120. For a list of operating ocean cables in 1875, which included 16 cables more than 1,000 miles long, see Prescott 1875.

84. Wilkinson 1896: 129–137.

85. For copious information on submarine cables in use, including international rates, see United States Hydrographic Office 1892.

86. Dibner 1959: 83.

87. This was alluded to in a company report published in "The Atlantic Telegraph," *NYT*, August 27, 1857.

88. In 1856, Field had queried Matthew Maury on cable design; he recommended a thinly insulated, light weight cable (Williams 1963: 234).

89. Bright also favored a thicker cable (Coates and Finn 1979: 31, note **), but he may not have pressed the case.

90. "From Europe," *SA*, September 25, 1865.

91. Coates and Finn (1979) list the many forecast benefits of the Atlantic telegraph.

92. See Coates and Finn 1979; Finn 1973; Headrick 1981; Hunt 1997; Nickles 2003.

Notes to Chapter 18

1. The definitive history of the Capitol is Allen 2001. On the dome, see Allen 1992. My background information on the Capitol extension project is from these sources. Large parts of this chapter are drawn, almost verbatim, from Schiffer 2008.

2. Allen 2001: 187.

3. Ibid.: 187.

4. On Brumidi and his art in the Capitol, see Murdock 1950; Wolanin 1998.

5. Schiffer et al. 2003: 230–232.

6. On the French invention, attributed to Villatte of Paris, see "New Mode of Lighting Street Lamps," *SA* 5 (1850): 364. On the Edinburgh installation, see "Gas Lighting by Electricity," *TE* 7 (1859): 356. For descriptions of some electric gas-lighting technologies in Europe, see Wise 1873.

7. "Another Use for the Telegraph Wires," *SA* 7 (1852): 406. Even earlier was a sketchy report on a British scheme for igniting the lamps on London streets, "Lighting Street Lamps by Electricity," *SA* 3 (1848): 356.

8. Born in Massachusetts, he appears as a New York resident in the federal census of 1860, but in 1870 is recorded as living in Washington, D.C; I have found no information on his later whereabouts.

9. The U.S. patent numbers for Gardiner's ore-processing machines are 9,640, 11,368, and 13,645. The ore crushing and pulverizing machine was brought to market by the Morgan Iron works of New York City; four machines were sold for $2,500 each on credit. In a petition to Congress in 1870, requesting a 7-year extension on the patent, Gardiner claimed that the development costs of several thousand dollars had not been recouped and that sales had been truncated by the Civil War. USNA, RG 233, Records of the U.S. House of Representatives, 41st Congress, Accompanying Paper Files, Box 11, file: "Samuel Gardiner, Jr." If, as Gardiner implied, he had fronted the development costs, it suggests that he could draw upon considerable financial resources. Although the House passed a bill (H.R. 1551) to extend the patent, the Senate did not concur (*U.S. House of Representatives Journal*, March 18, 1870, p. 490; *U.S. Senate Journal*, March 25, 1870, p. 410).

10. H. L. Hudson, undated manuscript, "History of the Galvanic and Magnetic Gas Lighting Inventions of the Capitol of the United States. USNA. RG-48, Entry 291, Box 3, "Dome-Lighting," p. 4–5. James Harlan, Secretary of the Interior, acknowledged receipt of this document, enabling it to be dated; Harlan to Hudson, December 28, 1865, USNA, Entry 295, Capitol Extension, letters sent, April 19, 1862–March 8, 1869, p. 152.

11. United States Census, 1860, New York Ward 17, District 9.

12. Draft patent application, USNA, RG-241, Records of the Patent and Trademark Office, Patented Files, Box 345, Patent No. 18,945.

13. Curiously, in questioning the application, the examiners did *not* cite the prior British patent of Charles Cowper, issued July 22, 1855 (no. 1732), for "lighting and extinguishing gas lights." Its gist is that "the valve or cock is opened by means of electro-magnetic arrangements, and at the same moment (or immediately afterwards) an electric spark is passed through the issuing

gas, or a fine platinum wire is ignited in the gas" (Great Britain Commissioner of Patents 1859: 583–584); a year later, a patent (no. 1775) was issued to Isham Baggs for a similar invention (pp. 587–590). Gardiner's patent, applied for on April 4, 1857, was issued on December 22, 1857; USNA, RG-241, Records of the Patent and Trademark Office, Patented Files, Box 345, Patent No. 18,945. By this time his invention had already received a British patent (no. 1060), issued in the name of William E. Newton, on April 14, 1857, despite the British precedents.

14. Norton deposition, December 2, 1857, USNA, RG-241, Records of the Patent and Trademark Office, Patented Files, Box 345, Patent No. 18,945.

15. *NYT*, May 28, 1857. Hudson (note 10 above) himself claimed that this trial was largely a failure.

16. "Galvanic Gas-lighter," *SA* 12 (1857): 320.

17. S. Marshall, manager of the Broadway Theatre, was deposed by the Patent Office. He stated, somewhat ambiguously, that Gardiner's system "was put in successful operation in May last." Marshall deposition, December 5, 1857, USNA, RG-241, Records of the Patent and Trademark Office, Patented Files, Box 345, Patent No. 18,945.

18. The patent numbers and dates are furnished in Schiffer 2008.

19. Patent No. 132,569 (1872). An example of this meter, very likely the patent model, is in the collections of the NMAH.

20. On this approach to explaining consumer behavior, see Schiffer 2000, Schiffer 2005b, and chapter 20 of the present work.

21. Hoping to entice the government into buying rights to use his patent, Gardiner argued in a petition to Congress that the gas lighter could be the basis of signal lights or adapted for detonating canons and submarine mines. The petition was rejected. *U.S. Senate Journal*, April 5, 1858, p. 314. Petition of Samuel Gardiner, Jr. USNA RG-46, Records of the U.S. Senate, 35th Congress, SEN 35A-H16, box 97.

22. General information about home lighting is from *NYT*, September 30, 1859. On the system in Gardiner's house, see "Gas-Lighting by Electricity," *The American Gas-Light Journal* 1 (1859): 62–63.

23. I have found no evidence that Gardiner was actually paid for this installation. Perhaps, like so many electrical firms later in the century, Gardiner installed it in the expectation of receiving favorable publicity as well as other government contracts. Because the old Senate chamber was about to be remodeling anyway, it carried low risk for the government.

24. Quotations in this paragraph are from "Gardiner's Galvanic Gas-Igniter," *SA* 2 (1860 n.s.): 133.

25. Wilson (1860) described his gas-lighter and presented a lengthy critique of Gardiner-type systems.

26. A Wilson system had been installed in the Cooper Institute, where it lit 300 burners in a large lecture hall: "The Polytechnic Association of the American Institute," *SA* 4 (1861, n.s): 75; Gustavus Miller recalled that it did not work well: "Lighting Gas by Electricity," *SA* 11 (1864): 342.

27. "Gas Lighting by the Electric-Spark," *The American Gas-Light Journal* 1 (1859): 90.

28. The patent (no. 40,468) for the exploding bullet was issued in 1863. Information in this paragraph is from www.inventors.about.com/library/inventors/blgun.htm and www.csa-dixie.com/csa/gazette/explosivemusketballs.htm.

29. On the political and technical difficulties of completing the extension project see Allen 2001; Meigs 2001.

30. Hazelton 1897: 60. This is also the source of the figures and dates in this paragraph.

31. Allen 2001: 335.

32. Thomas Walter to Gardiner, December 20, 1862. USNA, RG-46, Records of the U S. Senate, 42nd Congress, Sen 42A-H3, Petitions and Memorials, Box 104, "Gardiner and Glenn."

33. B. B. French to Gardiner, December 29, 1862, ibid. Gardiner's proposal is also in this file.

34. The general information about the dome lighting system here is liberally paraphrased from "Lighting up of the Capitol Dome," *SA* 14 (1866 n.s.): 97 and from "Gas Lighting by Electricity," *SA* 16 (1867 n.s.): 23. Knight (1876: 1315–1316) also discusses the project; for more recent mentions, see Allen 2001: 366; Myers 1978: 229.

35. Gardiner's invoice is in USNA, RG-48, Interior Department, Entry 291, Box 3, "Lighting the dome."

36. Gardiner to John P. Usher, February 11, 1864, USNA, RG-48, Entry 289, Interior Department, Letters received, Vol. 1, May 2, 1862–April 5, 1867.

37. B. B. French to James Harlan, June 30, 1865, USNA, RG-48, Entry 291, Interior Department, records relating to the extension of the U.S. Capitol, 1851–1872, Box 3 "Lighting of the Dome."

38. Allen 2001: 334.

39. James Harlan to Edward Clark, October 13, 1865, USNA, RG-48, entry 295, Interior Department, Capitol Extension, Letters Sent, April 19, 1862–March 8, 1869; Clark to Harlan, May 7, 1866, USNA, RG-48, Entry 291, Interior Department, records relating to the extension of the U.S. Capitol, 1851–1872, Box 3 "Lighting of the Dome."

40. These names and numbers are from affidavits, USNA, RG-48, Entry 291, Interior Department, records relating to the extension of the U.S. Capitol, 1851–1872, Box 3, "Lighting the dome."

41. Allen (2001: 366) and Frary (1969: 306) give 1,083 as the number of gas lamps in the dome.

42. U.S. patent number for the dial-plate switch: 57,697 (1866).

43. Harlan's estimate for the project, just before its completion, was $19,000. Harlan to Clark, October 13, 1865, USNA, RG-48, Entry 295, Interior Department, Capitol Extension, letters sent, April 19, 1862–March 8, 1869, p. 143. The report of the Harlan commission (see note 48 below) gave the cost of lighting the dome as $28,225.

44. On the salaries, see B. B. French, Report of the Commission of Buildings, October 30, 1866, USNA, RG-46, Records of the U.S. Senate, 42nd Congress, Sen 42A-H3, Petitions and Memorials, p. 11. On *Freedom* and Luetze's painting, see Allen 2001: 308, 315–316.

45. On the accusations, see 38th Congress, 1st Session, Senate Report no. 39.

46. Charles Page to James Harlan, undated but after November 1865. SIA, William S. Rhees Collection, RU 7081, Box 9, Folder 3, "Electricity."

47. This was misleading because Chester's $23 dollar estimate was for a battery differing in materials and construction. Charles Chester to Page, November 10, 1865, USNA, RG-48, Entry 291, Interior Department, records relating to the extension of the U.S. Capitol, 1851–1872, Box 3, "Lighting the Dome."

48. The report of the Harlan commission, solicited on October 21, 1865, is dated April 26, 1866. Untitled report, USNA, RG-48, Entry 291, Interior Department, records relating to the extension of the U.S. Capitol, 1851–1872, Box 3, "Lighting of the Dome."

49. It is not clear how the commission arrived at the $2,500 figure for materials, as contradictory data were already available. On the basis of three months of actual operation, Gardiner had estimated the yearly expenses of maintaining the dome-lighting system at $395 plus his own $2500 salary. Gardiner to Clark, March 23, 1866, unprovenanced letter, files of the Historian, Office of the Architect of the Capitol.

50. B. B. French, Report of the Commission of Buildings, October 30, 1866. USNA, RG-46, Records of the U.S. Senate, 42nd Congress, Sen 42A-H3, Petitions and Memorials, Box 104, "Gardiner and Glenn," p. 2.

51. Pike may have been Nicholas Pike, a naturalist and a telegraph enthusiast.

52. The commission's report was on pp. 12–13 of French's report (see above). Quotations in this paragraph are from p. 13. Away in Europe, Morse did not sign the report but concurred in its conclusions.

53. A copy of the contract, dated April 15, 1867, is in Architect of the Capitol, Curator's Office, RG-43, Series 43.2: Capitol Building Files, Capitol Extension, 1851–1875, Box 5, "Gas Lighting the Rotunda—Samuel Gardiner, 1867–1874."

54. "Lighting up of the Capitol Dome," *SA* 14 (1866 n.s.): 97.

55. On the falling out of Page and Morse, see Post 1976a: 164–170.

56. French to O. H. Browning, February 6, 1867, USNA, RG-48, Entry 291, Interior Department, records relating to the extension of the U.S. Capitol, 1851–1872, Box 3, "Lighting the Dome."

57. Edward Clark to Gardiner, March 25, 1873, Architect of the Capitol Letterbooks, Letters Sent, volume 16.

58. See note 10 above. Curiously, in 1869, Hudson along with A. W. Sharit and D. Lyman obtained a U.S. patent (no. 96,488) for an improved electromagnetic gas-lighter. Their complex system used chemical means, brought into play by an electromagnet, to ignite gas turned on by an electromagnet.

59. Edward Clark to Senator John Scott, January 11, 1873, Architect of the Capitol Letterbooks, Letters Sent, Vol. 16.

60. Information on Gardiner's claim is from the petition's supporting materials, U.S. National Archives, RG-46, Records of the U.S. Senate, 42nd Congress, Sen 42A-H3, Petitions and Memorials, Box 104, "Gardiner and Glenn."

61. *U.S. Senate Journal*, February 20, 1872, p. 259.

62. For example, this position appears on a signed letter to O. H. Browning, March 23, 1867, Architect of the Capitol, Curator's Office, RG-43, Series 43.2: Capitol Building Files, Capitol Extension, 1851–1875, Box 5, "Gas Lighting the Rotunda—Samuel Gardiner, 1867–1874."

63. Gardiner to Edward Clark, July 16, 1870, Architect of the Capitol, Curator's Office, RG-43, Series 43.2: Capitol Building Files, Capitol Extension, 1851–1875, Box 5, "Gas Lighting in House & Senate—Samuel Gardiner, 1870–1872."

64. Information in this paragraph is from A. L. Bogart to Edward Clark, September 15, 1870 and November 3, 1871; Edward Clark to J. Orr, April 16, 1872; Architect of the Capitol, Curator's Office, RG-43, Series 43.2: Capitol Building Files, Capitol Extension, 1851–1875, Box 5, "Gas Lighting in House & Senate—Samuel Gardiner, 1870–1872."

65. For a Western Electric advertisement, see *The Electrician and Electrical Engineer* 3 (1884): viii; see also Rhodes's offering of an "Electric Gas Burner," *Review of the Telegraph and Telephone* 1882: 283. Such apparatus was also marketed, probably in the mid 1880s, by the Electrical Supply Company of New York City (NMAH Library, undated Trade Catalog no. 9695. Curiously, the Senate in 1872 was already eager to try out another electrical lighting system, that of Dr. John Vansant (42nd Congress, Senate Bill 756).

66. Townsend 1873: 143–144. See also Ellis 1869: 77; Keim 1884: 83, 103.

67. Edison's acknowledged debt to gas lighting is covered by Israel (1998: 168–169) and by Friedel et al. (1986: 31–32).

68. Gardiner tried to sell this patent to the government. The Senate, in a unanimous resolution, directed the Secretary of the Navy to test the lamp "for making signal lights on vessels at sea and in harbors" and to report the findings at the next Congress. *U.S. Senate Journal*, June 12,

1858, p. 696. No action was taken. An example of this lamp, perhaps the patent model, is in the NMAH (catalog no. 251,226).

Notes to Chapter 19

1. On the dangers and follies of this scheme, see "Electric gas," *SA* 9 (1853): 130.

2. On Nollet's first magneto, see "Shepard's Patent Electro-Magnetic Heat, Light, and Motive-Power-Producing Machine," *MM* 54 (1851): 358, 362–364, 410–411. Shepard was Nollet's English patent agent.

3. For an illustration of Nollet's 1853 magneto, see King 1962b: 364. It is not clear whether Nollet knew about Woolrich's compound magnetos. In any event, as Page (1839b: 252) noted much earlier, the idea of a compound magneto was "obvious."

4. On Nollet and the Alliance Company, see Dredge (1882: 109, 117–118); Jarvis (1958: 181–182); King (1962b: 356). I rely mainly on King.

5. For an illustration of an Alliance magneto, see figure 19.2.

6. On the Holmes machine, see Holmes 1862 and "On Magneto-Electricity and Its Application to Lighthouse Purposes," *TE* 16 (1863): 337–338. On Faraday's impressions, see Faraday to George B. Airy, May 20, 1857 (Williams et al. 1971: 869–870). The London Science Museum possesses a later-model Holmes Magneto.

7. This constraint explains the armature's enormous size (Holmes 1862: 25–26). As the circumference of the armature increases, the speed at which the coils and magnets move relative to each other also increases, thereby augmenting the tension.

8. Faraday 1871: 62.

9. On the conflict between Trinity House and the Board of Trade, see MacLeod 1969: 15–16. The Dungeness lights had been extinguished by 1878, but Lizard now had two arc lights. On technical details of the Lizard installation, which employed six dynamos, see Higgs 1879a: 219–220.

10. On the Souter Point magnetos, see King 1962b: 362.

11. Reynaud 1876: 98–106. For technical details on the Alliance machines, see Le Roux 1868.

12. For example, "The Electric Light," *SA* 33 (1875): 35–36.

13. Henry to M. Day, December 13, 1872, *USNA*, Coast Guard, Lighthouse Service (RG-26), Letters Sent by the Light-House Board, Miscellaneous, Vol. 6.

14. Henry to Robert Doremus, January 20, 1869, SIA, Office of the Secretary, 1865–1891, Outgoing Correspondence, RU 33.

15. Elliot 1874: 12.

16. "Lights and Fog-Signals: European Systems and Improvements," *NYT*, August 31, 1874; "Light-Houses: Superiority of the European System," *NYT*, October 25, 1874.

17. Report of the Appropriations Committee, Forty-Third Congress, 2nd Session, U.S. Senate, Report no. 605 (1875), pp. 6–7.

18. Henry 1887: 370.

19. Parts of this section are drawn, almost verbatim, from Schiffer 2005b.

20. Despite Henry's opposition to an electric illuminant, an arc light was tried out during October 1867 in the tower of the Barge Office, a federal facility at the southern tip of Manhattan. This event was not a formal trial but a sales pitch by the Alliance Company seeking new markets. "Experiments with an Electric Light at the Battery," *SA* 17 (1867): 297; see also "New Electric Light," *NYT*, October 18, 1867.

21. Alglave and Boulard 1884: 389.

22. Ibid.: 390.

23. Ibid.: 393.

24. "What the Agent of the Line Says," *NYT*, December 2, 1873. By this time, all ships of that line had discontinued use of the electric light; see also "No Further Particulars Received—Probable Introduction of New Sea Lights," *NYT*, December 5, 1873.

25. "The Berlioz Electric Light," *SA* 19 (1868): 251; second and third quotations are from "The Berlioz Electric Light," *NYT*, October 1, 1968.

26. "Utility of the Electric Light," *SA* 11 (1864): 275.

27. "Fishing by Electric Light," *NYT*, July 30, 1865.

28. Gorman 1977: 526–527.

29. Nye 1998: 29.

30. Details on this demonstration are from "Experiment with Electric Light," *NYT* April 19, 1874.

31. Another useful source on applications of magneto arc lights is Guillemin and Thompson 1891.

32. "Electric Light," *SA* 27 (1872): 79.

33. Blasting and electromedicine continued to furnish large markets for small magnetos.

34. I use the term "generator" to include both magnetos and dynamos, as it was used after about 1880. For a brief history of generators, see Bowers 1982: 70–100.

35. Unless otherwise noted, information in the following discussion of dynamos is from Cardwell 1992, Dredge 1882, Jarvis 1958, King 1962b, Niaudet-Bréguet 1875, Schellen 1884, or "Recent Progress in Electromagnetism," *SA* 29 (1873): 351–352.

36. On Wilde's work, see Dredge 1882; Cardwell 1992; Gee 1920.

37. Wilde 1867. This machine received a very favorable write-up in "Electric Light—Wilde's Magneto-Electric Machine," *SA* 16 (1867): 248.

38. Wilde (1867: 81–83) himself was more concerned with resolving the apparent paradox that a small amount of magnetism (in the magneto) could be transformed into a much larger amount in the magneto's armature.

39. These numbers are from p. 5 of Gee 1920.

40. King (1962b: 376) mentions these applications.

41. "Recent Progress in Electromagnetism," *SA* 29 (1873): 351–352.

42. Inventors of the principle of self-excitation were Moses Farmer in the United States, Werner Siemens in Germany, and Charles Wheatstone, S. Alfred Varley, and Henry Wilde in England (King 1962b: 377–378). See also "Ordinary Meeting, February 19th, 1867," *Proceedings of the Literary and Philosophical Society of Manchester* 6: 103–107.

43. "Converting Electro-Magnetic Engine," *SA* 16 (1867): 216. For decades one could find generators with external exciters like the Wilde design, including Brush's ac dynamo, which used an external dc dynamo as the exciter; see "The Brush Electric Company's New Alternating-Current System," *Science* 44 (no. 338, 1889): 51–54. Indeed, they are still made today.

44. Many sources furnish images of the early dynamos and their immediate precursors (e.g., Dredge 1882; Guillemin 1891; King 1962b)

45. For a biography of Gramme, see Chavois 1963.

46. For a description of a Gramme machine, see Niaudet-Bréguet 1875. (The quoted passage is on p. 184.) On the first American report of the Gramme armature, see "New Magneto-Electric Machine," *SA* 26 (1872): 410. For a technical report on a Gramme lighting dynamo, see "The Gramme Magneto-Electric Machine," *SAS* 1 (1876): 257–258.

47. See, e.g., Hunt and Rudler 1878: 227.

48. See, e.g., Thompson 1886.

49. King 1962b: 379.

50. According to King (ibid.: 380–381), this dynamo technically incorporated self-excitation, but the means was curious: a small, built-in magneto supplied current for the field electromagnets.

51. For the numbers in this paragraph, see ibid.: 382–383. The efficiency of the electrometallurgy dynamos was calculated as follows: 1 horsepower is 746 watts, and 2 volts at 150 amps is 300 watts; 300 divided by 746 is 0.40 (40 percent).

52. Among those who mentioned the reversibility of motors and generators were Francis Watkins, Antonio Pacinotti, and Charles Siemens (MacLaren 1943: 89–90), Henry Paine ("Paine's Electric Engine," *SA* 10, 1855: 366), and Charles Page (1845a: 134). The French inventors of a rechargeable battery also made the same discovery: after the battery was charged by their Gramme dynamo, the latter began to rotate on its own. One wonders if this discovery was the immediate

inspiration for the Vienna demonstration. On the French discovery, see "Note on an Electro-Dynamic Experiment," *SA* 27 (1873): 245.

53. For example, Ayrton 1879a,b.

54. The definitive work on power transmission in the United States is Hunter and Bryant 1991.

55. "Transmission of Power at Rock Island Arsenal," *SA* 40 (1879): 117. See also Slattery 1990.

56. Higgs 1879b; Martin and Wetzler 1891.

57. Ayrton and Perry 1883: 3.

58. "Plowing by Electricity," *SA* 41 (1879): 41.

59. Ayrton 1879a: 571.

60. The dynamo for arc lighting, which could run at 900 rpm, "weighed 700 kg and measured 0.90 meter square by 0.65 meter high. There were 180 kg of copper wire on the electromagnets and 40 kg on the armature" (King 1962b: 384).

61. Source: http://edison.rutgers.edu/chron1.htm#76.

62. King 1962b: 385.

63. Ibid.: 385, for specifications of these machines.

64. Carré was turning out 6,500 feet of carbon rods per day. "Notes on Electrical Lighting," *SA* 40 (1879): 74.

65. Source: http://edison.rutgers.edu/chron1.htm#76.

Notes to Chapter 20

1. Large portions of this chapter are drawn, almost verbatim, from Schiffer 2005b.

2. Electric lights in 1879: in England, South Foreland (2), Souter Point, and Lizard (2); in France, la Hève (2) and Gris-Nez; elsewhere, Odessa and Port Said. The estimate of arc-lit lighthouses in 1895 is based on Findlay and Kettle 1896.

3. For additional details of this research process, see Schiffer 2000, 2005b.

4. These data are from Elliot 1874: 239. In the latter decades of the century, however, some French lighthouses employed hot air and petroleum engines to drive their generators (Blondel 1873: 50). The later American lighthouse at Hell Gate was said to consume 1,594 pounds of coal per night (Millis 1885: 163).

5. Elliot 1874: 72, 125. For English lighthouses constructed in the late 1870s, the cost differential was still great: the "first costs" of an oil-burning lighthouse were £7,400, versus at least £13,500 for the Lizard electrics; annual maintenance costs were, respectively, £793 and at least £1,523 (see Higgs 1879a: 222).

6. Elliot (1874: 233–236) quotes from a report on the French lights, which indicates the general reliability of the electrical system. See also Reynaud 1876: 101.

7. Faraday 1871: 63.

8. On annual operating expenses of the French lights, see Elliot 1874: 239. On the early British lights, see ibid.: 72, 239. Dynamos might have given electric lighting a slight advantage in operating costs in France, but in England they did not, as data from the Lizard lights demonstrated (Higgs 1879a: 221–223). However, the French continued to install magnetos, not dynamos, in their arc-lit lighthouses—even in the 1890s (Alglave and Boulard 1884: 291; Blondel 1893; Cadiat and Dubost 1896).

9. Reynaud 1871: 104–105.

10. Navigation manuals identified which lights were electric, sometimes emphasizing their "great brilliancy," as in King's (1878: 112) reference to the South Foreland lights.

11. On these disadvantages, see Douglass 1886: 11; Johnson 1890: 62–63.

12. See, e.g., the laudatory letters in Holmes 1862.

13. "Electric Lights in Lighthouses," *TE* 22 (1866): 32.

14. Elliot 1874: 247. At the same time that electric lights were being tried out in Europe and the United States, other fog- and haze-penetrating technologies were being widely adopted, including foghorns, sirens, whistles, and bells.

15. Findlay and Kettle 1896: 12–13.

16. On electrical enthusiasm, see Marvin 1998; Nye 1990. Simon (2004) sounds a discordant note, showing that electricity could also inspire anxiety and fear.

17. Here "political technology" denotes a wide range of functions that can serve a country's interests, especially in international relations; a political technology is not simply a symbol of political power (on the latter, see Mukerji 2003). Tagliacozzo (2005) places late-nineteenth-century lighthouses into the political context of competing Dutch and British empires in the Far East.

18. Allard 1876, 1881.

19. Allard 1881: 2 (my translation).

20. Ibid.: 3, my translation.

21. Ibid.: 38.

22. For these dates, see Lucas 1885: 16–17. The remaining arc lights had also been placed in prominent locations, mainly along the English Channel and the Atlantic coast.

23. Cordulack (2005) demonstrates the symbolic importance of the electric light to Parisians in the context of competition with the United States.

24. See "Lighthouse" in *Encyclopædia Britannica*, sixteenth edition.

25. Douglass 1886.

26. Ibid.: 40. A second committee reviewed the data and reaffirmed the conclusions of the original report, "Lighthouse Illuminants," *Science* 16 (1890): 267–269, 268.

27. Kenward 1893: 217. MacLeod (1969: 10–11) characterized English decisions on lighthouse illuminants: "In the end the question was usually decided on grounds of expense."

28. On the new British lights, see "Lighthouse," *Encyclopaedia Britannica*, sixteenth edition.

29. The difficulties of operating and maintaining arc lights also might have played a role in the retrenchment decision (Lucas 1885: 77). In 1900, France had around a dozen arc lights in lighthouses; seven were still operating in 1933—the only ones in the world—but these were gradually replaced by other illuminants (Schiffer 2005b).

30. Morton 1879.

31. On the electrical blasting at Hell Gate, see "Removal of the Hell Gate Rocks," *SA* 35 (1876): 214–216; "Success of the Hell Gate Explosion," *SA* 35 (1875): 226.

32. "The New Electric Light at Hell Gate Lighthouse," *SA* 50 (1884): 175, 178. Detailed sources are cited in note 70 of Schiffer 2005b. The 45-hp Brush No. 8 dynamo could power 65 lamps.

33. "The Hell Gate Electric Light to be Discontinued," *NYT*, October 24, 1886.

34. Using the example of lighthouse illuminants, Walker and Schiffer (2006) develop the implications of acquisition behavior as a manifestation of social power.

35. On the history and meanings of the Statue of Liberty, see Dillon and Kotler 1994; Lemoine 1986.

36. On lighting the Statue of Liberty, see Millis 1887. For a modern account, see Perrault 1986.

37. "Liberty's Torch Lighted," *NYT*, November 2, 1886.

38. *New York Times*, all 1886: "Liberty's Face Darkened," 5 November, p 8; "Lighting Liberty's Torch," 6 November, p. 8; "Cost of Liberty's Torch," 9 November, p. 2; "Liberty's Unlighted Torch," 10 November, p. 8; "The Bartholdi Statue Light " 16 November, p. 1; "Liberty's Torch," 17 November, p. 2; "Liberty to be Lighted," 18 November, p. 5; "Liberty Resumes her Task," 23 November, p. 2.

39. For technical details on the permanent installation, see Millis 1890.

40. For source material on the Gedney's Channel experiment on incandescent lighting, see Schiffer 2005b: 302, note 79.

41. The quotation is from Rodgers et al. 1889: 195. On the Sandy Hook Main Light, USNA, Treasury Department, Coast Guard, RG 26, Lighthouse Service, clipping file, district 3, box 6, "Sandy Hook Light Station, New Jersey."

42. Information on the Navesink light is from Ruth 1991.

43. On the Tino and Macquarie lighthouses, see "Lighthouse" in *Encyclopædia Britannica*, six-teenth edition. On Hanstholm, see "The Houstholm [*sic*] Electric Lighthouse," *Science* 14 (1889): 383.

44. For example, the Hanstholm light at two million candles was "the most powerful electric lighthouse in the world," "The Houstholm [*sic*] Electric Lighthouse," *Science* 14 (1889): 383; the Macquarie Lighthouse, which still stands, could produce "a light of 6,000,000 candles, the most powerful in the world at the time" (source: www.lighthouse.net.au/lights/NSW/Macquarie/Macquarie.htm#History). The Macquarie light had two de Méritens magnetos, one of which survives on site in the Powerhouse Museum.

Notes to Chapter 21

1. I use the term "corporate capitalism" to emphasize that capitalism existed long before the invention of the modern corporation.

2. The preceding biographical paragraphs are drawn, almost verbatim, from Schiffer et al. 1994: 10–11, 15. Material recycled from this source, which is scattered throughout this chapter and the next, was originally based mainly on Josephson 1959 and Friedel et al. 1986. New insights into Edison's electric-light project are furnished by Bazerman (1999). Israel 1998 is the definitive Edison biography, but Conot 1979 and Josephson 1959 remain useful. Millard (1990) treats Edison's approach to innovation. On the differing adoption patterns of electric light and power in Germany, Great Britain, and the United States during the late nineteenth and early twentieth centuries, see Hughes 1983. Marvin (1988), Nye (1990), Rose (1995), Schivelbusch (1989), and Simon (2004) treat the adoption, cultural meanings, and impacts of electric lighting. On the subsequent creation of "potent cultural images" around the Edison electric light, see Finn 1984.

3. For a rich account of this affair, see Simon 2004: 123–139. For other modern accounts, see Bazerman 1999: 128–129; Carlson 2003: 57–65; Hounshell 1980b; Israel 1998: 110–115; Susskind 1964.

4. "Weak Sparks," *SA* 33 (1875): 385.

5. "One of Mr. Edison's Curious Experiments," *SA* 34 (1875): 33.

6. "Etheric Force," *NYT*, December 3, 1875.

7. "Weak Sparks," *SA* 33 (1875): 385; "The Discovery of Another Form of Electricity," *SA* 33 (1875): 400–401; "The New Phase of Electric Force," *SA* 33 (1875): 401; the latter account was a firsthand description of the apparatus and observations of the effects it produced; the drawing of the set-up (see also Carlson 1991: 58) is amateurish, not executed in the usual *Scientific American* style.

8. "The 'Etheric' Force," *SA* 34 (1876): 36; "The 'Etheric' Phenomenon," *SA* 34 (1876): 116; the quotation is from the latter.

9. For biographical material on Beard, see Ellerbroek 1972; Simon 2004.

10. Beard letter, "The Nature of the Newly Discovered Force," *SA* 34 (1876): 57.

11. "Mr. Edison's Electric Discovery," *SA* 34 (1876): 17.

12. Edwin J. Houston, "Phenomena of Induction," *SAS* 1 (1876): 77–78. Susskind (1964: 36) claims that the 1871 experiments were actually done by Elihu Thomson. Carson 1991 is the definitive biography of Thomson.

13. Edison letter, "Mr. Edison's New Force," *SA* 34 (1876): 101. Those advocating an induction hypothesis include a letter signed "Electron": "Edison's New Force," *SA* 34 (1876) 69. P. H. Vander Weyde raised the possibility that the observed effects were produced by *electrostatic* induction: "The Nature of the Phenomena Discovered by Mr. Edison," *SA* 34 (1876): 89.

14. Dyer and Martin 1910, volume 2: 578.

15. On the work of Hertz, see Buchwald 1994.

16. Maxwell 1865.

17. Susskind (1964) discusses others who had reported these effects before Hertz, including Amos Dolbear and Joseph Henry (Moyer 1997: 172–176), but many others were not mentioned.

18. Israel 1998: 464.

19. On Edison's claim to be a scientist and on his relationship with academic scientists, see Hounshell 1980b; Israel 1998. In "Mr. Thomas A. Edison" (39, 1878: 5), *Scientific American* observed that "Mr. Edison may well pride himself as to his position in the world of science, standing, as he does, first among the inventors of the day."

20. Vanderbilt (1971) makes this point with particular reference to Edison's nickel-iron storage battery. On this project, see also Carlson 1988; Schiffer et al. 1994.

21. Wise 1885: ccclxxxvii–ccclxxxviii.

22. A lengthy and well illustrated article on regulators is 'Electric Lamps,' *SAS* 20 (1879): 2572–2577.

23. On the financing of the early electrical industry, see MacLaren 1943; McGuire 1990; Passer 1953.

24. Unless otherwise noted, information on Charles Brush and his lamp is from Jeffrey La Favre, "Charles Brush and the Arc Light" (http://www.lafavre.us/brush/brushbio.htm) or "Charles Francis Brush," *SAS* 18 (1884): 7286. See also Eisenman 1966: 511–512. On Brush's system, see Dredge 1882: 204–218. For technical details on dynamos of other early American manufacturers, see ibid.: 182–186, 224–234.

25. G. W. Stockly, "The Brush Electric Light" (letter) *SA* 40 (1879): 296.

26. "The Electric Light in Cleveland," *SA* 40 (1879): 329. I have found no evidence to contradict this claim.

27. The 80 percent figure is from p. 512 of Eisenman 1966.

28. Dredge 1882: 204.

29. The quotation is from p. 512 of Eisenman 1966. Carlson (1991: 133) mentions that this installation was not a true central station.

30. The technical argument was as follows (see also Vincenti 1995). The total power that could be transmitted, Ayrton noted, was equal to the voltage times the current, minus losses from resistance in the wires (today this law is often represented as $P = IE$, where P is power consumption in watts, I is current in amperes, and E is tension—electromotive force—in volts). According to a second law, the power consumed in a circuit equals the resistance times the current *squared* ($P = I^2R$, where R is resistance in ohms—this is known as "Joule's law"). Thus, to lessen the resistance losses *for a constant amount of power*, one had to reduce the current, which (by the first law, above) required an increase in tension.

31. Ayrton 1879a: 569.

32. Information and quotations in this paragraph are from Ayrton 1879b: 214.

33. Another early vision of the central station (battery-based) was articulated by M. Vergnes ("The Electro-Magnetic Engine," *SA* 12, 1857: 334). The term also appeared in an 1851 patent of François Dumont (Great Britain, Commissioner of Patents 1859: 216).

34. Information on the San Francisco central station is paraphrased from http://www.pge.com/field_work_projects/report_lights/light_history. See also Foster 1979: 72. Bright (1949: 32) recognized the San Francisco lighting plant as the "first central electric-generating station in America." Curiously, Israel (1998: 216) states that Edison's 1882 installation at Holborn Viaduct, in London, "was the first commercial central station in the world."

35. Ayrton 1879b: 213.

36. Unless otherwise noted, information on this project is from http://www.wabash.lib.in.us/city.html.

37. Source of anonymous quotation: http://www.wabash.lib.in.us/city.html.

38. "The City of Wabash, Ind., Now Entirely Lighted by Electric Light," *CT*, April 2, 1880, p. 12.

39. On the Jablochkoff candle, see "The Electric Light," *SA* 39 (1878): 326.

40. See Bright 1949: 29.

41. "The Cost of the Electric Light in Paris," *SA* 40 (1879): 164–165.

42. For comparative data from a Rouen experiment, see Alglave and Boulard 1884: 426–427.

43. On the Avenue de l'Opéra, see "The Electric Light in Paris," *SA* 40 (1879): 320; "The Cost of the Electric Light in Paris," *SA* 40 (1879): 164–165; Meyer (1972: 156) presents this example as a success. For other installations of the Jablochkoff candle, see Du Moncel 1882: 217–229;

Sprague 1878: 9–13; King 1962b: 393–407. King (1962b: 395) asserts, incorrectly, that the Jablochkoff candle solved the "problems of the subdivision of the electric light—that of placing several lights in the same circuit and that of reducing the intensity of the arc light." Although the Jablochkoff candle at 300 candles was dimmer than a regulated arc, it drew no less current. Thus, if one wanted to add lights to a circuit, the generating capacity had to be increased. King (1962b: 405–406) mentions—apparently without appreciating its implications—that the Siemens system for Jablochkoff candles came "in three sizes—4-, 8-, and 16-light machines that, with their exciters, required, respectively, 4, 7, and 13 hp." Subdivision was understood by most workers in the 1870s as a reduction in both light output and power consumption of each lamp (see chapter 22). In terms of power consumption, the Jablochkoff candle did not solve the subdivision problem.

44. "Progress of Electric Lighting in London," *SA* 42 (1880): 53; "Economy of the Electric Light," *SA* 41 (1879): 186.

45. Bright 1949: 29.

46. For these points I have drawn upon "The Electrical Department in the Mechanics' Fair, Boston, Mass.," *SA* 39 (1878): 289.

47. For biographical information on Edward Weston, see Woodbury 1949 and http://chem.ch .huji.ac.il/+eugeniik/history/weston.htm. Weston was best known in later years for his electrical measuring instruments.

48. Advertisement, *SA* 40 (1879): 397. A description of the Weston dynamos, "The Weston Dynamo-Electric Machine," *SA* 35 (1876): 150; a drawing of the Weston dynamo as employed in an electroplating firm is in "Electro-Plating Machinery," *SA* 37 (1877): 127, 130.

49. Meyer 1972: 157. For a description of the Wallace-Farmer lamp, which used slabs of carbon in an unusual configuration, see "The Wallace-Farmer Electric Light," *SA* 40 (1879): 54. On early American arc lighting companies, see Bright 1949: 30–34.

50. On the early history of incandescent lighting, see Bowers 1982; Bright 1949.

51. Other incandescent lamps of this period were invented by Hiram Maxim, Moses Farmer, and Joseph Swan. See Friedel et al. 1986: 114.

52. Unless otherwise noted, information about Sawyer and his lamp is from Wrege and Greenwood 1994, Bright 1949: 50–53, or Israel 1998: 188–190.

53. "The Sawyer-Man Electric Light," *SA* 40 (1879): 145. For an earlier account with a drawing of the light bulb, see "The Sawyer-Man Electric Lamp," *SA* 39 (1878): 351, 354–355.

Notes to Chapter 22

1. Hounshell 1980b: 613.

2. The previous paragraphs in this section are drawn, almost verbatim, from Schiffer et al. 1994: 15–17. On Edison's interactions with the press, see Bazerman 1999. I have drastically simplified

the complex business developments of 1875–1895. Carlson (1991) sheds much new light on this period.

3. Friedel et al. 1986: 7.

4. Source of quotation: ibid.: 8.

5. Source of quotation: ibid.: 13.

6. Bazerman 1999: 17.

7. On the financial arrangements, see Friedel et al. 1986: 21–22; Israel 1998: 173.

8. Ohm 1891.

9. Hounshell (1980b: 517, note 52) identifies George Barker as one who raised this issue. For an early statement of this erroneous design principle, see Jacobi 1841b: 153.

10. The Edison dynamo, and a closely related motor, were reported and illustrated in "Edison's Electrical Generator," *SA* 41 (1879): 242. The high efficiency of Edison's dynamo was disputed on theoretical grounds by Edward Weston in a letter, "Edison's Electrical Generator," *SA* 41 (1879): 276, as well as by Charles A. Seeley, "Edison's Electrical Generator," *SA* 41 (1879): 305. Francis Upton, Edison's resident physicist, refuted their arguments with gusto, "Edison's Electrical Generator," *SA* 41 (1879): 308, but the debate continued. Hounshell (1980b: 517, notes 40, 41) cites independent tests of Edison's dynamo that confirmed the high-efficiency claims (see also Friedel et al. 1986: 137). For a comparison of Edison's dynamo with German machines, see Du Moncel et al. 1884: 79–80. Thomson and Houston also argued in favor of a low-resistance armature (Carlson 1991: 86–87).

11. Friedel et al. 1986: 39–40.

12. Israel 1998: 184–185.

13. Higgs 1879a is a good source for early cost data on European arc light systems. See also Sprague 1878. The Sautter & Lemonnier factory, which built Fresnel lenses for lighthouses, replaced their gaslights with arc lights and enjoyed cost savings; "Illumination by the Electric Light," *SAS* 1 (1865): 599.

14. "Electric Illumination: Progress and Prospects of the New Light," *NYT*, June 15, 1879. A table demonstrated that large generators, like large steam engines, operated much more economically than small ones as measured by horsepower per 1,000 candles of light. According to Siemens, 15 times more light came from burning coal to power a dynamo than converting the coal into illuminating gas. "American History of the Electric Light," *SA* 40 (1879): 2.

15. "Progress of Electric Lighting," *SA* 40 (1879): 272.

16. Ibid.: 272.

17. "The Electric Light in Mining," *SA* 41 (1879): 24.

18. Meyer 1972: 156.

19. "The Electric Light in a Fish Market," *SA* 40 (1879): 354.

20. Alglave and Boulard 1884: 398.

21. On the Siemens system, see Sleeman 1880: 241.

22. "The Electric Light in Photography," *SA* 41 (1879): 47.

23. Quotations and information in this paragraph are from "Electric Illumination: Progress and Prospects of the New Light," *NYT*, June 15, 1879. On Maxim's arc lighting system, see "Electric Lighting," *SA* 39 (1878): 175, 178.

24. Preece 1879: 34. See also "Is the Subdivision of Electric Light a Fallacy? *SA* 40 (1879): 97; "Electric Lighting," *SA* 40 (1879): 138.

25. The quotation is from "Division of Electric Light," *SA* 42 (1879): 260; see also, "Progress of Artificial Illumination," *SA* 42 (1880): 16; "Practical Divisibility of the Electric Light," *SA* 40: 381, 389; "Subdivision of Electric Light," *MAB* 12 (1880): 14.

26. "The New Light," *Daily Constitution* (Atlanta), December 24, 1879; "The New Light," *CT*, December 24, 1879.

27. "Edison," *CT*, December 25, 1879.

28. "Two Serious Objections," ibid.: 3.

29. "Electrician Sawyer's Challenge to Electrician Edison," *CT*, December 24, 1879. In a later letter, Sawyer argued his priority for the paper-filament bulb while also mentioning that it could not burn for more than an hour. His carbon-rod lamp, he insisted, was superior. "Mr. Sawyer Questions the Novelty and the Practical Value of Edison's Discovery," *CT*, December 26, 1879.

30. "Edison's Gift," *CT*, December 25, 1879.

31. "Edison's Reply," *CT*, January 8, 1880, p. 3.

32. "Brush, the Cleveland Inventor, on Edison's Light," *CT*, December 27, 1879.

33. "A Scientific View of it: Prof. Henry Morton not Sanguine About Edison's Success," *NYT*, December 28, 1879.

34. Quoted on p. 187 of Israel 1998.

35. "The Edison Light," *Boston Daily Globe*, January 2, 1880, p. 1; "Edison and the Skeptics," *NYT*, January 4, 1880, p. 6.

36. "Edison's Latest Electric Light," *SA* 42 (1880): 19. Another useful early account is "Edison's Electric Light," *NYT*, December 28, 1879.

37. Friedel et al. 1986: 139.

38. Rossi 1880: 305.

39. Using the Edison electrical system as a case study, Vincenti (1995) emphasized the roles that fundamental laws and cost constraints play in the work of engineers. Contrast this with Usselman 1992.

40. These numbers are adapted from Friedel et al. 1986: 137.

41. Edison had decided on a high-resistance filament by early 1879, while still working on the platinum bulb (ibid.: 53).

42. Unsigned editorial, *Science* 2 (1881): 193.

43. On the *Columbia* installation, see Friedel et al. 1986: 140–144.

44. On sales of isolated plants, see Israel 1998: 214.

45. Ibid.: 206.

46. Ibid.: 198.

47. On these conflicts, see ibid., chapter 12.

48. Ibid.: 227–229.

49. Quotations are from ibid: 206, 207.

50. Information in this paragraph is from Foster 1979: 148–155.

51. "Mr. Sawyer Questions the Novelty and the Practical Value of Edison's Discovery," *CT*, December 26, 1879. Unless otherwise noted, information about Sawyer and his lamp is from Wrege and Greenwood 1994 or Bright 1949.

52. Usselman (1992) contrasts the approaches of Edison and Westinghouse to invention, organization, and commercialization.

53. On the Westinghouse system, see "The Westinghouse Alternating Current System of Electrical Distribution," *SA* 62 (1890): 113, 120–121.

54. Israel (1998: 219) mentions the 220-volt and 330-volt systems but later states that the "isolated and central station dynamos were 110 to 140 volts" (p. 326).

55. "Edison's Electric Light," *NYT*, December 28, 1879.

56. For modern discussions of the "war of the currents," see Friedel et al. 1986; Hughes 1983; Israel 1998; Jonnes 2003; Simon 2004.

57. Israel (1998: 325–326) notes that, early on, Edison had discounted alternating current on largely technical grounds, such as the lack of an ac motor, but such objections became increasingly untenable in the late 1880s.

58. I have compressed this episode. For detailed accounts, see Essig 2003; Simon 2004. On the changing social context of capital punishment in America, see Linders 2002.

59. "Electricity as an Executioner," *SA* 34 (1876): 16.

60. "Execution of Criminals," *NYT*, April 6, 1884.

61. The quotation is from "Death by Electricity, *NYT*, January 17, 1888. Additional information on the commission is from "Mr. Gerry on the Stand," *NYT*, July 19, 1889.

62. "The Abolition of Hanging," *NYT*, January 17, 1888.

63. Israel (1998: 329) maintains that it was "Edison's reputation as the nation's preeminent electrician" that carried the day in New York and, eventually, elsewhere.

64. "Executions by Electricity," *SA* 60 (1889): 181. On an earlier Edison demonstration, see "Experiments on Death by Electricity," *SA* 59 (1888): 368.

65. The term "electrocution" had been introduced by this time but was not yet in common use.

66. For a gruesome account, see "Far Worse Than Hanging," *NYT*, August 7, 1890.

67. On Edison's reaction, see ibid.: 1. On Westinghouse and his company, see "Westinghouse Is Satisfied," *NYT*, August 7, 1890.

68. The number 300 is from Sharlin 1961: 70. Sharlin places the triumph of alternating current in 1895, when it was used to transmit power 20 miles from generators at Niagara Falls to Buffalo. Sharlin also notes that dc systems were well suited for load leveling through the use of secondary batteries. Alternating current systems could be load leveled, but, in addition to batteries, they required rotary converters (i.e., motor-generator pairs).

69. This paragraph and the next contain material from Schiffer et al. 1994: 37–38. Finn (1984: 249) reports that there were 93,000 light bulbs at the Columbian Exposition. The Westinghouse ac system on display included a rotary converter supplying dc power for a traction motor (Sharlin 1961).

70. Ayrton's (1879a,b) arguments demonstrated that the voltage was crucial, not whether the electricity was ac or dc. In principle, either could be transmitted with low losses at high voltage. Alternating current systems required transformers to step up the voltage at the generator and step it down at the user's end; dc systems required rotary converters. As Hughes (1983) points out, when ac systems spread in the 1890s, older dc systems were coupled to them by rotary converters. It was Edison's commitment to low voltage not dc that doomed his system as local electric companies strove to service large areas.

71. On the founding of General Electric, see Carlson 1991.

72. On the battery project, see Carlson 1988; Schiffer et al. 1994.

73. As late as 1888, incandescent lighting cost consumers almost twice as much as gaslights of comparable candlepower (Foster 1979: 112).

References

Abel, F. A. 1884. Electricity Applied to Explosive Purposes. In *The Practical Applications of Electricity: A Series of Lectures Delivered at the Institution of Civil Engineers*. Institution of Civil Engineers, London.

Alglave, E., and J. Boulard. 1884. *The Electric Light: Its History, Production, and Applications*. Appleton.

Allan, Thomas. 1858. Electro-Magnetism as a Motive Power. *Mechanics' Magazine* 68: 316–322.

Allard, Émile. 1876. *Mémoire sur l'Intensité et la Portée des Phares*. Paris.

Allard, Émile. 1881. *Mémoire sur les Phares Électriques*. Paris.

Allen, William C. 1992. *The Dome of the United States Capitol: An Architectural History*. U.S. Government Printing Office.

Allen, William C. 2001. *History of the United States Capitol: A Chronicle of Design, Construction, and Politics*. U.S. Government Printing Office.

Ampère, André-Marie. 1820. De l'Action Mutuelle de Deux Courans Électriques. *Annales de Chimie et de Physique* 15: 59–76.

Ampère, André-Marie. 1826. *Théorie des Phénomènes Électro-Dynamiques, Uniquement Déduite de l'Expérience*. Paris.

Anderson, Chris. 2006. *The Long Tail: Why the Future of Business Is Selling Less of More*. Hyperion.

Appadurai, Arjun, ed. 1986. *The Social Life of Things: Commodities in Cultural Perspective*. Cambridge University Press.

Appleton. 1866. *Appleton's Dictionary of Machines, Mechanics, Engine Work, and Engineering*, volume 1.

Aschoff, V. 1984. *Geschichte der Nachrichtentechnik*. Berlin.

Ayrton, W. E. 1879a. Electricity as a Motive Power. *Nature* 20: 568–571.

Ayrton, W. E. 1879b. Electricity as a Motive Power. *Scientific American* 42: 213–214.

Ayrton, W. E., and John Perry. 1883. *Electro-Motors and Their Government*. London.

Bache, Alexander Dallas. 1832. Notice of Electro-Magnetic Experiments. *American Journal of Science and Arts* 22: 409–415.

Bain, Alexander. 1852. *A Short History of Electric Clocks, With Explanations of Their Principles and Mechanism, and Instructions for their Management and Regulation*. London.

Baird, Davis. 1994. Meaning in a Material Medium. In *Proceedings of the Biennial Meeting of the Philosophy of Science Association*.

Barlow, Peter. 1825. On the Laws of Electro-Magnetic Action, as Depending on the Length and Dimensions of the Conducting Wire, and on the Question, Whether Electrical Phenomena are due to the Transmission of a Simple or Compound Fluid. *Edinburgh Philosophical Journal* 12: 105–114.

Barnes, John S. 1869. *Submarine Warfare, Offensive and Defensive. Including a Discussion of the Offensive Torpedo System, its Effects upon Iron-Clad Ship Systems, and Influence upon Future Naval Wars*. Van Nostrand.

Bazerman, Charles. 1999. *The Languages of Edison's Light*. MIT Press.

Beardslee Magneto-Electric Co. 1863. *Directions for the Practical Working of Beardslee's Magneto-Electric Signal Telegraph*. New York.

Bearss, Edwin C. 1980. *Hardluck Ironclad: The Sinking and Salvage of the Cairo* (rev. ed). Louisiana State University Press.

Beauchamp, Kenneth. 2001. *History of Telegraphy*. Institution of Electrical Engineers, London.

Bektas, Yakup. 2001. Displaying the American Genius: The Electromagnetic Telegraph in the Wider World. *British Journal for the History of Science* 34: 199–232.

Bell, Louis. 1897. The Wonderful Expansion in the Use of Electric Power. *Engineering Magazine* 12: 630–642.

Bernal, J. D. 1965. *Science in History*, volume 2: *The Scientifc and Industrial Revolutions*. 3rd ed. MIT Press.

Biagioli, Mario. 2006a. *Galileo's Instruments of Credit: Telescopes, Images, Secrecy*. University of Chicago Press.

Biagioli, Mario. 2006b. From Print to Patents: Living on Instruments in Early Modern Europe. *History of Science* 139–186.

Bijker, Wiebe E. 1995. *Of Bicycles, Bakelites, and Bulbs: Toward a Theory of Sociotechnical Change*. MIT Press.

Bijker, Wiebe E., Thomas P. Hughes, and Trevor J. Pinch, eds. 1987. *The Social Construction of Technological Systems: New Directions in the Sociology and History of Technology*. MIT Press.

Blondel, André. 1893. On the Electric Machinery and Arc Light of Lighthouses: Experiments Made by the Lighthouse Department of France. International Maritime Congress, London Meeting.

Bowers, Brian. 1982. *A History of Electric Light and Power*. Peter Peregrinus.

Brett, J. W. 1857. On the Submarine Telegraph. *The Engineer* 3: 321–322.

Briggs, Charles F., and Augustus Maverick. 1858. *The Story of the Atlantic Telegraph*. Fudd & Carleton.

Bright, Arthur A., Jr. 1949. *The Electric-Lamp Industry*. Macmillan.

Bright, Charles. 1908. *The Life Story of Charles Tilston Bright, Civil Engineer, with which is Incorporated the Story of the Atlantic Cable, and the First Telegraph to India and the Colonies*. Archibald Constable.

Bright, Charles T. 1860/61. Synopsis of the Surveys of the *Fox*, under the Command of Capt. Allen Young, F.R.G.S. *Proceedings of the Royal Geographic Society of London* 5 (2): 70–79.

Brown, Sanborn C. 1979. *Count Rumford, Physicist Extraordinary*. Greenwood.

Bruce, Robert V. 1987. *The Launching of Modern American Science*. Cornell University Press.

Bruno, Leonard C. 1997. *Science and Technology Firsts*. Gale.

Buchwald, Jed Z. 1994. *The Creation of Scientific Effects: Heinrich Hertz and Electric Waves*. University of Chicago Press.

Byrne and Spon. 1874. *Spons' Dictionary of Engineering, Civil, Mechanical, Military, and Naval*, volume 2. Spon.

Cadiat, E., and L. Dubost. 1896. *Traité Pratique d'Électricité Industrielle*. Paris.

Callan, Nicholas J. 1837. A Description of the Most Powerful Electro-Magnet Yet Constructed. *Annals of Electricity, Magnetism, and Chemistry* 1: 376–378.

Caneva, Kenneth L. 1974. Georg Simon Ohm. In *Dictionary of Scientific Biography*. Scribner.

Cardwell, Donald. 1976. Science and Technology: The Work of James Prescott Joule. *Technology and Culture* 17: 674–687.

Cardwell, Donald. 1989. *James Joule: A Biography*. Manchester University Press.

Cardwell, Donald. 1992. On Michael Faraday, Henry Wilde, and the Dynamo. *Annals of Science* 49: 479–487.

Carlson, W. Bernard. 1988. Thomas Edison as a Manager of R&D: The Case of the Alkaline Storage Battery, 1898–1915. *IEEE Technology and Society Magazine* 7 (4): 4–12.

Carlson, W. Bernard. 1991. *Innovation as a Social Process: Elihu Thomson and the Rise of General Electric*. Cambridge University Press.

Carrier, Elba O. 1965. *Humphry Davy and Chemical Discovery*. Watts.

Carter, Samuel, III. 1968. *Cyrus Field: Man of Two Worlds*. G.P. Putnam's Sons.

Cavicchi, Elizabeth. 2006. Nineteenth-Century Developments in Coiled Instruments and Experiences with Electromagnetic Induction. *Annals of Science* 63: 319–361.

Chaiklin, Seth, and Jean Lave, eds. 1993. *Understanding Practice: Perspectives on Activity and Context*. Cambridge University Press.

Chandler, Alfred D., Jr. 1977. *The Visible Hand: The Managerial Revolution in American Business.* Harvard University Press.

Channing, William F. 1849. *Notes of the Medical Application of Electricity.* Boston.

Channing, William F. 1852. The Municipal Electric Telegraph; Especially its application to Fire Alarms. *American Journal of Science and Arts* 13 (n.s): 58–79

Channing, William F. 1855. The American Fire-Alarm Telegraph. In Ninth Annual Report, Board of Regents, Smithsonian Institution.

Chavois, Louis. 1963. *Histoire Merveilleuse de Zénobe Gramme, Inventeur de la Dynamo.* Librairie Scientifique et Technique, Paris.

Children, John George. 1809. An Account of Some Experiments, Performed with a View to Ascertain the Most Advantageous Method of Constructing a Voltaic Apparatus, for the Purposes of Chemical Research. *Philosophical Transactions of the Royal Society* 99: 32–38.

Christopher, Milbourne, and Maurine Christopher. 1996. *The Illustrated History of Magic.* Heinemann.

Clark, Latimer. 1868. *An Elementary Treatise on Electrical Measurement for the Use of Telegraph Inspectors and Operators.* Spon.

Clarke, E. M. 1835. On a New Phenomenon of Magneto-Electricity. *London and Edinburgh Philosophical Magazine* 6: 169–170.

Clarke, E. M. 1836. Description of E. M. Clarke's Magnetic Electrical Machine. *London and Edinburgh Philosophical Magazine* 9: 262–266.

Clarke, E. M. 1837. Reply of Mr. E. M. Clarke to Mr. J. Saxton. *London and Edinburgh Philosophical Magazine* 10: 455–459.

Coates, Vary T., and Bernard Finn.1979. *A Retrospective Technology Assessment: Submarine Telegraphy.* San Francisco Press.

Committee on Electrical Standards. 1913. *Reports of the Committee of Electrical Standards Appointed by the British Association for the Advancement of Science.* Cambridge University Press.

Conot, Robert. 1979. *Thomas A. Edison: A Streak of Luck.* Da Capo.

Cooper, Carolyn C. 1991. *Shaping Invention: Thomas Blanchard's Machinery and Patent Management in Nineteenth-Century America.* Columbia University Press.

Cordulack, Shelley W. 2005. A Franco-American Battle of the Beams: Electricity and the Selling of Modernity. *Journal of Design History* 18: 147–166.

Cowan, Ruth Schwartz. 1997. *A Social History of American Technology.* Oxford University Press.

Coyne, Terrence E. 1995. The Hoosac Tunnel: Masschusetts' Western Gateway. *Historical Journal of Massachusetts* 23: 1–20.

Crosse, Andrew. 1857. *Memorials, Scientific and Literary, of Andrew Crosse, the Electrician.* London.

Crowther, J. G. 1937. *Famous American Men of Science*. Norton.

Culley, R. S. 1878. *A Handbook of Practical Telegraphy*, seventh edition. Longmans, Green, Reader, and Dyer.

Daniell, J. F. 1836a. On Voltaic Combinations. *Philosophical Transactions of the Royal Society* 126: 107–124.

Daniell, J. F. 1836b. Additional Observations on Voltaic Combinations. *Philosophical Transactions of the Royal Society* 126: 125–129.

Daniell, J. F. 1837. Further Observations on Voltaic Combinations. *Philosophical Transactions of the Royal Society* 127: 141–160.

Davenport, Thomas. 1851. Autobiography of Thomas Davenport. Manuscript on file, Vermont Historical Society, Montpelier.

Davenport, Walter R. 1929. *Biography of Thomas Davenport, the "Brandon Blacksmith," Inventor of the Electric Motor*. Vermont Historical Society.

Davenport, Willard G. 1900. Thomas Davenport: Inventor of the Electric Motor. *Proceedings of the Vermont Historical Society*, pp. 59–81.

Davis, Daniel, Jr. 1838. *Davis's Descriptive Catalog of Apparatus and Experiments*. Marden & Kimball.

Davis, Daniel, Jr. 1842. *Davis's Manual of Magnetism*. Boston.

Davis, Daniel, Jr. 1848. *Catalogue of Apparatus, to Illustrate Magnetism, Galvanism, Electro-Dynamics, Electro-Magnetism, Magneto-Electricity, and Thermo-Electricity*. Boston.

Davis, Daniel, Jr. 1851. *Book of the Telegraph*. Boston.

Davis, Daniel, Jr. 1857. *Davis's Manual of Magnetism, Including Galvanism, Magnetism, Electro-Magnetism, Electro-Dynamics, Magneto-Electricity, and Thermo-Electricity*. twelfth edition. Palmer and Hall.

Davy, Humphry. 1808. The Bakerian Lecture: on Some New Phenomena of Chemical Changes Produced by Electricity, Particularly the Decomposition of the Fixed Alkalies, and the Exhibition of the New Substances Which Constitute their Bases; and on the General Nature of Alkaline Bodies. *Philosophical Transactions of the Royal Society* 98: 1–44.

Davy, Humphry. 1812. *Elements of Chemical Philosophy*, volume 1, Part 1. London.

Davy, Humphry. 1816. On the Fire-Damp of Coal Mines, and on Methods of Lighting the Mines so as to Prevent its Explosion. *Philosophical Transactions of the Royal Society* 106: 1–22.

Dawson, Keith. 1976. Electromagnetic Telegraphy: Early Ideas, Proposals, and Apparatus. *History of Technology* 1: 114–141.

de Cogan, D. 1985. Dr. E. O. W. Whitehouse and the 1858 Trans-Atlantic Cable. *History of Technology* 10: 1–15.

de Guili, Italio. 1932. *Submarine Telegraphy: A Practical Manual*. Isaac Pitman & Sons.

Dibner, Bern. 1959. *The Atlantic Cable*. Burndy Library, Norwalk, Connecticut.

Dibner, Bern. 1961. *Oersted and the Discovery of Electromagnetism.* Burndy Library, Publication No. 18. Norwalk, Connecticut.

Dibner, Bern. 1964. *Alessandro Volta and the Electric Battery.* F. Watts.

Dillon, Wilton S., and Neil G. Kotler, eds. 1994. *The Statue of Liberty Revisited.* Washington.

Dobres, Marcia-Anne. 2000. *Technology and Social Agency.* Blackwell.

Dobres, Marcia-Anne, and C. R. Hoffman, eds. 1999. *The Social Dynamics of Technology.* Smithsonian Institution Press.

Douglas, Mary, and Baron Isherwood. 1996. *The World of Goods: Towards an Anthropology of Consumption.* Routledge.

Douglass, James N. 1886. Address to the Mechanical Science Section of the British Association. Birmingham.

Dray, Philip. 2005. *Stealing God's Thunder: Benjamin Franklin's Lightning Rod and the Invention of America.* Random House.

Dredge, James, ed. 1882. *Electric Illumination*, volume 1. London.

Du Boff, Richard B. 1980. Business Demand and the Development of the Telegraph in the United States, 1844–1860. *Business History Review* 54: 459–479.

Du Boff, Richard B. 1984. The Telegraph in Nineteenth-Century America: Technology and Monopoly. *Comparative Studies in Society and History* 26: 571–586.

Du Moncel, Théodose. 1859. *Revue des Applications de l'Électricité en 1857 et 1858.* Paris.

Du Moncel, Théodose. 1867. *Notice sur l'Appareil d'Induction Électrique de Ruhmkorff.* Paris.

Du Moncel, Théodose. 1880. *L'Éclairage Électrique.* Paris.

Du Moncel, Théodose. 1882. *Electric Lighting.* Routledge.

Du Moncel, Théodose, Frank Geraldy, and C. J. Wharton. 1883. *Electricity as a Motive Power.* Spon.

Du Moncel, Théodose, C. Herz, A. Guerout, F. Geraldy, O. Kern, E. Sartiaux, P. Clemenceau, and C.-C. Soulages. 1884. *Études sur l'Exposition Internationale d'Électricité de Munich.* Paris.

Dyer, Frank L., and Thomas C. Martin. 1910. *Edison, His Life and Inventions.* Two volumes. Harper & Brothers.

Edgerton, David. 2007. *Technology and Global History Since 1900.* Oxford University Press.

Edmondson, T. 1834. The Rotating Armatures. *American Journal of Science and Arts* 26: 205–206.

Eisenman, Harry J. 1966. The Brush Double-Arc Lamp. *Technology and Culture* 7: 511–512.

Electro-Magnetic Association. 1837. *Electro-Magnetism. History of Davenport's Invention of the Application of Electro-Magnetism to Machinery.* New York.

Ellerbroek, W. C. 1972. Dr. George M. Beard: Pioneer in Psychosomatic Medicine. *Psychosomatics* 13: 57–60.

Elliot, George H. 1874. *Report of a Tour of Inspection of European Light-house Establishments, Made in 1873*. Washington.

Ellis, John B. 1869. *The Sights and Secrets of the National Capitol*. United States Publishing Co.

Emmet, John P. 1833. A New Mode of Developing Magnetic Galvanism, by Which May Be Obtained, Shocks, Vivid Sparks and Galvanic Currents from the Horse-Shoe Magnet. *American Journal of Science and Arts* 24: 78–86.

Essig, Mark. 2003. *Edison and the Electric Chair: A Story of Light and Death*. Walker.

Ewbank, Thomas. 1973 [1850]. The Motors: Chief Levers of Civilization. In *The New American State Papers. Science and Technology*, volume 1, pp 435–449. Scholarly Resources.

Fahie, John J. 1884. *A History of Electric Telegraphy, to the Year 1837*. London.

Faraday, Michael. 1834. On the Magneto-Electric Spark and Shock, and on a Peculiar Condition of Electric and Magneto-Electric Induction. *London and Edinburgh Philosophic Magazine* 5: 349–354.

Faraday, Michael. 1854. On Electric Induction—Associated Causes of Current and Static Effects. *Philosophical Magazine* 7: 197–208.

Faraday, Michael. 1952a [1855]. On Some New Electro-Magnetical Motions and on the Theory of Magnetism. In *Great Books of the Western World*, ed. R. Hutchins. Encyclopædia Britannica.

Faraday, Michael. 1952b [1855]. Electro-Magnetic Rotation Apparatus. In *Great Books of the Western World*, ed. R. Hutchins. Encyclopædia Britannica.

Faraday, Michael. 1952c [1839]. Experimental Researches in Electricity, First Series. In *Great Books of the Western World*, ed. R. Hutchins. Encyclopædia Britannica.

Faraday, Michael. 1952d [1839]. Experimental Researches in Electricity, Third Series. In *Great Books of the Western World*, ed. R. Hutchins. Encyclopædia Britannica.

Faraday, Michael. 1952e [1839]. Experimental Researches in Electricity, Seventh Series. In *Great Books of the Western World*, ed. R. Hutchins. Encyclopædia Britannica.

Faraday, Michael. 1952f [1839]. Experimental Researches in Electricity, Eighth Series. In *Great Books of the Western World*, ed. R. Hutchins. Encyclopædia Britannica.

Faraday, Michael. 1871 [1861]. On the Condition of Lights, Buoys, and Beacons. In *Extracts Republished for the Use of the U.S. Light-House Establishment*, by Her Britanic Majesty's Commissioners. U.S. Government Printing Office.

Farrar, Wilfred V. 1974. William Sturgeon. In *A Biographical Dictionary of Scientists*, second edition, ed. T. Williams. Halsted.

Ferguson, James. 1778. *An Introduction to Electricity in Six Sections*, third edition. London.

Field, Henry M. 1866. *History of the Atlantic Telegraph*. Scribner.

Field, Henry M. 1893. *The Story of the Atlantic Telegraph*. Scribner.

Findlay, Alexander G., and William R. Kettle. 1896. *The Lighthouses of the World, and Coast Fog Signals*, 36th edition. London.

Finn, Bernard S. 1973. *Submarine Telegraphy: The Grand Victorian Technology*. Science Museum, London.

Finn, Bernard S. 1975. Heinrich Daniel Rühmkorff. In *Dictionary of Scientific Biography*. Scribner.

Finn, Bernard S. 1976. William Sturgeon. In *Dictionary of Scientific Biography*. Scribner.

Finn, Bernard S. 1984. The Incandescent Electric Light. In *Bridge to the Future*, ed. M. Latimer, B. Hindle, and M. Kranzberg. New York Academy of Sciences.

Fischer, Claude S. 1992. *America Calling: A Social History of the Telephone to 1940*. University of California Press.

Foote, George A. 1952. Sir Humphry Davy and His Audience at the Royal Institution. *Isis* 43: 6–12.

Forman, Paul. 2007. The Primacy of Science in Modernity, of Technology in Postmodernity, and of Ideology in the History of Technology. *History and Technology* 23: 1–152.

Foster, Abram J. 1979. *The Coming of the Electrical Age to the United States*. Arno.

Foster, G. C. 1866. On the Electrical Principles of the Atlantic Telegraph. *Popular Science Review* 5: 416–428.

Foucault, Michel. 1973. *The Order of Things*. Pantheon.

Foucault, Michel. 1977. *Discipline and Punish*. Pantheon.

Fox, Robert. 1969. James Prescott Joule (1818–1889). In *Mid-Nineteenth-Century Scientists*, ed. J. North. Pergamon.

Franklin, Benjamin. 1986. *Benjamin Franklin: The Autobiography and Other Writings*. New York.

Franklin, Benjamin. 1996 [1769]. *Experiments and Observations on Electricity*. New York.

Franklin Institute. 1885. International Electrical Exhibition, 1884, Report of the Examiners of Section XXIV: Electro-Dental Apparatus.

Frary, Ihna T. 1969. *They Built the Capitol*. Books for Libraries Press.

Frazier, Arthur H. 1975. *Joseph Saxton and His Contributions to the Medal Ruling and Photographic Arts*. Smithsonian Institution Press.

Friedel, Robert. 2007. *A Culture of Improvement: Technology and the Western Millennium*. MIT Press.

Friedel, Robert, Paul Israel, and Bernard S. Finn. 1986. *Edison's Electric Light: Biography of an Invention*. Rutgers University Press.

Friedlander, Amy. 1995. *Natural Monopoly and Universal Service: Telephones and Telegraphs in the U.S. Communications Infrastructure, 1837–1940*. Corporation for National Research Initiatives.

Gage, Simon H. 1908. The Origin and Development of the Projection Microscope. *Transactions of the American Microscopical Society* 28: 5–60.

Gale, Leonard D. 1875. Historic Facts Concerning Morse's Electro-Magnetic Telegraph. In *Memorial of Samuel Finley Breese Morse, Including Appropriate Ceremonies of Respect at the National Capitol, and Elsewhere*. U.S. Government Printing Office.

Galison, Peter. 2003. *Einstein's Clocks, Poincaré's Maps: Empires of Time*. Norton.

Galvani, Luigi. 1953 [1791]. *Commentary on the Effects of Electricity on Muscular Motion*. Burndy Library, Norwalk, Connecticut.

Gassiot, J. P. 1859. On the Application of Electrical Discharges from the Induction Coil to the Purposes of Illumination. *Proceedings of the Royal Society of London* 10: 432.

Gavarret, Jules. 1861. *Télégraphie Électrique*. Victor Masson et Fils, Paris.

Gee, Brian. 1991. Electromagnetic Engines: Pre-Technology and Development Immediately Following Faraday's Discovery of Electromagnetic Rotations. *History of Technology* 13: 41–72.

Gee, Brian. 1993. The Early Development of the Magneto-Electric Machine. *Annals of Science* 50: 101–133.

Gee, George E. 1885. *The Silversmith's Handbook, Containing Full Instructions for the Alloying and Working of Silver*. Crosby Lockwood.

Gee, W. W. 1920. Henry Wilde. *Memoirs and Proceedings of the Manchester Literary and Philosophical Society* 63 (5): 1–16.

Gell, Alfred. 1998. *Art and Agency: Anthropological Theory*. Clarendon.

Gernsheim, Helmut, and Alison Gernsheim. 1969. *The History of Photography from the Camera Obscura to the Beginning of the Modern Era*. McGraw-Hill.

Giedion, Siegfried. 1969 [1948]. *Mechanization Takes Command. A Contribution to Anonymous History*. Norton.

Gooday, G. J. N. 2004. *The Morals of Measurement: Accuracy, Irony and Trust in Late Victorian Electrical Practice*. Cambridge University Press.

Gooding, David. 1990. Mapping Experiment as a Learning Process: How the First Electromagnetic Motor Was Invented. *Science, Technology, and Human Values* 15: 165–201.

Gordon, John S. 2003. *A Thread Across the Ocean: The Heroic Story of the Transatlantic Cable*. Perennial.

Gore, George. 1884. *The Art of Electro-Metallurgy, Including all Known Processes of Electro-Deposition*. Longmans, Green.

Gorman, Mel. 1977. Electric Illumination in the Franco-Prussian War. *Social Studies of Science* 7: 525–529.

Great Britain, Commissioner of Patents. 1859. *Patents for Inventions. Abridgments of Specifications Relating to Electricity and Magnetism, Their Generation and Applications*. London.

Great Britain, Commissioner of Patents. 1874. *Patents for Inventions. Abridgments of Specifications Relating to Electricity and Magnetism and Their Generation and Applications: Part II—A.D. 1858–1866*, second edition. London.

Greenslade, Thomas B., Jr. 1986. Apparatus for Natural Philosophy: Barlow's Wheel. *Rittenhouse* 1 (1): 26–28.

Griglietta, C., compiler. 1838. *Electro-Magnetism. A Brief Essay or Informal Lecture on Electro-Magnetism, With a Full Description of Models of Davenport's Machines, as now Exhibited in New York; and at the Masonic Hall, Philadelphia.*

Grove, William K. 1845. De l'Application de la Puissance Incandescente de l'Électricité Voltaïque à l'Éclairage des Mines. *Archives de l'Électricité* 5: 547–550.

Grove, William K. 1849. On Heat and Electric Light. *Scientific American* 4: 211.

Guillemin, Amédée, and Sylvanus P. Thompson. 1891. *Electricity and Magnetism.* Macmillan.

Hachette, Jean. 1832. Nouvelle Construction d'une Machine Électro-Magnétique. *Annales de Chimie et de Physique* 50: 322–324.

Hachette, Jean. 1833. Chemical Action and Decomposition of Water, Produced by Electrical Induction. *American Journal of Science and Arts* 24: 142–145.

Hacking, Ian. 1983. *Representing and Intervening: Introductory Topics in the Philosophy of Natural Science.* Cambridge University Press.

Hacking, Ian. 1999. *The Social Construction of What?* Harvard University Press.

Hague, Douglas B., and Rosemary Christie. 1975. *Lighthouses: Their Architecture, History and Archaeology.* Gomer.

Hamilton, William. 1838. Report on Prof. Morse's Electro-Magnetic Telegraph. *Journal of the Franklin Institute* 25: 106–108.

Hare, Robert. 1832. On the Application of Galvanic Ignition in Rock Blasting. *American Journal of Science and Arts* 21: 139–141.

Hare, Robert. 1834. Description of a Process, and an Apparatus, for Blasting Rocks by Means of Galvanic Ignition. *American Journal of Science and Arts* 26: 352–360.

Hare, Robert. 1840. *Brief Exposition of the Science of Mechanical Electricity, or Electricity Proper; Subsidiary to the Course of Chemical Instruction in the University of Pennsylvania.*

Harrison, Joseph L. 1891. *The Great Bore: A Souvenir of the Hoosac Tunnel.* North Adams, Massachusetts.

Hayden, Brian. 1998. Practical and Prestige Technologies: The Evolution of Material Systems. *Journal of Archaeological Method and Theory* 5: 1–55.

Hazelton, George C., Jr. 1897. *The National Capitol: Its Architecture, Art, and History.* New York.

Headrick, Daniel R. 1981. *The Tools of Empire: Technology and European Imperialism in the Nineteenth Century.* Oxford University Press.

Hearder, J. N. 1865. *An Account of Some Experiments with the Electric Light, to Test its Value for Nocturnal Military Operations.* Devonshire Association of Science and Art.

Hearn, Chester G. 2002. *Tracks in the Sea: Matthew Fontaine Maury and the Mapping of the Oceans.* International Marine/McGraw-Hill.

Hearn, Chester G. 2004. *Circuits in the Sea: The Men, the Ships, and the Atlantic Cable*. Praeger.

Henry, Joseph. 1831a. On the Application of the Principle of the Galvanic Multiplier to Electro-Magnetic Apparatus, and also to the Developement of Great Magnetic Power in Soft Iron. *American Journal of Science and Arts* 19: 400–408.

Henry, Joseph. 1831b. On a Reciprocating Motion Produced by Magnetic Attraction and Repulsion. *American Journal of Science and Arts* 20: 340–343.

Henry, Joseph. 1832. On the Production of Currents and Sparks of Electricity from Magnetism. *American Journal of Science and Arts* 22: 403–408.

Henry, Joseph. 1835. Facts in Reference to the Spark, &c. from a Long Conductor Uniting the Poles of a Galvanic Battery. *American Journal of Science and Arts* 28: 331.

Henry, Joseph. 1839. Contributions to Electricity and Magnetism. *Annals of Electricity, Magnetism, and Chemistry* 4: 281–310.

Henry, Joseph. 1887. Researches in Sound, in Relation to Fog-Signalling. In Scientific Writings of Joseph Henry, Smithsonian Institution, Miscellaneous Collections No. 30 (part II), p. 370–510.

Henry, Joseph. 1935 [1874]. Biographical Memoir of Joseph Saxton. In *Joseph Saxton*, ed. J. Pendleton. Reading, Pennsylvania.

Henry, Joseph. 1981 [1838]. Review of "Report of the Committee on Naval Affairs, to Whom Was Referred the Memorial of Henry Hall Sherwood" In *The Papers of Joseph Henry*, volume 4: *Janury 1838–December 1840, The Princeton Years*, ed. N. Reingold, A. Molella, and M. Rothenberg. Smithsonian Institution Press.

Henry, Joseph, and Philip Ten Eyck. 1831. An Account of a Large Electro-Magnet, made for the Laboratory of Yale College. *Journal of American Science and Arts* 20: 201–203.

Hicks, Brian, and Schuyler Kropf. 2002. *Raising the Hunley: The Remarkable History and Recovery of the Lost Confederate Submarine*. Ballantine Books.

Higgs, Paget. 1879a. *The Electric Light in its Practical Applications*. London.

Higgs, Paget. 1879b. *Electric Transmission of Power: Its Present Position and Advantages*. Soon.

Hill, Forest G. 1960. Formative Relations of American Enterprise, Government and Science. *Political Science Quarterly* 75: 400–419.

Hindle, Brooke. 1981. *Emulation and Invention*. Norton.

Hindle, Brooke, and Steven Lubar. 1986. *Engines of Change: The American Industrial Revolution, 1790–1860*. Smithsonian Institution Press.

Hirshfeld, Alan. 2006. *The Electric Life of Michael Faraday*. Walker.

Hodgins, Eric, and F. Alexander Magoun. 1932. *Behemoth: The Story of Power*. Doubleday, Doran.

Holmes, A. Bromley. 1884. *Practical Electric Lighting*, second edition. London.

Holmes, Frederick H. 1862. *Holmes' Magneto-Electric Light, as Applicable to Lighthouses*. London.

Hospitalier, E., and Julius Maier. 1883. *The Modern Applications of Electricity.* Two volumes. Appleton.

Hounshell, David A. 1980a. Public Relations or Public Understanding?: The American Industries Series in *Scientific American. Technology and Culture* 21: 598–593.

Hounshell, David A. 1980b. Edison and the Pure Science Ideal in 19th-Century America. *Science* 207: 612–617.

Hounshell, David A. 1984. *From the American System to Mass Production, 1800–1932: The Development of Manufacturing Technology in the United States.* Johns Hopkins University Press.

Hubbard, Geoffrey. 1965. *Cooke and Wheatstone and the Invention of the Electric Telegraph.* Routledge and Kegan Paul.

Huggins, William. 1864. On the Spectra of Some of the Chemical Elements. *Philosophical Transactions of the Royal Society* 154: 139–160.

Hughes, Thomas P. 1983. *Networks of Power: Electrification in Western Society, 1880–1930.* Johns Hopkins University Press.

Hugill, Peter J. 1999. *Global Communications Since 1844: Geopolitics and Technology.* Johns Hopkins University Press.

Hunt, Bruce J. 1991. Michael Faraday, Cable Telegraphy and the Rise of Field Theory. *History of Technology* 13: 1–19.

Hunt, Bruce J. 1994. The Ohm Is Where the Art Is: British Telegraph Engineers and the Development of Electrical Standards. *Osiris* 9, second series: 48–63.

Hunt, Bruce J. 1996. Scientists, Engineers and Wildman Whitehouse: Measurement and Credibility in Early Cable Telegraphy. *British Journal for the History of Science* 29: 155–169.

Hunt, Bruce J. 1997. Doing Science in a Global Empire: Cable Telegraphy and Electrical Physics in Victorian England. In *Victorian Science in Context*, ed. B. Lightman. University of Chicago Press.

Hunt, Robert. 1851. *Elementary Physics, an Introduction to the Study of Natural Philosophy.* Reeve and Benham.

Hunt, Robert. 1854. *A Manual of Photography*, fourth edition. Griffin.

Hunt, Robert. 1860., ed. *Ure's Dictionary of Arts, Manufactures, and Mines*, fifth edition. Longman, Green, Longman, and Roberts.

Hunt, Robert, and F. W. Rudler. 1878. *Ure's Dictionary of Arts, Manufactures, and Mines*, seventh edition. Longmans, Green.

Hunter, Louis C. 1985. *A History of Industrial Power in the United States*, volume 2: *Steam Power.* University Press of Virginia.

Hunter, Louis C., and Lynwood Bryant. 1991. *A History of Industrial Power in the United States*, volume 3: *The Transmission of Power.* MIT Press.

Hutchins, E. 1995. *Cognition in the Wild*. MIT Press.

Iles, George. 1904. *Flame, Electricity and the Camera*. J. A. Hill.

Israel, Paul. 1992. *From Machine Shop to Industrial Laboratory: Telegraphy and the Changing Context of American Invention, 1830–1920*. Johns Hopkins University Press.

Israel, Paul. 1998. *Edison: A Life of Invention*. Wiley.

Jacob, Margaret C. 1997. *Scientific Culture and the Making of the Industrial West*. Oxford University Press.

Jacobi, Moritz. 1837. On the Application of Electro-Magnetism to the Moving of Machines. *Annals of Electricity, Magnetism, and Chemistry* 1: 408–415.

Jacobi, Moritz. 1841a. *Galvanoplastik: or the Process of Cohering Copper into Plates, or Other Given Forms, by Means of Galvanic Action on Copper Solutions*. Manchester.

Jacobi, Moritz. 1841b. On the Principles of Electro-Magnetic Machines. *Annals of Electricity, Magnetism, and Chemistry* 6: 152–159.

Jahns, Patricia. 1961. *Matthew Fontaine Maury and Joseph Henry: Scientists of the Civil War*. Hastings House.

James, Frank. 1999. "The Civil-Engineer's Talent": Michael Faraday, Science, Engineering and the English Lighthouse Service, 1836–1865. *Transactions of the Newcomen Society* 70: 153–160.

James, Frank. 2000. Michael Faraday and Lighthouses. In *British Social and Economic History, 1850–1870*, ed. I. Inkster et al. Ashgate.

Jarvis, C. Mackechnie. 1958. The Generation of Electricity. In *A History of Technology*. Volume 5: *The Late Nineteenth Century*, ed. C. Singer, E. Holmyard, A. Hall, and T. Williams. Clarendon.

Jelved, Karen, Andrew D. Jackson, and Ole Knudsen, eds. 1998. *Selected Scientific Works of Hans Christian Oersted*. Princeton University Press.

Johnson, Arnold B. 1890. *The Modern Light-House Service*. Washington.

Jones, A. V., and R. P. Tarkenter. 1992. *Electrical Technology in Mining. The Dawn of a New Age*. Peter Peregrinus.

Jones, Thomas P. 1837. Remarks by the editor. *Journal of the Franklin Institute* 20: 342–343.

Jonnes, Jill. 2004. *Empires of Light: Edison, Tesla, Westinghouse, and the Race to Electrify the World*. Random House.

Josephson, Matthew. 1959. *Edison: A Biography*. McGraw-Hill.

Joule, James P. 1838. Description of an Electro-Magnetic Engine. *Annals of Electricity, Magnetism, and Chemistry* 2: 122–123.

Joule, James P. 1839a. Investigations in Magnetism and Electro-Magnetism. *Annals of Electricity, Magnetism, and Chemistry* 4: 131–135.

Joule, James P. 1839b. Description of an Electro-Magnetic Engine. *Annals of Electricity, Magnetism, and Chemistry* 4: 204–205.

Joule, James P. 1840a. On Electro-Magnetic Forces. *Annals of Electricity, Magnetism, and Chemistry* 4: 474–481.

Joule, James P. 1840b. On Electro-Magnetic Forces. *Annals of Electricity, Magnetism, and Chemistry* 5: 187–198.

Joule, James P. 1842. On a New Class of Magnetic Forces. *Annals of Electricity, Magnetism, and Chemistry* 8: 219–224.

Joule, James P. 1884. *The Scientific Papers of James Prescott Joule.* Physical Society of London.

Keim, de B. Randolph. 1884. *Keim's Illustrated Hand-Book. Washington and Its Environs,* nineteenth edition. Washington.

Keithley, Joseph F. 1999. *The Story of Electrical and Magnetic Measurements from 500 BC to the 1940s.* IEEE Press.

Keller, Charles M., and Janet D. Keller. 1996. *Cognition and Tool Use: The Blacksmith at Work.* Cambridge University Press.

Kendall, Amos. 1854. The American Electro-Magnetic Telegraph. *Shaffner's Telegraph Companion* 1 (2): 73.

Kenward, J. 1893. Some Notes on Lighthouse Apparatus. *Science* 21: 216–218.

Keyser, Paul T. 1993. The Purpose of the Parthian Galvanic Cells: A First-Century A.D. Electric Battery Used for Analgesia. *Journal of Near Eastern Studies* 52: 81–98.

Khan, B. Zorina. 2005. *The Democratization of Invention: Patents and Copyrights in American Economic Development, 1790–1920.* Cambridge University Press.

King, James W. 1878. *The Pilots Handbook for the English Channel,* eighth edition. London.

King, Moses. 1883. *King's Handbook of Boston.* Boston.

King, W. James. 1962a. The Development of Electrical Technology in the 19th Century: 1. The Electrochemical Cell and the Electromagnet. *United States National Museum Bulletin* 228: 233–271.

King, W. James. 1962b. The Development of Electrical Technology in the 19th Century: 3. The Early Arc Light and Generator. *United States National Museum Bulletin* 228: 334–407.

King, W. James. 1962c. The Development of Electrical Technology in the 19th Century: 2. The Telegraph and the Telephone. *United States National Museum Bulletin* 228: 272–333.

Kingery, W. D., ed. 1996. *Learning from Things: Method and Theory in Material Culture Studies.* Smithsonian Institution Press.

Kinney, Thomas A. 2004. *The Carriage Trade: Making Horse-Drawn Vehicles in America.* Johns Hopkins University Press.

Kinsey, Thomas A. 2004. *The Carriage Trade: Making Horse-Drawn Vehicles in America.* Johns Hopkins University Press.

Knight, Edward H. 1876. *Knight's American Mechanical Dictionary,* volume 2. Hurd and Houghton.

Koestler, Arthur. 1964. *The Act of Creation*. Macmillan.

LaMotta, Vincent M., and Michael Brian Schiffer. 2001. Behavioral Archaeology: Towards a New Synthesis. In *Archaeological Theory Today*, ed. I. Hodder. Polity.

Landes, David S. 1969. *The Unbound Prometheus: Technological Change and Industrial Development in Western Europe from 1750 to the Present*. Cambridge University Press.

Langdon, William E. 1877. *The Application of Electricity to Railway Working*. Macmillan.

Lansing, J. Stephen. 1991. *Priests and Programmers: Technologies of Power in the Engineered Landscape of Bali*. Princeton University Press.

Larkin, Oliver W. 1954. *Samuel F. B. Morse and American Democratic Art*. Little, Brown.

Latour, Bruno, and Steve Woolgar. 1979. *Laboratory Life: The Social Construction of Scientific Facts*. Sage.

Lave, Jean, and Etienne Wenger. 1991. *Situated Learning: Legitimate Peripheral Participation*. Cambridge University Press.

Layton, Edwin. 1971. Mirror-Image Twins: The Communities of Science and Technology in 19th-Century America. *Technology and Culture* 12: 562–580.

Lemoine, Bertrand. 1986. *La Statue de la Liberté*. Brussels.

Lemonnier, P. 1992. *Elements for an Anthropology of Technology*. University of Michigan Museum of Anthropology.

Lemonnier, P., ed. 1993. *Technological Choices: Transformation in Material Cultures Since the Neolithic*. Routledge.

Leroi-Gourhan, André. 1993 [1964]. *Gesture and Speech*. MIT Press.

Le Roux, F. P. 1868. *Les Machines Magneto-Électriques Françaises et l'Application de l'Électricité a l'Éclairage des Phares*. Paris.

Licht, Walter. 1995. *Industrializing America: The Nineteenth Century*. Johns Hopkins University Press.

Linder, Annulla. 2002. The Execution Spectacle and State Legitimacy: The Changing Nature of the American Execution Audience, 1833–1937. *Law and Society Review* 36: 607–656.

Lindley, David. 2004. *Degrees Kelvin: A Tale of Genius, Invention, and Tragedy*. Joseph Henry Press.

Lipton, Leah. 1981. William Dunlap, Samuel F. B. Morse, John Wesley Jarvis, and Chester Harding: Their Careers as Itinerant Portrait Painters. *American Art Journal* 13 (3): 34–50.

Lucas, Félix. 1885. *Les Machines Magnéto-Électriques et l'Arc Voltaïque des Phares*. Paris.

Lundeberg, Philip K. 1974. *Samuel Colt's Submarine Battery: The Secret and the Enigma*. Smithsonian Institution Press.

Mabee, Carleton. 1969. *The American Leonardo: A Life of Samuel F. B. Morse*. Octagon Books.

MacLaren, Malcolm. 1943. *The Rise of the Electrical Industry During the Nineteenth Century*. Princeton University Press.

MacLeod, Roy M. 1969. Science and Government in Victorian England: Lighthouse Illumination and the Board of Trade, 1866–1886. *Isis* 60: 4–38.

Mangin, Arthur. 1866. *Le Feu du Ciel: Histoire de l'Électricité et de ses Principales Applications*. 3rd ed. Tours.

Marland, E. A. 1964. *Early Electrical Communication*. Abelard-Schuman.

Martin, Thomas, ed. 1932. *Faraday's Diary*. volume 1, Sept. 1820–June 11, 1832. G. Bell & Sons.

Martin, T. Commerford, and Joseph Wetzler. 1891. *The Electric Motor and Its Applications*, third edition. W. J. Johnston.

Marvin, Carolyn. 1988. *When Old Technologies Were New: Thinking about Electrical Communication in the Late 19th Century*. Oxford University Press.

Maury, Richard L. 1915. *A Brief Sketch of the Work of Matthew Fontaine Maury During the War 1861–1865*. Whittet & Shepperson.

Maxwell, James Clerk. 1865. A Dynamical Theory of the Electromagnetic Field. *Philosophical Transactions of the Royal Society of London* 155: 459–512.

Mayr, Otto. 1971. Feedback Mechanisms in the Historical Collections of the National Museum of History and Technology. *Smithsonian Studies in History and Technology*, No. 12.

McCracken, Grant. 1988. *Culture and Consumption*. Indiana University Press.

McGuire, Patrick. 1990. Money and Power: Financiers and the Electric Manufacturing Industry, 1878–1896. *Social Science Quarterly* 71: 510–530.

McMahon, A. Michal. 1984. *The Making of a Profession: A Century of Electrical Engineering in America*. IEEE Press.

McMillan, Walter G. 1891. *A Treatise on Electro-Metallurgy*. Griffin.

McMillan, Walter G., and W. R. Cooper. 1910. *A Treatise on Electro-Metallurgy*, third edition. Griffin.

Meigs, Montgomery C. 2001. *Capitol Builder: The Shorthand Journals of Montgomery C. Meigs, 1853–1859, 1861*. U.S. Government Printing Office.

Merrill, John P. 1874a. Lecture on Galvanic Batteries and Electrical Machines, as Used in Torpedo Operations. Part I: Galvanic Batteries. U.S. Torpedo Station, Newport, Rhode Island.

Merrill, John P. 1874b. Lecture on Galvanic Batteries and Electrical Machines, as Used in Torpedo Operations. Part II. Frictional and Magneto-Electric Machines. U.S. Torpedo Station, Newport, Rhode Island.

Mertens, Joost. 1998. From the Lecture Room to the Workshop: John Frederic Daniell, the Constant Battery, and Electometallurgy around 1840. *Annals of Science* 55: 241–261.

Merton, Robert K. 1968. The Matthew Effect in Science: The Reward and Communication Systems of Science are Considered. *Science* 159: 56–63.

Meskell, Lynn. 2004. *Object Worlds in Ancient Egypt: Material Biographies Past and Present.* Berg.

Meyer, Herbert W. 1972. *A History of Electricity and Magnetism.* Burndy Library, Norwalk, Connecticut.

M'Gauley, Prof. 1867. On the Applicability of the Electric Light to Lighthouses. *Intellectual Observer* 11: 325, 327.

Michalowicz, Joseph C. 1948. Origin of the Electric Motor. *Electrical Engineering* 67: 1035–1040.

Millard, Andre. 1990. *Edison and the Business of Innovation.* Johns Hopkins University Press.

Miller, Daniel. 1995a. Consumption and Commodities. *Annual Review of Anthropology* 19: 453–505.

Miller, Daniel. 1995b. *Acknowledging Consumption.* Routledge.

Miller, Daniel, ed. 2005. *Materiality.* Duke University Press.

Millis, John. 1885. Appendix 2: Report upon the Hell Gate Electric Light, Hallet's Point, N.Y., and On Progress of Experiments on Electricity as an Illuminant for Light-Houses. *U.S. Treasury Department, Report of the Light-House Board*, pp. 159–173.

Millis, John. 1887. The Installation of the Electric Light Plant of the Statue of Liberty Enlightening the World. *U.S. Treasury Department, Report of the Light-House Board*, pp. 117–127.

Millis, John. 1890. Appendix 3: Illumination of the Statue of "Liberty Enlightening the World." *U.S. Treasury Department, Report of the Light-House Board*, pp. 227–237.

Molella, Arthur P. 1976. The Electric Motor, the Telegraph, and Joseph Henry's Theory of Technological Progress. *Proceedings of the IEEE* 64: 1273–1278.

Molella, Arthur P. 1984. At the Edge of Science: Joseph Henry, 'Visionary', and the Smithsonian Institution. *Annals of Science* 41: 445–461.

Molella, Arthur P., and Nathan Reingold. 1973. Theorists and Ingenious Mechanics: Joseph Henry Defines Science. *Science Studies* 3: 323–351.

Molella, Arthur P., Nathan Reingold, Marc Rothenberg, Joan F. Steiner, and Kathleen Waldenfels, eds. 1980. *A Scientist in American Life: Essays and Lectures of Joseph Henry.* Smithsonian Institution Press.

Moll, Gerard. 1830. Electro-Magnetic Experiments. *Edinburgh Journal of Science, New Series* 3 (6): 209–218.

Moll, Gerard. 1831. Electro-Magnetic Experiments. *American Journal of Science and Arts* 19: 329–337.

Morse, Edward L., ed. 1973 [1914]. *Samuel F. B. Morse: His Letters and Journals*, volume. 2. Da Capo.

Morse, Samuel F. B. 1855. The Electro-Magnetic Telegraph. A Defence Against the Injurious Deductions Drawn from the Deposition of Prof. Joseph Henry. *Shaffner's Telegraph Companion* 2: 6–96.

Morse, Samuel F. B. [and Taliaferro Shaffner]. 1855. History of Morse's Telegraph. *Shaffner's Telegraphic Companion* 2: 399–416.

Morton, Henry. 1879. Report on Experiments with Machines for Producing Electric Light. In *U.S. Treasury Department, Report of the Light-House Board*, pp. 88–136.

Morton, Henry. 1895a. Engineering Fallacies. *Cassier's Magazine* 7: 200–211.

Morton, Henry. 1895b. Engineering Fallacies. *Cassier's Magazine* 7: 428–439.

Morus, Iwan R. 1988. The Sociology of Sparks: An Episode in the History and Meaning of Electricity. *Social Studies of Science* 18: 387–417.

Morus, Iwan R. 1998. *Frankenstein's Children: Electricity, Exhibition, and Experiment in Early-Nineteenth-Century London.* Princeton University Press.

Moyer, Albert E. 1997. *Joseph Henry: The Rise of an American Scientist.* Smithsonian Institution Press.

Mukerji, Chandra. 2003. Intelligent Uses of Engineering and the Legitimacy of State Power. *Technology and Culture* 44: 655–676.

Mumford, Lewis. 1963 [1934]. *Technics and Civilization.* Harcourt Brace Jovanovich.

Murdock, Myrtle C. 1950. *Constantino Brumidi: Michelangelo of the United States Capitol.* Monumental.

Myers, Denys P. 1978. *Gaslighting in America: A Guide for Historic Preservation.* U.S. Department of the Interior, Heritage Conservation and Recreation Service.

Napier, James. 1876. *A Manual of Electro-Metallurgy: Including the Applications of the Art to Manufacturing Processes*, fifth edition. Griffin.

Navy Department, Naval History Division. 1971. *Civil War Naval Chronology, 1861–1865.* Washington.

New York City Mercantile Manufacturers. 1857. *Business Directory for the Year Ending May 1st, 1857.* West, Lee, and Bartlett.

Newman, J. 1836. *A Catalogue of Philosophical Instruments Manufactured and Sold by J. Newman, Philosophical Instrument Maker, by Appointment, to the Royal Institution of Great Britain.* London.

Niaudet-Bréguet, Alfred. 1875. Gramme's Magneto-Electric Machines. *Telegraphic Journal* 3: 184–186, 196–200, 223–226.

Niaudet-Bréguet, Alfred. 1880. *Elementary Treatise on Electric Batteries.* Wiley.

Nickerson, Edward N. 1885. *Joseph Henry and the Magnetic Telegraph. An Address Delivered at Princeton College, June 16, 1885.* Scribner.

Nickles, David P. 2003. *Under the Wire: How the Telegraph Changed Diplomacy.* Harvard University Press.

Nielsen, Axel. 1995. Architectural Performance and the Reproduction of Social Power. In *Expanding Archaeology*, ed. J. Skibo et al. University of Utah Press.

Noad, Henry M. 1857. *Manual of Electricity. Part 2: Magnetism and the Electric Telegraph*. George Knight.

Noble, Dennis L. 1997. *Lighthouses and Keepers: The U.S. Lighthouse Service and Its Legacy*. Naval Institute Press.

Nye, David. 1990. *Electrifying America: Social Meanings of a New Technology*. MIT Press.

Nye, David. 1994. *American Technological Sublime*. MIT Press.

Nye, David. 1998. *Narratives and Spaces: Technology and the Construction of American Culture*. Columbia University Press.

O'Brien, Michael J., T. D Holland, R. J. Hoard, and G. L. Fox. 1994. Evolutionary Implications of Design and Performance Characteristics of Prehistoric Pottery. *Journal of Archaeological Method and Theory* 1: 259–304.

Ohm, Georg S. 1891. *The Galvanic Circuit Investigated Mathematically*. Van Nostrand.

Oudshoorn, Nelly, and Trevor Pinch. 2003. *How Users Matter: The Co-Construction of Users and Technology*. MIT Press.

Pacey, Arnold. 1992. *The Maze of Ingenuity: Ideas and Idealism in the Development of Technology*, second edition. MIT Press.

Page, Charles G. 1839a. On Electro-Magnetism, as a Moving Power. *American Journal of Science and Arts* 35: 106–111.

Page, Charles G. 1839b. Magneto-Electric and Electro-Magnetic Apparatus and Experiments. *American Journal of Science and Arts* 35: 252–268.

Page, Charles G. 1845a. New Electro-Magnetic Engine. *American Journal of Science and Arts* 49: 131–135.

Page, Charles G. 1845b. Axial Galvanometer, and Double Axial Reciprocating Engine. *American Journal of Science and Arts* 49: 136–142.

Page, Charles G. 1867. *History of Induction. The American Claim to the Induction Coil and its Electrostatic Developments*. Washington.

Page, Charles G. 1973a [1850]. Report to William A. Graham, Secretary of the Navy, 30 August 1850. In *The New State Papers: Science and Technology*, volume 13: *Special Studies*. Scholarly Resources.

Page, Charles G. 1973b [1851]. Report to William A. Graham, Secretary of the Navy, 28 November 1851. In *The New State Papers: Science and Technology*, volume 13: *Special Studies*. Scholarly Resources.

Paine, Elijah. 1838. Thomas Davenport and Myrick W. Nelson vs. Edwin Williams. Bill of Complaint, Chancery Court, State of New York.

Palmer, Edward. 1840. *Palmer's New Catalogue*. London.

Pancaldi, Guiliano. 2003. *Volta: Science and Culture in the Age of Enlightenment*. Princeton University Press.

Pantalony, David, Richard L. Kremer, and Francis J. Manasek. 2005. *Study, Measure, Experiment: Stories of Scientific Instruments at Dartmouth College*. Terra Nova.

Passer, Harold C. 1953. *The Electrical Manufacturers, 1875–1900: A Study in Competition, Entrepreneurship, Technical Change, and Economic Growth*. Harvard University Press.

Patterson, Elizabeth C. 1975. Mary Fairfax Greig Somerville. In *Dictionary of Scientific Biography*. Scribner.

Pavlova, S. A. 1968. *Electrodeposition of Metals: A Historical Survey*, ed. S. Pogodin. Israel Program for Scientific Translations, Jerusalem.

Pepper, John H. 1869. Some Experiments with the Great Induction Coil at the Royal Polytechnic. *Scientific American* 21: 275.

Pera, Marcello. 1992. *The Ambiguous Frog: The Galvani-Volta Controversy on Animal Electricity*. Princeton University Press.

Perrault, Carole L. 1986. Liberty Enlightening the World. *The Keeper's Log*, spring: 1–15 and summer: 6–18.

Perry, Milton. 1965. *Infernal Machines: The Story of Confederate Submarine and Mine Warfare*. Louisiana State University Press.

Pfaffenberger, B. 1992. Social Anthropology of Technology. *Annual Review of Anthropology* 21: 491–516.

Pielke, Roger A., Jr. 2007. *The Honest Broker: Making Sense of Science in Policy and Politics*. Cambridge University Press.

Pixii, Hippolyte. 1833? *Nouveaux Appareils Électro-Magnétiques*. Paris.

Pixii, Père et Fils. 1835. *Catalogue des Principaux Instrumens de Physique, Chimie, Optique, Mathématiques et Autres, à l'Usage des Sciences*. Paris.

Pixii, Père et Fils. 1845. *Catalogue des Principaux Instrumens de Physique, Chimie, Optique, Mathématiques et Autres, à l'Usage des Sciences*. Paris.

Pope, Franklin L. 1869. *Modern Practice of the Electric Telegraph: A Handbook*. Russell Brothers.

Pope, Franklin L. 1891a. The Inventors of the Electric Motor—I. With Special Reference to the Work of Thomas Davenport. *The Electrical Engineer* 11: 1–5.

Pope, Franklin L. 1891b. The Inventions of Thomas Davenport. *Transactions of the American Institute of Electrical Engineers* 8: 93–97.

Post, Robert C. 1972. The Page Locomotive: Federal Sponsorship of Invention in Mid-19th-Century America. *Technology and Culture* 13: 140–169.

Post, Robert C. 1974. Electro-Magnetism as a Motive Power: Robert Davidson's Galvani of 1842. *Railroad History, Bulletin* 130: 5–22.

Post, Robert C. 1976a. *Physics, Patents, and Politics: A Biography of Charles Grafton Page*. Science History Publications.

Post, Robert C. 1976b. "Liberalizers" versus "Scientific Men" in the Antebellum Patent Office. *Technology and Culture* 17: 24–54.

Post, Robert C. 1976c. Stray Sparks from the Induction Coil: The Volta Prize and the Page Patent. *Proceedings of the IEEE* 64: 1276–1279.

Post, Robert C. 2003. *Technology, Transport, and Travel in American History*. American Historical Association.

Preece, William H. 1879. The Electric Light. *London, Edinburgh, and Dublin Philosophical Magazine and Journal of Science* 7 (Fifth Series): 29–34.

Preece, William H., and James Sivewright. 1891. *Telegraphy*, ninth edition. Longmans, Green.

Prescott, George B. 1860. *History, Theory, and Practice of the Electric Telegraph*. Ticknor and Fields.

Prescott, George B. 1875. Ocean Telegraphy II. *Scientific American* 32: 40–41.

Prescott, George B. 1977. *Electricity and the Electric Telegraph*. Appleton.

Prescott, George B. 1888. *Electricity and the Electric Telegraph* seventh edition. Two volumes. Appleton.

Pursell, Carroll W. 1976. Testing a Carriage: The "American Industry Series" of Scientific American. *Technology and Culture* 17: 82–92.

Pursell, Carroll W. 1995. *The Machine in America: A Social History of Technology*. Johns Hopkins University Press.

Redfield, W. C. 1833. Notices of American Steamboats. *American Journal of Science and Arts* 23: 311–318.

Redner, Harry. 1987. *The Ends of Science: An Essay in Scientific Authority*. Westview.

Reid, James D. 1879. *The Telegraph in America. Its Founders Promoters and Noted Men*. Derby Brothers.

Reingold, Nathan. 1972. Joseph Henry. In *Dictionary of Scientific Biography*. Scribner.

Reingold, Nathan. 1991. *Science, American Style*. Rutgers University Press.

Reingold, Nathan, Stuart Pierson, and Arthur P. Molella, eds. 1972. *The Papers of Joseph Henry*, volume 1: *December 1797–October 1832, The Albany Years*. Smithsonian Institution Press.

Reingold, Nathan, Arthur P. Molella, and Michele L. Aldrich, eds. 1975 *The Papers of Joseph Henry*, volume 2: *November 1832–December 1835, The Princeton Years*. Smithsonian Institution Press.

Reingold, Nathan, Arthur P. Molella, and Marc Rothenberg, eds. 1979. *The Papers of Joseph Henry*, volume 3: *January 1836–December 1837, The Princeton Years*. Smithsonian Institution Press.

Reingold, Nathan, Arthur P. Molella, and Marc Rothenberg, eds. 1981. *The Papers of Joseph Henry*, volume 4: *January 1838–December 1840, The Princeton Years*. Smithsonian Institution Press.

Research Publications, Inc. 1980. *Early Unnumbered United States Patents, 1790–1836: Index and Guide to the Microfilm Edition.* Research Publications.

Reynaud, Léonce. 1871 [1864]. *Memoire upon the Light-House Illumination of the Coasts of France.* Washington.

Reynaud, Léonce. 1876 [1864]. *Memoir Upon the Illumination and Beaconage of the Coast of France.* U.S. Government Printing Office.

Ritchie, William. 1833. Experimental Researches in Electro-Magnetism and Magneto-Electricity. *Philosophical Transactions of the Royal Society* 123: 313–321.

Robertson, Charles J. 2006. *Temple of Invention: History of a National Landmark.* Scala.

Robertson, Robert, and Harold Hartley. 1933. Michael Faraday and Electro-Chemistry. *Proceedings of the Royal Institution of Great Britain* 27: 332–356.

Rodgers, Frederick, et al. 1889. Appendix 1: Electric-Lighted Buoys in Gedney's Channel Harbor. *U.S. Treasury Department, Report of the Light-House Board*, pp. 185–197.

Rose, Mark H. 1995. *Cities of Light and Heat: Domesticating Gas and Electricity in Urban America.* Pennsylvania State University Press.

Rosenberg, Nathan. 1970. Economic Development and the Transfer of Technology: Some Historical Perspectives. *Technology and Culture* 11: 550–575.

Rossi, Auguste. 1880. Cyprien M. Tessié du Motay. *Journal of the American Chemical Society* 2: 305–314.

Rothenberg, Marc, and Kathleen W. Dorman, eds. 1998. *The Papers of Joseph Henry*, volume 8: *January 1850–December 1853, The Smithsonian Years.* Smithsonian Institution Press.

Rothenberg, Marc, Kathleen W. Dorman, and Frank R. Millikan, eds. 2002. *The Papers of Joseph Henry*, volume 9: *January 1854–December 1857, The Smithsonian Years.* Science History Publications.

Rothenberg, Marc, Kathleen W. Dorman, and Frank R. Millikan, eds. 2004. *The Papers of Joseph Henry*, volume 10: *January 1858–December 1865, The Smithsonian Years.* Science History Publications.

Routledge, Robert. 1879. *Discoveries and Inventions of the Nineteenth Century.* George Routledge and Sons.

Rowbottom, Margaret, and Charles Susskind. 1984. *Electricity and Medicine: History of Their Interaction.* San Francisco.

Ruth, Kim M. 1991. The Twin Lights of Navesink. *The Keeper's Log*, fall: 2–7.

Sabine, Robert. 1869. *The History and Progress of the Electric Telegraph. With Descriptions of Some of the Apparatus*, second edition. Van Nostrand.

Salsbury, Stephen. 1988. The Emergence of an Early Large-Scale Technical System: The American Railroad Network. In *The Development of Large Technical Systems*, ed. R. Mayntz and T. Hughes. Westview.

Saxton, Joseph. 1836. Mr. J. Saxton on his Magneto-Electrical Machine; with Remarks on Mr. E. M. Clarke's Paper in the Preceding Number. *London and Edinburgh Philosophical Magazine* 9: 360–365.

Scharf, J. Thomas. 1887. *History of the Confederate States Navy from its Organization to the Surrender of its Last Vessel.* Rogers & Sherwood.

Scharlott, Bradford. 1986. The Telegraph and the Integration of the U.S. Economy: The Impact of Electrical Communications on Interregional Prices and the Commercial Life of Cincinnati. Ph.D. dissertation, University of Wisconsin, Madison.

Scharlott, Bradford. 2004. Communication Technology Transforms the Marketplace: The Effect of the Telegraph, Telephone, and Ticker on the Cincinnati Merchants' Exchange. *Ohio History* 113: 4–17.

Schellen, Heinrich. 1850. *Der Electromagnetische Telegraph.* Vieweg.

Schellen, Heinrich. 1884. *Magneto-Electric and Dynamo-Electric Machines: Their Construction and Practical Application to Electric Lighting and the Transmission of Power.* Van Nostrand.

Schiffer, Michael Brian. 1976. *Behavioral Archeology.* Academic Press.

Schiffer, Michael Brian. 1987. *Formation Processes of the Archaeological Record.* University of New Mexico Press.

Schiffer, Michael Brian. 1991. *The Portable Radio in American Life.* University of Arizona Press.

Schiffer, Michael Brian. 1992. *Technological Perspectives on Behavioral Change.* University of Arizona Press.

Schiffer, Michael Brian. 1993. Cultural Imperatives and Product Development: The Case of the Shirt-Pocket Radio. *Technology and Culture* 34: 98–113.

Schiffer, Michael Brian. 1996. Some Relationships Between Behavioral and Evolutionary Archaeologies. *American Antiquity* 61: 643–662.

Schiffer, Michael Brian. 2000. Indigenous Theories, Scientific Theories and Product Histories. In *Matter, Materiality and Modern Culture*, ed. P. Graves-Brown. Routledge.

Schiffer, Michael Brian, ed. 2001a. *Anthropological Perspectives on Technology.* University of New Mexico Press.

Schiffer, Michael Brian. 2001b. The Explanation of Long-Term Technological Change. In *Anthropological Perspectives on Technology*, ed. M. Schiffer. University of New Mexico Press.

Schiffer, Michael Brian. 2002. Studying Technological Differentiation: The Case of 18th-Century Electrical Technology. *American Anthropologist* 104: 1148–1161.

Schiffer, Michael Brian. 2003. Properties, Performance Characteristics and Behavioral Theory in the Study of Technology. *Archaeometry* 45: 169–172.

Schiffer, Michael Brian. 2004. Studying Technological Change: A Behavioral Perspective. *World Archaeology* 36: 579–585.

Schiffer, Michael Brian. 2005a. The Devil is in the Details: The Cascade Model of Invention Processes. *American Antiquity* 70: 485–502.

Schiffer, Michael Brian. 2005b. The Electric Lighthouse in the Nineteenth Century: Aid to Navigation and Political Technology. *Technology and Culture* 46: 275–305.

Schiffer, Michael Brian. 2007. Some Thoughts on the Archaeological Study of Social Organization. In *Archaeological Anthropology: Perspectives on Method and Theory*, ed. J. Skibo, M. Graves, and M. Stark. University of Arizona Press.

Schiffer, Michael Brian. 2008. A Cognitive Analysis of Component-Stimulated Invention: Electromagnet, Telegraph, and the Capitol Dome's Gas Lighter. *Technology and Culture* 49, no. 2: 376–398.

Schiffer, Michael Brian, Tamara C. Butts, and Kimberly K. Grimm. 1994. *Taking Charge: The Electric Automobile in America*. Smithsonian Institution.

Schiffer, Michael Brian, Kacy Hollenback, and Carrie Bell. 2003. *Draw the Lightning Down: Benjamin Franklin and Electrical Technology in the Age of Enlightenment*. University of California Press.

Schiffer, Michael Brian, and Andrea Miller. 1999. *The Material Life of Human Beings: Artifacts, Behavior, and Communication*. Routledge.

Schiffer, Michael Brian, and James M. Skibo. 1987. Theory and Experiment in the Study of Technological Change. *Current Anthropology* 28: 595–622.

Schiffer, Michael Brian, and James M. Skibo. 1997. The Explanation of Artifact Variability. *American Antiquity* 62: 27–50.

Schivelbusch, Wolfgang. 1989. *Disenchanted Light: The Industrialization of Light in the Nineteenth Century*. University of California Press.

Schweigger, Johann. 1821. Noch Einige Wort über Diese Neuen Elektromagnetischen Phänomene. *Neues Journal für Chemie und Physik* 1: 35–41.

Shaffner, Taliaferro P. 1859. *The Telegraph Manual*. Pudney & Russell.

Shaffner, Taliaferro P. 1859/60. Communication with America, via the Faroes, Iceland, and Greenland. *Proceedings of the Royal Geographical Society of London* 4 (3): 101–108.

Sharlin, Harold I. 1961. The First Niagara Falls Power Project. *Business History Review* 35: 59–74.

Sharlin, Harold I. 1970. Peter Barlow. In *Dictionary of Scientific Biography*. Scribner.

Sharlin, Harold I. 1979. *Lord Kelvin: The Dynamic Victorian*. Pennsylvania State University Press.

Sherman, Roger. 1988. Charles Page, Daniel Davis, and Their Electromagnetic Apparatus. *Rittenhouse* 2 (6): 34–47.

Shiers, George. 1977. *The Electric Telegraph: An Historical Anthology*. Arno.

Silliman, Benjamin. 1837. Notice of the Electro-Magnetic Machine of Mr. Thomas Davenport, of Brandon, near Rutland, Vermont. *American Journal of Science and Arts* 32 (1): Appendix (separately paginated).

Silverman, Kenneth. 2003. *Lightning Man: The Accursed Life of Samuel F. B. Morse*. Knopf.

Simon, Linda. 2004. *Dark Light: Electricity and Anxiety from the Telegraph to the X-Ray*. Harcourt.

Singer, George J. 1814. *Elements of Electricity and Electro-Chemistry*. London.

Skibo, James M. 2001. Understanding Artifact Variability and Change: A Behavioral Framework. In *Anthropological Perspectives on Technology*, ed. M. Schiffer. University of New Mexico Press.

Skibo, James M., and Michael B. Schiffer. 2008. *People and Things: A Behavioral Approach to Material Culture*. Springer.

Skibo, James M., William H. Walker, and Axel E. Nielsen, eds. 1995. *Expanding Archaeology*. University of Utah Press.

Slattery, Thomas J. 1990. An Illustrated History of the Rock Island Arsenal and Arsenal Island, Parts One and Two. U.S. Army Armament, Munitions and Chemical Command. Historical Office, Rock Island, Illinois.

Sleeman, C. W. 1880. *Torpedoes and Torpedo Warfare: Containing a Complete and Concise Account of the Rise and Progress of Submarine Warfare*. Griffin, Portsmouth. England.

Slight, Julian. 1849. *A Narrative of the Loss of* Royal George, *at Spithead, August, 1782; Tracey's Attempt to Raise Her in 1783; Her Demolition and Removal by Major-General Pasley's Operations*. Portsea.

Smee, Alfred. 1841. *Elements of Electro-Metallurgy, or The Art of Working in Metals by the Galvanic Fluid*. London.

Smith, Crosbie, and M. Norton Wise. 1989. *Energy and Empire: A Biographical Study of Lord Kelvin*. Cambridge University Press.

Smith, Cyril S. 1974. Reflections on Technology and the Decorative Arts in the Nineteenth Century. In *Technological Innovation and the Decorative Arts*, ed. I. Quimby and P. Earl. University of Virginia Press.

Smith, Merritt Roe. 1985. *Military Enterprise and Technological Change: Perspectives on the American Experience*. MIT Press.

Smithsonian Board of Regents. 1858. Report of the Special Committee of the Board of Regents on the Communication of Professor Henry. *Annual Report of the Board of Regents of the Smithsonian Institution for 1857*, p. 88–117. Washington.

Snelling, Henry H. 1849. *The History and Practice of the Art of Photography*. G. P. Putnam.

Société d'Histoire de la Poste et de France Télécom en Alsace. 1992. *Le Pantélégraphe de l'Abbé Caselli* (exhibition catalog). Société d'Histoire de la Poste et de France Télécom en Alsace, Strasbourg.

Sömmerring, Samuel. 1811. Bemerkungen über Herrn Prem. Lieut. C.J.A Prätorius Aufsatz: über die Unstatthaftigkeit der Elektrischen Telegraphen für Weite Fernen. *Annalen der Physik* 39: 478–482.

Spencer, Thomas. 1840. *Instructions for the Multiplication of Works of Art in Metal, by Voltaic Electricity.* Glasgow.

Spon, Edward. 1874. Boring and Blasting. In *Spons' Dictionary of Engineering, Civil, Mechanical, Military, and Naval.* Spon.

Sprague, John T. 1875. *Electricity: Its Theory, Sources, and Applications.* Spon.

Sprague, John T. 1878. *Electric Lighting: Its State and Progress, and Its Probable Influence upon the Gas Interests.* Spon.

Staite, G. H., compiler. 1882. *Staite's Electric Light, 1846–1853.* Chester.

Staiti, Paul J. 1989. *Samuel F. B. Morse.* Cambridge University Press.

Standage, Tom. 1999 [1998]. *The Victorian Internet.* Berkley.

Staudenmaier, John M. 1985. *Technology's Storytellers: Reweaving the Human Fabric.* Society for the History of Technology and MIT Press.

Stauffer, David M. 1906. *Modern Tunnel Practice.* Engineering News Publishing Co.

Stauffer, Robert C. 1953. Persistent Errors Regarding Oersted's Discovery of Electromagnetism. *Isis* 44: 307–310.

Stauffer, Robert C. 1957. Speculation and Experiment in the Background of Oersted's Discovery of Electromagnetism. *Isis* 48: 33–50.

Steffens, Henry John. 1979. *James Prescott Joule and the Concept of Energy.* Science History Publications.

Stevenson, Thomas. 1871. *Lighthouse Illumination, Being a Description of the Holophotal System and of Azimuthal Condensing and Other New Forms of Lighthouse Apparatus,* second edition. Edinburgh.

Stotherd, Richard H. 1872. Electrical Ignition of Explosives. *Journal of the Society of Telegraph Engineers* 1: 209–220.

Sturgeon, William. 1826. Account of an Improved Electro-Magnetic Apparatus. *Annals of Philosophy* 12: 357–361.

Sturgeon, William. 1830. *Recent Experimental Researches in Electro-Magnetism, and Galvanism.* London.

Sturgeon, William. 1834. Account of Some Magneto-Electrical Experiments Made with the Large Magnet at the Exhibition Room, Adelaide Street. *London and Edinburgh Philosophical Magazine* 5: 376–377.

Sturgeon, William. 1835. Explanatory Facts. *London and Edinburgh Philosophical Magazine* 6: 231–234.

Sturgeon, William. 1836. Description of an Electro-Magnetic Engine for Turning Machinery. *Annals of Electricity, Magnetism, and Chemistry* 1: 75–78.

Sturgeon, William. 1839. Historical Sketch of the Rise and Progress of Electro-Magnetic Engines for Propelling Machinery. *Annals of Electricity, Magnetism, and Chemistry* 3: 429–437.

Sturgeon, William. 1850. *Scientific Researches, Experimental and Theoretical, in Electricity, Magnetism, Galvanism, Electro-Magnetism, and Electro-Chemistry*. Bury.

Suchman, Lucy. 1983. *Plans and Situated Actions: The Problem of Human-Machine Communication*. Cambridge University Press.

Susskind, Charles. 1964. Observations of Electromagnetic-Wave Radiation Before Hertz. *Isis* 55: 32–42.

Tagliacozzo, Eric. 2005. The Lit Archipelago: Coast Lighting and the Imperial Optic in Insular Southeast Asia, 1860–1910. *Technology and Culture* 46: 306–328.

Taylor, William B. 1879a. *A Memoir of Joseph Henry. A Sketch of His Scientific Work*. Philadelphia.

Taylor, William B. 1879b. *An Historical Sketch of Henry's Contribution to the Electro-Magnetic Telegraph with an Account of the Origin and Development of Prof. Morse's Invention*. U.S. Government Printing Office.

Thompson, Robert L. 1972 [1947]. *Wiring a Continent. The History of the Telegraph Industry in the United States, 1832–1866*. Arno.

Thompson, Sylvanus P. 1886. *Dynamo-Electric Machinery: A Manual for Students of Electrotechnics*. Spon.

Thomson, William. 1855. On the Theory of the Electric Telegraph. *Proceedings of the Royal Society* 7: 382–399.

Thomson, William. 1856a. On Practical Methods for Rapid Signalling by the Electric Telegraph. *Proceedings of the Royal Society* 8: 299–303.

Thomson, William. 1856b. On Practical Methods for Rapid Signalling by the Electric Telegraph (Second Communication). *Proceedings of the Royal Society* 8: 303–307.

Thomson, William. 1884. *Mathematical and Physical Papers*, volume 2. Cambridge University Press.

Toumey, Christopher P. 1991. Modern Creationism and Scientific Authority. *Social Studies of Science* 21: 681–699.

Townsend, George, A. 1873. *Washington, Outside and Inside*. James Betts.

Trescott, Martha M. 1981. *The Rise of the American Electrochemicals Industry, 1880–1910*. Greenwood.

Tyndall, John. 1874. On the Atmosphere as a Vehicle of Sound. *Transactions of the Royal Society* 164: 183–244.

United States Hydrographic Office. 1892. Submarine Cables. Bureau of Navigation, Navy Department.

Urquhart, J. W. 1882. *Electro-Motors: A Treatise on the Means and Apparatus Employed in the Transmission of Electrical Energy and Its Conversion into Motive Power*. William T. Emmott, Manchester.

Usselman, Steven W. 1992. From Novelty to Utility: George Westinghouse and the Business of Innovation during the Age of Edison. *Business History Review* 66: 251–304.

Vail, Alfred. 1845. *The American Electro Magnetic Telegraph*. Lea & Blanchard, Philadelphia.

Vail, J. Cummings. 1914. *Early History of the Electro-Magnetic Telegraph, from Letters and Journals of Alfred Vail*. Hine Brothers.

Vander Weyde, P. H. 1886. Some Early Experiences with Electric Motors. *Electrical World* 7: 234.

Vanderbilt, Byron M. 1971. *Thomas Edison, Chemist*. American Chemical Society.

Varley, Cromwell F. 1862. On the Relative Speed of the Electric Wave through Submarine Cables of Different Lengths, and a Unit of Speed for Comparing Electric Cables by Bisecting the Electric Wave. *Proceedings of the Royal Society* 12: 211–216.

Vergnes, M. 1857. *Electricity Considered in Its Application to Motive Power*. New York.

Vincenti, Walter G. 1995. The Technical Shaping of Technology: Real-World Constraints and Technical Logic in Edison's Electrical Lighting System. *Social Studies of Science* 25: 553–574.

Volta, Alessandro. 1800. On the Electricity Excited by the Mere Contact of Conducting Substances of Different Kinds. *Philosophical Transactions of the Royal Society* 90: 403–431.

von Urbanitzky, A.R., R. Wormell, and R. M. Walmsley. 1890. *Electricity in the Service of Man: A Popular and Practical Treatise on the Applications of Electricity in Modern Life*. Cassell.

Wahl, William H. 1871. A Large Induction Coil. *Journal of the Franklin Institute* 91: 212–216.

Walker, Charles V. 1841. *Electrotype Manipulation: Being the Theory and Plain Instructions in the Art of Working in Metals, by Precipitating Them from Their Solutions, Through the Agency of Galvanic or Voltaic Electricity*, second edition. George Knight and Sons.

Walker, William H., and Michael B. Schiffer. 2006. The Materiality of Social Power: The Artifact-Acquisition Perspective. *Journal of Archaeological Method and Theory* 13: 67–88.

Wallace, Anthony F. C. 2003. *The Social Context of Innovation: Bureaucrats, Families, and Heroes in the Early Industrial Revolution, as Foreseen in Bacon's* New Atlantis. University of Nebraska Press.

Walters, Ronald G., ed. 1997. *Scientific Authority and Twentieth-Century America*. Johns Hopkins University Press.

Warner, Deborah J. 1994. Compasses and Coils: The Instrument Business of Edward S. Ritchie. *Rittenhouse* 9: 1–24.

Watkins, Francis. 1828. *A Popular Sketch of Electro-Magnetism, or Electro-Dynamics*. London.

Watkins, Francis. 1835. Observations on Mr. Sturgeon's letter contained in the Lond., and Edinb. Phil. Mag. for November 1834. *London and Edinburgh Philosophical Magazine* 6: 239.

Watkins and Hill. 1836. *A Descriptive Catalogue of Optical, Mathematical, Philosophical, and Chemical Instruments and Apparatus*. London.

Watt, Alexander. 1860. *Electro-Metallurgy Practically Treated*. John Weale.

Weiss, George. 1926. *The Lighthouse Service: Its History, Activities and Organization*. Johns Hopkins University Press.

Wheatstone, Charles, and Frederick A. Abel 1861. *Report to the Secretary for War, on the Results of Investigations, Conducted at Woolwich and Chatham, on the Application of Electricity from Different Sources to the Explosion of Gunpowder*. Her Majesty's Stationery Office.

White, Samuel S. 1995 [1876]. *Samuel S. White Catalogue of Dental Instruments and Equipment*. Norman.

Whitehouse, E. O. Wildman. 1857. Experiments on the Retardation of Electric Signals, Observed in Submarine Conductors. *The Engineer* 3: 62–63.

Whitehouse, E. O. Wildman. 1858. Atlantic Telegraph Instruments. *Mechanics' Magazine* 69: 395.

Wideman, John C. 1993. *The Sinking of the USS Cairo*. University Press of Mississippi.

Wilde, Henry. 1867. Experimental Researches in Magnetism and Electricity. *London, Edinburgh, and Dublin Philosophical Magazine and Journal of Science* 34: 80–104.

Wilkinson, H. D. 1896. *Submarine Cable Laying and Repairing*. The Electrician Publishing Co.

Williams, Frances L. 1963. *Matthew Fontaine Maury: Scientist of the Sea*. Rutgers University Press.

Williams, L. Pearce. 1965. *Michael Faraday: A Biography*. Basic Books.

Williams, L. Pearce. 1970. André-Marie Ampère. In *Dictionary of Scientific Biography*. Scribner.

Williams, L. Pearce. 1971. Michael Faraday. In *Dictionary of Scientific Biography*. Scribner.

Williams, L. Pearce. 1974. Hans Christian Oersted. In *Dictionary of Scientific Biography*. Scribner.

Williams, L. Pearce. 1983. What Were Ampère's Earliest Discoveries in Electrodynamics? *Isis* 74: 492–508.

Williams, L. Pearce, Rosemary FitzGerald, and Oliver Stallybrass, eds. 1971. *The Selected Correspondence of Michael Faraday*. Cambridge University Press.

Wilson, Archibald. 1860. Patent Gas Lighter. *Journal of the Franklin Institute* 34: 385–389.

Window, F. R. 1857. On Submarine Electric Telegraphs. *The Engineer* 3: 44.

Winner, Langdon. 1985. Do Artifacts Have Politics? In *The Social Shaping of Technology*, ed. D. MacKenzie and J. Wajcman. Open University Press.

Winner, Langdon. 1986. *The Whale and the Reactor*. University of Chicago Press.

Wise, W. Lloyd. 1873. On Gas-Lighting by Electricity, and Means for Lighting and Extinguishing Street and Other Lamps. *Telegraphic Journal and Electrical Review* 1: 122–124.

Wise, W. Lloyd. 1885. Appendix II: Classified Index of English Patents. In *Electric Illumination*, volume 2, ed. J. Dredge. London.

Wolanin, Barbara A., compiler. 1998. *Constantino Brumidi: Artist of the Capitol*. U.S. Government Printing Office.

Woodbury, David O. 1949. *A Measure for Greatness: A Short Biography of Edward Weston*. McGraw-Hill.

Wrege, Charles D., and Ronald G. Greenwood. 1994. William E. Sawyer and the Rise and Fall of America's First Incandescent Electric Light Company, 1878–1881. www.h-net.org./~business/bhcweb/publications/BEHprint/v013/p0031-p0049.pdf.

Wright, Lewis. 1897. *The Induction Coil in Practical Work: Including Röntgen X-Rays*. Macmillan.

Wylde, James, ed. 1881–2. *The Industries of the World*. Two volumes. London Printing and Publishing.

Zöllner, Julius. 1865. Die Kräfte der Natur und ihre Benuzung. Leipzig.

Index

Abel, Frederick, 130
Adams, Henry, 185
Adelaide Gallery, 54–58
Agamemnon (ship), 228–230
Allan, T., 193
Allard, Émile, 275, 276
Alliance Company, 256, 257, 260–262
Alternating current, 314–316
American Association for the Advancement
    of Science, 180
American Electric Manufacturing Company,
    280
American Institute, 105, 139
Ampère, André-Marie, 23, 24, 27, 43, 52, 154
Anglo-American Telegraph Company, 234
Apps, A., 158, 159
Arago, François, 23, 51, 52
Atlantic Cable Company, 226–230
Authority, scientific, 8, 9
Ayrton, William, 291, 292

Babbage, Charles, 73
Bache, Alexander Dallas, 69, 74, 113
Badger, George, 123
Bain, Alexander, 193, 195, 196, 203
Ballard, J. A., 193
Barlow, Peter, 43–45
Barlow's law, 44, 45
Batteries, 13–16, 33, 75
  constant, 210
  Daniell, 76, 77

medical, 158
  submarine, 121–129
Beard, George, 286
Beardslee, George, 88, 89, 167
Becquerel, Antoine, 213
Bell, Alexander Graham, 152, 270
Bentley, E. M., 311, 312
Benton, Thomas Hart, 165, 166
Berzole light, 178, 184–186
Berryman, O. H., 224
Blake, J. P., 183
Block signaling, 200–202
Blossom, Levi, 253, 254
Brady, Mathew, 138, 215
Brett, John, 227
Bright, Charles Tilson, 193, 225, 227
Broadway Theatre, 243, 244
Browne, Charles, 132, 133
Brush, Charles, 289–292, 308
Brush Dynamo Company, 309
Brush Electric Company, 290, 305
Bunsen, Robert Wilhelm, 77, 160
Burleigh, Charles, 132

Cables, 139–142, 221–230
California Electric Light Company, 291, 292
Caselli, Giovanni, 203–205
Cebrián, John, 305
Centennial Exhibition (1876), 269, 270
Channing-Farmer system, 198, 199
Channing, William, 197–199

Chappe, Claude, 41
Children, John George, 17
Chronometry, 193–196
Civil War, 127–130, 138, 167, 245, 246
Clark, Latimer, 225, 236
Clarke, Edward, 59–62, 73, 248
Coils, 24, 27, 158–162
Colt, Samuel, 121–129, 139
Commercialization, 5, 6, 52–54, 59–63, 72,
    73, 79–82, 85–90, 99–117, 146–149, 152–
    154, 172–174, 285
Commission des Phares, 207, 208
Communication technology, 146–149,
    152–154
Commutator, 58, 59
Compagnie Transatlantique, 260
Compound magnet and electrotome, 157,
    158
Condenser, 158
Cook, Ransom, 70–72, 105–116
Cooke, William, 91, 138, 200
Cooper, Peter, 172, 224
Cornell, Ezra, 141, 222
Coulomb, Charles-Augustin, 23
Credit, allocation of, 62, 150, 151
Crosse, Andrew, 79

Daguerre, Louis, 137, 138
Daguerreotype process, 137, 138, 214, 215
Dalton, John, 162
Dana, James Freeman, 95
Daniell, John Frederic, 73–77, 82
Davenport, Thomas, 37, 63–72, 105–116
Davidson, Robert, 155
Davis, Daniel, Jr., 156, 157, 192
Davy, Humphrey, 15–19, 25, 26, 52
Deception, scientific, 175–190
Depth sounder, 193
Detonation, 119–135
Dixon, Joseph, 183
Draper, John, 123
Draper, William, 138
Duboscq, Jules, 210, 217, 220

Du Moncel, Théodose, 194
Du Motay, Cyprien Tessié, 307, 309
Dynamos, 147, 174, 255, 266–270, 289–
    291, 295, 296, 300, 301, 304, 305,
    311

Eaton, Amos, 37
Edison Electric Light Company, 309–312
Edison, Thomas Alva, 1, 154, 270, 283–288
Electrical devices and vehicles
  annunciator, 192
  chairs, 313, 314
  gas lighter, 241–246
  locomotive, 167–169
  mines, 121–129
  organ, 196
  torpedo boat, 193
  trolleys, 312
Electricity
  animal, 13, 14
  contact, 13, 14
  frictional, 12–14
  static, 12
  voltaic electricity, 13, 14, 80
Electrodeposition, 80–90
Electromagnetic Association, 105–109
Electromagnetic radiation, 288
Electromagnetic regulator, 194
Electromagnetic theory, 288
Electromagnetism, 19–29
Electromagnets, 28, 29, 32–37, 191, 192
Electromechanical devices, 206
Electrometallurgy, 78–90
Electroplating, 80–90, 84
Electrostatic machines, 158
Electrostatics, 12
Electrotherapy, 59, 60
Electrotome, 157
Electrotyping, 82–84
Elitism, 62, 65, 112
Elkington, George, 82, 85
Elkington, Henry, 82, 85
Elliott, George, 259, 260, 274

Ellsworth, Annie, 142, 144
Ellsworth, Henry, 99
Engines
  electro axial, 161, 162
  electromagnetic, 162, 163, 169–172
  steam, 270
Everett, William, 229
Ewbank, Thomas, 169, 185
Explosives, 119–135

Faraday (ship), 235
Faraday, Michael, 24–27, 49–52, 73, 78, 119,
    212, 213, 225, 236, 237, 257, 258, 273,
    274
Farmer, Moses, 198, 199, 305
Ferguson, James, 27
Ferris, Charles, 139, 140
Field, Cyrus, 222–238
Filaments, 309
Flagler, D. W., 268
Fisher, James, 139, 142
Fishing, 261
Forces
  electromagnetic, 162–164
  etheric, 286–288
  unity of, 21, 22, 49, 162–164
Franco-Prussian War, 216
Franklin, Benjamin, 11, 12, 27, 31
French, Benjamin, 247–251
French Lighthouse Commission, 258, 259
Fresnel, Augustin, 73, 208
Froment, Paul-Gustav, 164, 165, 203
Fulton, Robert, 122, 123
Fuses, 132, 133

Gale, Leonard, 99, 141, 185
Galvani, Luigi, 13
Galvanism, 13–16, 79
Galvanometers, 24, 25, 229
Gardiner, Samuel, Jr., 167, 241–254
Gassiot, John, 159
Geissler tube, 159, 160
General Electric Corporation, 315, 316

Generators
  electromagnetic, 49–62
  electrostatic, 156 157
Geometer, 74
Gisborne, Frederick, 223
Goff, E. H., 280
Gramme, Zénobe, 265–268
Great Eastern (ship), 232–237
Grove, William, 77, 210, 213
Gutta percha, 222, 223
Gutta-Percha Company, 222, 227

Hachette, Jean, 52
Hall, Thomas, 201
Hansen, W., 193
Hare, Robert, 26, 112, 120–122, 131
Harlan, James, 248–252
Heap, D. P., 230
Hearder, J. H., 216
Helix, 24, 27, 158–162
Hell Gate Light, 134, 278–280
Henry, Joseph, 9, 31–40, 45–50, 64, 55, 68,
    69, 73, 74, 77, 91, 137–139, 148–151, 156,
    159, 172, 173, 180, 185, 208, 213 223,
    230, 249, 250
Hertz, Heinrich, 287, 283
Hoe & Co., 196, 197
Holmes, Frederick, 256, 257
Hoosac Tunnel, 131–135
Houston, Edwin, 287
Hudson, L. A., 241–246
Hunt, Robert, 169 173
Hydro Electric Light, 178–186

Induction, 24, 27, 158–162, 287
Industrialization, 85–90, 99, 100, 105–117
Invention, 5, 38–40
Inventors, 172–175

Jablochkoff, Paul, 292
Jablochkoff candles, 292, 293
Jackson, Charles, 97
Jacobi, Moritz, 79, 81, 155, 161, 268

Joint-stock companies, 72, 73, 105–109,
    283
Jones, Thomas, 112
Joule, James Prescott, 162–164, 192

Kendall, Amos, 148
Kirchhoff, Gustav, 150
Knight, Edward, 250
Knight, W. H., 311, 312

Lanterns
  Gardiner-Blossom, 153, 154
  magic, 215, 216
Liebig, Justus von, 170
Light-House Board, 207, 280–282
Lighthouses, 73, 178, 207–213, 255–260,
    271–282
Lighting
  arc, 210–220, 259, 260, 271–275, 289–297,
    303–306, 309
  of Capitol dome, 239, 240, 247–253
  electric, 219–221, 273
  gas, 288
  incandescent, 281, 294, 297–316
  magneto-powered, 256–270
  of steamships, 260, 261
Lundeberg, Philip, 122, 126

MacLaren, Malcolm, 174
Magnetic Telegraph Company, 148
Magnetos, 52–62, 86, 167, 255
  Beardslee, 88, 89
  Holmes, 257, 258
  Nollet-Van Malderen, 258, 259
  Woolrich, 86–88
Marconi, Guglielmo, 287
Marine technologies, 121–129, 192, 193, 220,
    224, 260–262
Mascher, J. F., 170
Mathiot, George, 181–183
Matteuci, Carlo, 78
Matthew effect, 62
Maury, Matthew, 129, 130, 224

Maxwell, James Clerk, 288
Mechanics' Magazine, 175, 176, 212
Medical technology, 59, 60
Meigs, Montgomery, 246
Menlo Park, New Jersey, 285
Military technology, 103, 119–130, 216,
    304
Molera, Eusebius, 305
Molera-Cebrián lighting system, 305, 306
Moll, Gerard, 33, 34
Morse, O'Reilly v., 150, 151
Morse, Samuel F. B., 17, 63–65, 91–103, 115–
    117, 123, 137–152, 223, 227, 228, 236,
    237, 250, 251
Morse Code, 142–145
Morton, Henry, 278, 305, 308, 309
Motors, 70–72, 109–112, 171, 172, 178, 187,
    188, 268–270
  of Davenport and Cook, 70–72
  designs of, 155, 156
  reciprocating, 161, 162
  rocking-beam, 37, 38
  rotary, 67
Mowbray, George, 132
Munn & Co., 175–178

Napier, James, 170
Napier press, 114, 115
Napoléon III, 217, 219, 256
Naturphilosophie, 21, 22, 49
Nautical technologies, 121–129, 192, 193,
    220, 224, 260–262
Nelson, Myrick, 114
Newton, Isaac, 33
New York, Newfoundland, and London
    Telegraph Company, 224
Niagara (ship), 228–230
Nitroglycerin, 132, 133
Nollet, Floris, 256
North Atlantic Telegraph Company, 232

Oerstad, Hans Christian, 21–23
Ohm, Georg, 301

Ohm's Law, 45, 301
O'Reilly, Henry, 150, 151
O'Reilly v. Morse, 150, 151
Orton, J. W., 178

Pacinotti, Antonio, 265
Page, Charles Grafton, 113, 121, 147, 155–171, 249, 250
Paine, Henry, 178–190
Paine Electromagnetic Engine Company, 187, 188
Palmer, Edward, 84, 155
Pantelegraph, 203, 204
Pasley, Charles William, 119–121
Patent agencies, 175–178
Patent law, 175, 179, 180
Patent processes, 60–62, 81, 82, 102, 149–152, 301
Pearl Street District, 1, 310, 311
Penfield, Allen, 36, 37
Perkins, Jacob, 54
Phonographs, 299, 300
Photography, 137, 138, 214, 215, 304
Pike, Nicolas, 250
Pixii et Fils, 52–54, 58, 61, 62
Platinum bulb, 302, 303
Polarization, 75, 76
Practicality, 2, 3, 7, 12, 19, 42
Preece, William, 305
Pretsch, Paul, 85
Printing industry, 81–84, 109, 114, 193, 196, 197, 203–205
Pure science, 112

Railroad industry, 199–202, 219, 220
Railroad switch and alarm, 201, 202
Ritchie, Edward, 158, 159
Robert-Houdin, Jean-Eugéne, 196
Roe, George, 291, 292
Roosevelt, Hilborne, 196
Rowland, Henry, 187, 188
Rühmkorff, Heinrich, 160, 161
Rutter, John, 194

Safety industry, 194, 197–199
Safety lamp, 18, 19
Salvá, Don Francisco, 41, 42
Sawyer, William, 286, 297, 306–308, 312
Sawyer-Man Company, 312
Saxton, Joseph, 54, 62
Schweigger, Johann, 32
Scientific American, 170, 171, 175–184 197, 212, 213, 221, 244, 264, 265, 309
Security industry, 193, 194
Serrell, James, 141
Shaffner, Taliaferro, 232, 250
Sherwood, Henry, 74
Siemens, Werner, 265, 311
Silliman, Benjamin, 33–35, 47, 106, 107
Simpson, James, 249
Singer, George, 17, 18, 78
Smee, Alfred, 81, 82
Smith, Cyril Stanley, 79
Smith, Francis, 141, 142
Smith, J. E., 138
Solenoid, 24, 27, 158–162
Somerville, Mary, 107
Sömmerring, Samuel, 42, 43
Southeastern Railway, 134
Speedwell Iron Works, 99, 100
Spencer, Thomas, 79–81
Staite, W. Edwards, 210–213
Stanley, William, Jr., 312
Statue of Liberty, 280, 281
Steam power, 174
Steinheil, Karl, 91
Sternberg, George, 194
Stockly, George, 289, 290
Sturgeon, William, 27–29, 32, 54, 58, 59, 62, 86–90, 106, 155, 162
Submarine Battery Company, 123
Submarine Cable Company, 224
Submarine light, 220
Surveying, 145

Taft, Timothy, 36, 37
Tatham, Benjamin, 141

Taylor, Thomas, 192
Telegraph Construction & Maintenance
    Company, 232
Telegraph poles, 142
Telegraphs
  Atlantic, 221, 222
  fire alarm, 197–199
  railroad, 200–202
  semaphore-type, 41
Telegraph system, 97–103
Telephone, 152, 153, 270
Ten Eyck, Philip, 35, 45, 46
Tesla, Nikola, 312
Thermo-regulator, 194
Thompson, Benjamin, 15, 18, 154
Thomson, Elihu, 287
Thomson, William, 222, 225–229, 232–237
Thomson-Houston, 312, 316
Totten, Joseph, 126
Trinity House, 207, 257
Turner, Edward, 67, 68

Unio-directive discharger, 58, 59
Upshur, Abel, 123
U.S. Light-House Board, 259, 260
U.S. Patent Office, 39, 70, 99, 100, 175
U.S. Senate chamber, 245

Vail, Alfred, 91, 99–102, 139–142, 147,
    150
Vander Weyde, P. H., 172, 187, 188, 304
Van Malderen, Joseph, 256
Van Rensselaer, Stephen, 68, 69
Varley, Cromwell, 194
Varley, Samuel, 194
Volta, Alessandro, 13, 14

Wallace, William, 300
Walter, Thomas, 239, 246–248
Watkins, Francis, 58
Watson, Joseph, 170
Watt, Alexander, 82
Watt, James, 162

Western Union, 147, 148, 270
Westinghouse, George, 283, 284, 312–316
Weston, Edward, 293, 294, 301
Weston Dynamo Machine Company, 294
Wheatstone, Charles, 73–77, 91, 102, 138,
    154
Whitehouse, E. O. Wildman, 222, 225, 227,
    229, 236, 237
Wideman, John, 129
Wilcox, J. W., 83, 84
Wilde, Henry, 263–265
Wilkins, William, 126
Williams, Edwin, 114
Wilson, Archibald, 245, 252, 253
Woolrich, John, 86–88
Wright, E., 182, 183